Builder's Guide to Accounting

═Revised═

By
Michael C. Thomsett

Craftsman Book Company
6058 Corte del Cedro Carlsbad, CA 92009

Foreword

This reference is written for builders, contractors, subcontractors, and anyone in related specialty trades. It has been geared to the construction business, and the problems builders have in maintaining a good set of useful books. Company sizes range from one-man operations all the way up to some of the world's largest corporations. The smaller the business, the more need there is for economical record-keeping.

A business operating out of a small office on a tight budget often can't afford full-time bookkeeping and accounting services. In view of the competition for contracts, the small operation which has the lowest possible overhead will be the one best able to bid in line with larger companies, win contracts, and make a profit. But just like an outfit which wastes material and manhours, the operation that depends too much on expensive professional services cannot compete with an efficient shop.

This is a reference book. It provides step-by-step suggestions for keeping books and records, and won't burden the reader with unnecessary terminology. The examples given are explained in detail and are accompanied by illustrations meant to show the way toward simplifying what has to be done to maintain a good set of books. This book is not written for accountants. It is meant to help the owner or manager of a construction company save money and time by getting involved in much of his own bookkeeping.

The information can be applied immediately. Without revamping your books all at once, you should be able to streamline gradually and still have all the information that you and your accountant need.

The ideas explored here can be adapted to the needs of your operation. Every business is different, and no one method will work for everyone. This information should be interpreted your way and used for your own, unique requirements. With the variety of things your accountant has been trained to do, he should be pleased when you work with him to modify your records and books to make his job and yours as easy as possible.

At first, you will probably need to spend some time learning how to keep records the way you need and want. You should have little difficulty developing the bookkeeping ability needed to use your financial information to your own benefit. In time, you'll only need a few hours each week. You'll be a better-rounded businessman when you're involved in the development of your own financial information.

You don't need an accounting background to keep your own in-house books, only the desire to be involved, to increase your understanding, and cut your overhead. When you can look at a statement or report and use the valuable information in it effectively, you'll have added a new dimension to your management skills. Then you can comfortably leave the tax planning and financial statement preparation to your accountant — your financial subcontractor.

Library of Congress Cataloging-in-Publication Data

Thomsett, Michael C.
  Builders guide to accounting.

  Includes index.
  1. Construction industry--Accounting. I. Title.
HF5686.B7T47 1987 657'.869 87-6817

First edition   © 1979 Craftsman Book Company
Second edition © 1987 Craftsman Book Company
Eighth printing 1992

ISBN 0-934041-18-0

Contents

Foreword 2

1 Why Keep Records? 5
   Accounting System Components,
   An Overview 5
   Double Entry Bookkeeping 12

Section I. Sales and Accounts Receivable
2 Accounting Methods 18
   Percentage of Completion Method 18
   Completed Contract Method 20
   Combined Accounting Methods 21

3 Cash and Charge Sales 28
   Recording Cash Sales 28
   Charge Sales 29
   Accounts Receivable 33

4 Managing Receivables 42
   Average Length Study 43
   Reversing an Unfavorable Trend 44
   Collection Notes 46

5 Bad Debt Procedures 49
   Journal Entries For Bad Debts 50
   Estimating Bad Debts 50

6 Sales Records and Cash Budgeting 56
   Cash Receipts Summaries 57
   Cash Receipts Projections 58
   Calculating Profits By Job 60

7 Sales Planning 63
   Yield and Risk Standards 64
   Income Statements by Product 65
   Market Assumptions 67

8 Planning For Profits 68
   Profit and Loss Statements 69
   Profit and Loss Analysis by Job 70
   Successful and Unsuccessful Volume .. 72
   Management Guidelines 74

Section II. Costs and Expenses
9 Check Writing and Recording 77
   Maintaining the Check Book 77
   Keeping a Check Register 82
   Writing Checks to "Cash" 83
   Handling Voided Checks 85
   Pegboard Check Registers 85
   Paying Accounts Payable 87
   Multiple Businesses 91
   Computerized Check Writing 92

10 Accounting For Materials 94
   Purchase Orders 95
   Inventory 98
   Valuing Inventories 98
   Analyzing Material And Inventory ... 100

11 Payroll Accounting 102
   Time Cards 105
   Payroll Checks 106
   Earnings Record 108
   Payroll Register 109
   Accounting For Tax Liabilities 110
   Payroll Tax Limits 114
   Payroll Forms 115

12 Overhead Expenses 118
   Variable and Fixed Expenses 119
   Budgeting Expenses 120
   General Ledger Accounts 122
   Ratios 124
   Recording Expenses 126

13 Equipment Records 130
    Classifying Fixed Assets 131
    Controls on Purchasing Equipment ...134
    Unit Cost of Equipment 134
    Depreciation 137
    Depreciation Methods 139
    Amortization 144
    Sales and Trade-Ins 145

14 Cash Budgeting 147
    Testing The Cash Position 148
    Preparing a Cash Budget 149
    Figuring a Breakeven Point 152
    Cash Controls 153

15 Cost and Expense Records 158
    Scheduling Costs And Expenses.....160
    Performance Standards 162
    Control of Deferred Costs 163

16 Accounting For Costs and Expenses 166
    Direct Labor................167
    Material and Subcontracts........169
    Overhead 169
    Job Cost Ledger Card 170
    Accounting For Completed
    Contract Expenses 171

17 Petty Cash Funds 174
    Imprest System 174
    Controlling Petty Cash 175

18 Balancing The Checking Account 178
    Account Balancing Checklist 182
    How to Construct a Balance 184
    Old Outstanding Checks 184

19 Accounting For Estimates 186
    Job Completion Schedules 186
    The Estimate 186
    Allocating Fixed Overhead 189

Section III. Financial Statements
20 Recording Before The Event 193
    Accounts Receivable 194
    Prepaid Expenses 194
    Deferred Assets 197
    Provision Accounts 197

21 Financial Statements 199
    Relationship Between Statements ...200
    The Trial Balance 202
    Supplementary Schedules 202
    Balance Sheet Accounts 202
    Income Statement Accounts 205

22 Using Financial Information 207
    Setting Standards 208
    Enforcing Standards 209
    Controlling Cash Flow 210
    Short Term Goals 211

23 Financial Ratios 212
    Balance Sheet Ratios 213
    Combined Ratios 214
    Income Account Ratios.........216
    Presenting Ratios 217
    Comparative Ratio Analysis 218

24 Putting Together a Statement 220
    General Ledger Posting 221
    The Trial Balance 224
    Balance Sheet 226
    Income Statement 227
    Closing The Books 227

25 Comparative Period Statement 230
    Comparative Balance Sheets 230
    Comparative Income Statement ...230
    Monthly Income Statements 234
    Using Comparative Data 235
    Comparison With Budgets 236

26 Restatements By Accounting Methods ..237
    Cash Accounting 238

27 Statements By Job 242
    Plotting Job Progress 243
    Long Range Job Planning 245
    Levels of Control 246

28 Statements For Loan Applications 248
    Presenting Data to Lenders 250
    Small Business Administration 251
    The Prospectus Format 254

Appendix 256
    The Chart of Accounts 257
    Johnson Construction Company —
    Complete Financial Statements 263
    Expanding the Account System ...273
    Automating Your Accounting.....279
    Income Tax Planning..........283
    Blank Forms...............286

Index 299

Chapter 1

Why Keep Records ?

Builders, like other businessmen, have a very real need for keeping books. You are engaged in a complex industry, and a good set of records is as necessary as any other modern piece of equipment.

You know that in figuring the cost of a job, a major part of your bid is for labor. Imagine how few contracts you'd win if your labor was twice as expensive as everyone else's. The same holds true with a bookkeeping method. The less time you have to take, or pay someone else to take, the cheaper it is. That, and the goal of maximum efficiency, is the purpose of a bookkeeping method. If you can reach that goal and still have everything you need, you'll have a good set of books. And with it, you'll definitely save time and money.

Keeping records is required by law. As a result, many builders think of their bookkeeping as a necessity, and nothing more. But if you can make good use of the financial information in your books, you'll be a more informed manager. The more insight you develop, the more effective you'll become. This is why extensive records were maintained in the United States long before income tax even existed.

A set of books should tell you everything you need to know in the least complicated way possible. The fewer papers you have to push, the better. A method should contain just the right degree of information—no more and no less than you really need. This guideline is one often forgotten by businessmen—accountants included.

A method that gives you too little information is not going to save you time. Sooner or later, you'll have to go back and fill in the gaps. Imagine working from a blueprint only half completed or a set of specs with missing pages. If you don't have it all, you're wasting your time.

A method with unneeded features is going to be time-consuming and expensive. Whether you do the work yourself or pay someone else, it's costing you. A good method doesn't have any fat. The following is an overview of a good accounting system for builders. Each component will be discussed in detail later in this guide.

SALES

Keeping track of sales gives you and your

accountant a basis for backing up what is reported for sales tax (if your state levies a sales tax). Sales records are also needed for some insurance reports, Census Bureau reports, and other forms you have to provide to various governmental agencies. For example, some states collect a tax based on hauling and freight operations income. A sales journal breaks out this type of sale and supports what is claimed on a filed report. This support feature is important throughout a set of books, to verify reported information.

The sales journal is a detailed record of all your income from operations. This detail is needed for financial statements, and is a good way for you to see how you're doing compared to last year, last month, or last week.

The comparison of income to past performance is very important to you in planning. No matter what direction your business takes, you should have an idea of your probable cash flow—how much cash will be coming in, and when? You cannot make complete plans unless you have a well-based idea of your future income potential.

Good sales records are also helpful in finding the right market for your services. Because there are so many different kinds of contracting markets, you must know which ones will allow you to compete and still have profits. You must develop an understanding of the financial effects of providing or not providing a particular kind of service. For example, you might think you're making money in one type of work. Good records will help you determine exactly how your company is doing in that area. The result will help you decide which types of jobs are best for your company.

Sales records should provide information on specific jobs as well. There is more than one way to account for income, and picking the accounting method to use for your work can have drastic effects on your financial statements and tax liability. Rather than leaving such decisions to your accountant, you will be able to discuss the alternatives with him when you understand the methods and their consequences.

RECEIVABLES

Naturally, you want to know who owes you money. Your method of keeping track of receivables is important to the accuracy of your billings, your collections, bad debts, and financial statements. Census Bureau reports for many of the trades require information on accounts receivable.

Many builders run into their single largest bookkeeping problem when it comes to keeping track of receivables. This is because of the volume of record-keeping that is required and the investment in time needed to keep up a customer ledger. Finding a good method isn't necessarily easy. A lot of seemingly good ideas for saving time actually result in a greater workload, while providing you with less control. It is easy to come up with halfway measures that call for duplication of effort. Not only will you waste a lot of time writing down the same information two or even three times, but you soon lose touch with the real purpose for the work.

Accounts receivable can be controlled effectively with the proper accounting method. Contractors run into problems with receivables because there are special considerations such as retainages. These withheld amounts should be accounted for separately from the normal trade receivables.

When you do work before you are paid, it is inevitable that some portion of your receivables will never be collected. By understanding the direction your bad debts are going, you can both budget for them and take the steps necessary to tighten up on collection procedures. This could reduce your exposure to bad debt losses. Surprisingly, many builders are very casual about doing work on credit. If your volume increases over a short period of time, it may be important to look at bad debt statistics critically. You might discover that the percentage of total receivables that go bad is increasing.

By checking historical information on receivables and bad debts, a rising trend may be discovered. But without detailed accounting information you can not have a full picture of the direction you're going. A general familiarity with your own business affairs is not enough to serve as a dependable guideline for making important decisions.

CHECKS

Knowing where your money is spent is important for controlling direct costs and overhead. Within a procedure for controlling expenses, you need to come up with a good way to analyze costs by job. Your billings, which are based on actual spendings, require 100% accuracy.

The check register is a listing of all checks. It breaks expenses into account categories. It is important to write up this record each month, or even each week. It would be foolish to wait until the end of the year to construct a check register for the previous twelve months. Considering the number of checks most businesses write in a year, there would be no hope for any effective control this way. Good controls require timely information. Seeing from month to month where you are spending money will give you timely control information. Your accountant needs updated information to post your general ledger. Your analysis of budgeted and actual expenses must show a total amount each month. The check register is essential for balancing a bank account as well.

Like most good sets of records, a check register should be flexible. It can be set up to provide a fool-proof method for checking the accuracy of your math. This requires listing all checks twice—once in a "total" column and once in a category of accounts column, such as "materials" or "office supplies." This is not a duplication of effort. A double listing gives totals for both all expenses and expenses by category.

To provide yourself with direct cost records by job, design the check record to fit your own needs. Many builders make copies of invoices and place them in a job file, or go through the check register and recopy the information. This is duplication of effort. Both of these methods accumulate incomplete information and both are time-consuming. Breaking down your check register by jobs saves time and makes your check register a primary control tool.

The check register is also a primary source of other information you need to compile. Your check register can be arranged in many ways. The method you use should fit your business. Your books and records, if efficient and flexible, can be worth the investment you make in them many times over.

PAYROLL

Proper payroll records show you labor costs by job and provide information essential for union reports and payroll tax returns. Many of these records are required by law. Your records, in addition to being useful, should be so designed to meet the legal requirement. Otherwise, your time will be spent in going through and reorganizing your records every time a payroll report is due.

A necessary part of payroll bookkeeping is organization. Breaking out information by job can be difficult if a set of books has to be reconstructed for each different kind of information you need. Here again, you can avoid the cumbersome and lengthy paperwork so often seen in payroll methods.

Too many builders make no attempt to find out their actual labor costs by job. Their billings are based on a fixed rate per hour worked by each classification of employee. This method has obvious shortcomings. A truly useful and efficient method computes the actual cost at the same time the payroll is prepared and checks are made up. All payroll (except overhead payroll like the office staff) should be assigned to one or more job categories. That way your direct cost of labor is known immediately. Job categories should include not only those for large bid contracts, but for various one-time work, shop, maintenance, or idle time.

BANK ACCOUNT

Don't depend on your bank to keep your check book straightened out. Balance your bank account every month to be sure of the cash you really have. Balancing a bank account is not difficult, although many builders think it is. The secret of breezing through a bank account is having a good method and understanding exactly how your method works.

Having your accountant balance your bank account is poor practice and an unnecessary expense. No one knows your check book as well as you do. After all, if you add up your own bank deposits, prepare your own checks and figure all the math, who is better suited to prove the balance? And if you make yourself responsible for this job, the side benefit is that you are likely to work more accurately. You'll be the one to find those same mistakes at the end of the month.

Some builders hardly ever balance their bank accounts. This is a dangerous practice. You may suddenly discover that you don't have nearly as much cash as your balance claims. Errors can accumulate over the months or years. By the time you discover this problem, it could be too late to correct it without embarrassment. There is no "good" time to be out of funds. You might discover, too, that you have more than you thought in your bank account. While this might be a nice surprise, it is a less common discovery. Do not assume that the bank statement's ending balance is a true amount. Unless you balance the account, you don't know the real balance. The bank statement doesn't reflect any checks it hasn't received.

It doesn't always occur to business owners to check the statement for bank errors. But everyone should do this. Banks, and their computers, are run by humans who can and do make mistakes. It is up to the business owner (and bank customer) to discover these errors. Do yourself a favor, and spend a little time in checking someone else's work. You may be lucky and never find a mistake. If that happens, stay with that bank. You might not know how lucky you really are!

You need a thorough, step-by-step procedure to follow. Such a procedure is outlined in this guide. If all steps in the procedure are gone through completely, the bank account will balance every time. The only way to learn is to do it yourself. Once you are used to it, the whole procedure will become quite routine.

CASH ON HAND

Most builders need a cash fund for the little expenses that come up from day to day: C.O.D. deliveries, postage due, coffee and donuts, a newspaper—expenses too small to write out a check for. But you may be amazed at how much cash can be spent in one month from a small office fund.

Setting up a controlled cash fund, called "petty cash," is the best way to control these disbursements. In a petty cash system, you "vouch" for all expenses by replacing cash removed with a slip of paper which explains the reason for the expense or lists an account number and gives the amount.

This fund should have enough cash to suit your needs. If you are running out often, the fund is too small. But keeping too large an amount of cash around the office is not a good idea. So find the amount of petty cash that is right for your business.

You should be able to add all the cash and the vouchers in the cash box at any time and arrive at the petty cash fund balance. You can expect occasional minor over and short problems, but a well-controlled fund will always be in balance. Document fully all of your expenses—even the petty ones—so you can take the full tax deduction you deserve.

EQUIPMENT RECORDS

Few builders can operate without having some kind of machinery and equipment. Everything from carpentry tools to multi-axle trucks are needed in many contracting businesses. The money you spend on tools and equipment is most likely the largest investment you will make. Good equipment records supply information for financial statements, property taxes, capital gains and losses, motor vehicle reports, and depreciation. Your equipment records can also supply direct cost information, if the data is organized and available.

Your records should allow you to figure out the hourly cost of owning and operating your equipment. Each job can then be charged appropriately for its share of the equipment cost. This is essential for determining the profit or loss you are experiencing on each of your jobs.

Your equipment cost will have an effect on how you plan for the future. Will you lease or buy heavy equipment? How much use will you get from a piece of equipment and will that use justify the investment it will require? When you prepare bids on future contracts, you must know how much per hour to charge for operation of equipment and machinery for those projects.

OVERHEAD EXPENSES

The fixed, necessary expenses of being in business are often considered to be uncontrollable: rent, telephone and utilities, shop and office supplies, insurance. You should have some way to plan for these expenses because planning can help you control your overhead.

Control of your overhead is important when estimating future jobs and will often make the difference between profit and loss. Your record of past overhead expenses is the best indication of what you can expect in the future. A good budget for general (overhead) expense will serve as an important guideline for month to month planning. Trying to stay within a realistic budget will almost always lower actual expenses and increase profits. Builders who don't budget their overhead invariably spend more on overhead expenses than those who examine and control their overhead.

ESTIMATING RECORDS

Job estimates are often prepared in a hurry and under pressure. But estimates require both attention to detail and accuracy. You have only a limited amount of time to familiarize yourself with the proposal and come up with a profitable but competitive bid. The method you use to figure labor time, material cost, overhead, and profit must guarantee accuracy. Otherwise, your bid may be successful but still result in an unwanted loss.

Following a set procedure on every bid will save time and assure a higher degree of accuracy. Use this same formula for every bid to make the entire estimate fall together smoothly. Document your costs so you have both historical data and current knowledge. This is the best way to assure yourself that the bid is going to yield a reasonable profit.

To have this confidence in your estimating, your books and records have to be organized to supply the kind of accurate information you need in a hurry. The best estimators use carefully prepared past cost records to back up their conclusions.

RECORDS BY JOB

If, like many contractors, you work on several jobs at one time, you can only know your success if you keep cost records for each job. Your right to progress payments depends upon these records. Accuracy is required to support your charges.

Use by-the-job records to examine the cash flow and profit you have. Your method should be complete and consistent. Since you are the one person most in touch with your operation, you are best qualified to keep these kinds of records. And you are the person most likely to get useful information from them.

Compare the costs for each job with your estimate. Modify your planning and estimating expectations accordingly. Doing this will help you develop valuable historical information which you can use in the future.

You may find that your profits from one job are being eroded by two others, even though you thought you were doing well on all three contracts. A builder can't tell how he's doing on each of several jobs without good cost records.

ACCRUALS

Cash received and paid out in a business does not make up the total financial picture. Billings you have mailed out are income, even before the cash is taken in. And the bills you owe are direct costs or operating expenses, even before the checks are written. These billed accounts are very real parts of your financial picture at any time. Your books and records are only summaries of cash which has changed hands. The true and complete picture of your operation has to include adjusting entries. These are called accruals.

It is generally a good idea to let your accountant handle these entries, as well as the actual general ledger recordings of all your financial business. But you should understand what is accrued, and why, so that you can set up records which will help your accountant establish accurate accruals. He can't make the correct entries unless you have good records. Accruals are estimated much too often. An understatement of true accruals must be made if your records can't establish support for the actual numbers.

You don't have to train yourself to become a professional accountant to know how to break out the significant figures in your books. This book will help you increase the quality of all financial records your company maintains.

RATIOS

Financial and operating ratios are very useful to builders and contractors who understand their meaning. These ratios tell the month to month trend of your business. These trends are useful and revealing financial statements themselves.

It is difficult for you to learn everything you need to know about your business by looking at the monthly numbers. Ratios allow you to divorce yourself from dollars and cents and follow instead the trend of your business. Ratios help you to find the good and the bad situations and show the relative health of your business. It takes only minutes to figure your own financial and operating ratios but they can speak volumes about how you are doing each month.

Builder's Guide To Accounting

FINANCIAL STATEMENTS

Most businessmen don't know how to read financial statements. But nowhere is the skill of comprehending a statement so important as it is in construction. Builders who don't understand simple accounting practice and the usefulness of financial information are ripe for failure in their business.

You must make effective use of your own internal reports and involve yourself directly in record-keeping. But you should also be able to understand your own financial statement. This includes knowing the categories of the statement—where the numbers come from, what the account titles mean, and what the classifications on the statement are. You have an accountant, so you don't have to master the mechanical skills of double-entry bookkeeping to gain this knowledge. You don't want to become an accountant, but you do want to be able to read the statement you are paying for.

Preparing financial statements is a routine chore when the books and records are complete and efficient. But a financial statement can be useful in many ways. For instance, a statement which compares this year's performance to last year's is valuable in finding out whether your management skills are producing profits and whether your financial health is improving or declining. Another type of statement shows what your income, costs, expenses, and profits are by comparing it to previous data using different accounting methods. Statements of income or loss can be prepared by job so you can compare the value of different types of work and different size contracts.

Current statements are essential for performance bonds and loan applications. An impressively complete statement is more likely to result in a needed loan being granted, simply because more than enough information is available to answer any questions a banker might want to ask.

Small Business Administration (S.B.A.) loans require extensive historical and financial information. Your past records must be able to serve this purpose. Complete applications involve less processing delay.

USING A COMPUTER

As small business computers become affordable and software (programs) get better, more and more builders have automated their accounting routines.

A number of versatile and easy-to-use accounting programs are on the market, as well as a good variety of affordable hardware. You can decide to put only some functions on a computer, or the entire system.

A specialized program will provide you with good mathematical controls, and help avoid the struggle with balancing the books from one month to another. In some cases, a good program makes the job easier to understand and manage; in others, it's simply easier to store a lot of information on a disk.

Before deciding to automate, analyze the number of transactions you handle every month. Are there enough to justify getting a computer? If not, you should know that a computer will not replace human effort, nor will it clear up accounting problems you are now having.

Automation is designed to manage information. If that's your goal, then a computer will save you money every month. But some builders have regretted the decision to automate because they didn't understand how a computer works or the advantage of using a computer.

It isn't necessary to put all of your accounting functions on a computer. Some fully integrated programs are designed to manage everything for you. But more affordable, specialized routines can help you manage key heavy-volume areas such as accounts receivable, purchasing, payroll, or job costing.

The traditional argument, often promoted by the manufacturers of computers themselves, is "you must get a computer in order to remain competitive." That may be true when time becomes a critical factor. But for the daily routine, it isn't always the case.

Buying a computer always requires an investment in time as well as money. It takes time to master any new accounting system. Unless the program ultimately saves you a great deal of time, you won't recapture your investment.

Be sure before you buy that you have a critical need for accounting on an automated system. If you approach the problem from that point of view, you're less likely to buy something you don't really need. Appendix D has some suggestions that may help you decide if a computer is needed in your office.

INCOME TAXES

Tax reform has complicated the way builders keep their books and records. The Tax Reform Act of 1986 was the largest overhaul in tax history. And there is more to come.

From 1981 to 1986, there was a major new tax bill every year. This means accounting rules are likely to change every year: long-range policies can't be set with any dependability, and estimating after-tax net income is virtually impossible.

Your accounting system is affected by changing tax rules. Rules for accounting methods are more complex than ever, and any planning you do today may change next year or the year after.

Deciding to incorporate or operate as a sole proprietorship or partnership is going to be affected by tax law. Tax rates have been shifted, adjusted, and scheduled by a three-year phase-in under the 1986 law.

Many business tax breaks have been discontinued as well. You can no longer claim the investment tax credit, once a major benefit of investing in equipment and machinery. Capital gains are no longer taxed at favorable rates. And depreciation schedules have been lengthened.

To further complicate your accounting, many of the provisions that were changed under the 1986 tax law are sure to be revised in the future. An overhaul of that scope is never the end of the story. Within two months after passage of the bill, members of Congress were discussing the next step in tax reform.

Here are some of the changes I wouldn't be surprised to see:

- A reinstatement of capital gains tax rates and the investment tax credit

- Changes in depreciation rates

- Modification of both individual and corporate tax rates

Since part of your accounting is forecasting, the changing face of taxation definitely affects your future profitability. At the same time, the uncertainty of future legislation means you cannot depend on this year's rules being in effect one, two or three years from now.

Even assuming that the present rules remain intact, it's not easy to figure out what a phased-in change will mean in terms of future tax liabilities.

I'll suggest a solution: Keep efficient, complete records of all transactions and do solid planning for the immediate and foreseeable future. Find and use a reliable, professional tax expert. This may be your accountant or a tax specialist.

A book you should find helpful is *Contractor's Year-Round Tax Guide*. An order form bound into the back of this manual shows how to get one.

It isn't possible to keep yourself up to date on all the provisions of the tax law as they affect you. So use professional advice, and plan and forecast according to what the professionals think is coming up.

THE CHART OF ACCOUNTS

The chart of accounts is a numbered listing of the accounts on your general ledger and operating statement. It tells at a glance the financial categories you maintain. Numbered accounts simplify your recordkeeping and provide a shorthand method of applying costs and income.

There is a common and fairly standard order for listing accounts within categories. While each business has its own unique needs, most use a few basic accounts. Construction contractors, more than most other businesses, have a large number of account categories that are unique to the industry. You should understand how the chart of accounts is organized. As you become more aware of the utility and potential for your own records, you might even want to create additional categories which can help you keep track of certain key expenses.

Within one set of books and within one

company, it is possible to have integrated accounts for use in a variety of jobs kept under separate accounting methods. A good chart of accounts is descriptive enough to keep these types of figures separate from each other and yet supply important overall information necessary for reports.

THE SYSTEM IN REVIEW

Any good set of books will yield the financial statements and reports needed for insurance companies, governmental agencies, and bankers. There may be special considerations in your business which justify major changes in the way you keep your books. By the time you finish this book you will see many changes that would improve the utility of your business records. First, the way you keep records should be unique, designed just for you and your particular operation. Second, it should be as flexible as possible to allow for increases in volume and sudden changes in requirements. Third, there should be no room in your system for unneeded work.

You may come to a point where no manual system is good enough to take care of your payroll or your receivables. At that point, do you hire more bookkeepers and accountants? Do you struggle on with what you have, and hope for the best? Do you get in touch with a small computer service? How do you decide? And whatever your decision is, can the cost be justified? This manual is intended to help you make decisions like this.

Just as some jobs could require over-investment in machinery or too large a labor force for you to manage, your bookkeeping system should suit your individual needs.

A review of your system is needed periodically. How often depends on you. Your opinion of a review's importance will be reflected in the efficiency of your operation. Your own personality, then, is a major factor in any review.

FORMS

Rather than constantly recopying or redesigning the forms you use for working up information or preparing your own information reports, you need a set of good, usable and practical forms. They will almost certainly be developed by trial and error over a period of time. The blank forms at the end of this book will help you improve the forms you already have. By comparing the blank forms to the corresponding accounting examples, you will learn how forms are developed. Then you can apply your new understanding to your own business and come up with your own forms. When you have created a form that you know will be useful for a long time, type it up and have copies made.

Only you can decide what kind of records you want to have in your construction business. But keep in mind that your accountant depends on your books and records for just about everything he does for you. Sales, property, payroll, and income tax returns, financial statements, and reports all are based on the information you provide. A complete system will provide all the information your accountant needs, at the time it is most needed. The most valuable service your accountant provides, his advice, can only be based on what he sees in your books and records. Make sure that what he sees is an accurate reflection of what your business has done.

DOUBLE-ENTRY BOOKKEEPING

The most common modern bookkeeping method is called *double-entry bookkeeping*. It is doubtful that any alternative method can be found to provide the same degree of bookkeeping control yet remain as simple to use. Training and experience are required to fully master double-entry bookkeeping, and you shouldn't expect to become sidetracked into a bookkeeping career just to manage your operation's affairs. But it is always smart and a good business practice to know every aspect of your operation, especially the books, records, and management systems. The following overview of the double-entry method should help you better understand the rest of this book and give you enough bookkeeping knowledge to let you maintain your own books and records.

Double-entry bookkeeping is so called because every transaction requires two entries—one debit and one credit. Two entries provide an important control throughout the bookkeeping documents that is not available with any other system. A debit is an entry made to the left side of an account, and a credit is an entry made to the right side. Debits and credits will always equal each other. The total of all accounts will be zero if the general ledger is completed accurately. In other words, the debits (pluses) and the credits (minuses) of account transactions will cancel each other out when the accounts are added up.

Some accounts normally have debit balances, and some normally have credit balances. A complete summary of typical accounts and their balance types is listed below.

ASSETS:	Debit	Credit
Cash	XXX	
Accounts Receivable	XXX	
Bad Debt Reserve		XXX
Fixed Assets	XXX	
Depreciation Reserve		XXX
LIABILITIES:		
Accounts Payable		XXX
Taxes Payable		XXX
Notes Payable		XXX
NET WORTH:		
Capital Stock		XXX
Retained Earnings		XXX
SALES:		
Income Accounts		XXX
COST OF GOODS SOLD:		
Materials	XXX	
Labor	XXX	
EXPENSES:		
Operating Expense Accounts	XXX	

Journal entries are created to change the balances in various accounts. A good example of a journal entry is shown below. Here, a builder wants to show the effect of bank charges on his bank account. He must reduce his cash account by the amount of the monthly charge. He must also increase his expense account for bank charges.

	Debit	Credit
Bank Charge Expense	$4.82	
Cash in Bank		$4.82

The debit will increase an expense account (expense accounts are usually debit-balance accounts) and decrease the balance of cash (also a debit-balance account).

All entries to the general ledger—the record that summarizes all business operations—are made from journal-type entries consisting of equal debits and credits. These entries are readily noticed in a general journal. But entries from a cash receipts journal require a different format. One column represents the total of cash received. This becomes a debit (increase) to the cash account. Other columns, dividing the total into appropriate categories, are for credits to income, accounts receivable, and sales tax accounts. Debits and credits will always balance to zero.

A cash disbursements journal (cash paid out) has several columns, as well. One column represents the total decreases to cash. This total decrease is entered on the cash account as a credit. Distribute the total decreases to various cost and expense accounts as debits. Again, the total of all debits and credits in the cash disbursement journal will be zero.

Figure 1-1 shows the traditional style of ledger page. Debits are posted on the left side and credits on the right. The balance in an account—the net remaining amount when credits are subtracted from debits—is written under the last amount posted on one side or other of the account. Which side this is depends upon which side has the highest total amounts. If the debits are greater than the credits, record the net total on the debit (left) side. If credits are greater, record the net balance on the credit (right) side. Recording on the right side would result in a "credit balance," or negative account total. Income accounts, liabilities, and net worth accounts usually contain net credit balances.

Figure 1-2 shows the result of posting to a single general ledger account from several sources. The T-account—so named because of the letter "T" created by the lines—is used here to demonstrate this. The T-account is simply a way of separating debits and credits on a worksheet and can be used to estimate the results of business before financial statements have been prepared. In the figure, the Cash in Bank account has received three entries. The debit (increase) came from the cash receipts journal and reflects all cash collected in one month. The decreases come from the cash disbursement journal (showing the total of checks issued in one month) and the general journal (where bank service charges were recorded).

Ledger Sheet

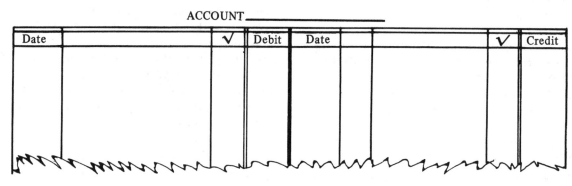

Figure 1-1

'T' Account

```
              CASH IN BANK
              Debit      Credit
Balance Forward   1,486.80
Cash Receipts    22,401.60
Cash Disbursements          21,844.62
General Journal                  4.82
Balance Forward   2,038.96
```

Figure 1-2

Several of the controls found in double-entry bookkeeping can be explained using the T-account summary as an example. First, the beginning balance and the ending balance of a cash account let you know what is in the bank. These balances should agree with reconciled totals from the monthly statement the bank sends. Any errors or omissions by either the bank or your bookkeeper become apparent when the reconciled balance is compared to the general ledger account balance.

The general ledger will be *out of balance* if a posting error has been made in it or if the math has not been carried through correctly. In other words, the grand total of all debits less all credits will not be zero. An error has been made when this is the case, and the error must be found before financial statements can be prepared. No confidence can be placed in a general ledger that is out of balance. You or the bookkeeper can be certain that all posting has been done correctly only when the general ledger credits and debits have been added to get zero. Of course, entries may be posted to the wrong accounts; diligence is still required in the posting process. No method can ever eliminate human error, but double-entry bookkeeping does provide the best controls against mathematical errors. Throughout this book, many additional controls will be discussed in relation to the many aspects of double-entry accounting and bookkeeping procedures.

The general ledger should not be burdened with large numbers of detail accounts, analysis and budgetary controls, or excessive detail of any kind. Secondary or subsidiary ledgers and accounts should be kept apart from the general ledger for these control details.

Financial statements are reports which express information summarized from the general ledger in a format designed to pass on information rather than to control a large number of detailed transactions. Financial statements reveal the status, progress, and control you have exercised, and the degree of business that has been generated as a result. These statements are only as accurate, informative, and concise as are the books of original entry (journals) and the summary document (the general ledger).

The balance sheet is one of three main financial statements. It is a listing of properties (assets), debts (liabilities), and ownership value (net worth). The term "balance sheet" refers to: (1) the listing of account balances (or net totals when credits are subtracted from debits), and (2) proof that all debits listed on the balance sheet are equal to all credits. The basic formula which defines the balance sheet is:

Assets = Liabilities *plus* Net Worth

There are two parts to the balance sheet: 1) listing of all assets and a total of those accounts, and 2) a listing of liabilities and net worth and a total of those accounts. Both totals (assets and liabilities/net worth) will be the *same number*. Therefore, the two sides will balance.

The income statement, also called Statement of Profit and Loss or the Summary of Operation, lists income, direct costs, operating expenses, and profits. While the balance sheet lists the balances in asset, liability, and net worth accounts *as of* a specific date, the income statement reports the results of operations within a defined period (such as one month, one quarter, or one year).

The statement of cash flows, also called the Statement of Provisions and Uses of Funds, summarizes the management of cash during a specified period of time. It shows how the net profits of a business have been used—payment of liabilities, buying new assets, or distribution to owners—in the course of business. It also shows the source of funds. Funds may come from operations (profits), from the sale of assets, or from outside loans. Owners may contribute additional capital, resulting in an increase of funds. The cash flow statement is useful in judging the degree of management control you exercise over your funds.

The double-entry system forces the user to perfect his entries before issuing accurate statements. If the general ledger is out of balance, the interrelationship of the three statements will be off. These relationships are summarized below.

1. Assets must equal the total of liabilities and net worth shown on the balance sheet.

2. The net income or loss must equal the increase or decrease to retained earnings shown as part of the net worth on the balance sheet.

3. The increase or decrease in funds shown on the statement of cash flows must equal the change in current asset and liability accounts. Current assets less current liabilities at the end of the period, *less* current assets less current liabilities at the beginning of the period, must equal the increase or decrease in funds.

The balance sheet reflects the value of a business in terms of properties and debts. Those properties owned by the builder are subject to debts related to them. At the same time, some portion of those assets are truly owned. If debts, or liabilities, represent 60% of assets, and net worth 40%, a different conclusion must be drawn than if liabilities represent 95% of assets, and owner equity only five percent.

This comparison of net worth to liabilities is called a *ratio*. Ratios are explained in more detail in Chapter 23. The financial statements provide information for several useful ratios needed by bankers considering loan applications and by the builder himself in assessing his own financial condition. Accurate financial statements and the ratios drawn from them help control costs and expenses, inventory levels, cash flow and accounts receivable.

The income statement shows total sales, direct costs, operating expenses, and profits within a period of time. Prepare this statement on a comparison basis to get the most value from it. Last year's income compared to this year's will reveal the good and bad trends in the business and the increase or decrease in volume, expenses, and profits.

The statement of cash flows, which is often wrongly excluded from the set of financial statements, is in many ways the most valuable report you can have. Control of cash in the construction business is crucial to success. Many builders have problems in this area, and poorly designed cash procedures or lack of cash flow planning altogether can cripple an operation.

To build an accounting system that provides accurate financial statements, you need a series of documents. At the same time, the entire procedure should run smoothly. Otherwise, the office bogs down in its own paperwork, and nothing is completed in time to be of use. And the accuracy and availability of information suffers in a poorly designed accounting system.

Figure 1-3 summarizes the relationship of the various reports, controls, and documents contained in an accounting system. Understanding the workings of these stages will help the builder to appreciate the value of his double-entry system, and the resulting value of information he can have in a well-organized and thought-out plan of operation.

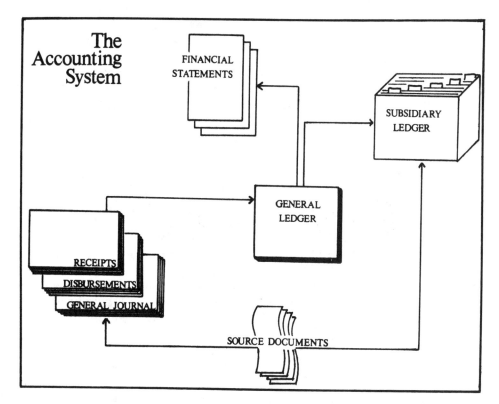

Figure 1-3

The general ledger must be a summary document containing the minimum amount of information needed to produce maximum information in financial statements.

The cash receipts and cash disbursements journals supporting the general ledger must contain math control features, report information legibly, and provide the detail needed for analysis of accounts. The general journal must also be controllable and concise.

Many secondary systems are required to control the several general ledger accounts. Among these is a subsidiary control for accounts receivable. Most builders have a large volume of business transacted on credit. You must be able to render accurate statements to customers, control the level of receivables and bad debts, and set up reserves for future losses.

Source documents—invoices, purchase orders, receipts—must be handled efficiently to support the accounting journals and ledgers. Filing systems and methods for labeling and identifying source documents must be practical and time-saving.

The apparent complexity of the double-entry system will clear up if you can relate to the results rather than the mechanical processes. However, you need to recognize the need for controls over those processes in order to produce concise reports. Without controls, no system of keeping books will yield a dependable, accurate, and profitable financial statement. And no profitable decisions can be made based on misleading information.

The terminology of accounting can be confusing and misleading. Unfortunately, most accountants and others who work intimately in the accounting field must, from necessity, communicate in the common language of accounting. This alienates the unexposed builder and keeps him from fully understanding the meaning of his own books and records.

This book has been designed to introduce concepts to the builder that are useful in a practical way. Specialized terminology that could confuse and mislead the reader has been avoided whenever possible. You are interested in understanding accounting from a builder's point of view, not from an accountant's. This text is not intended to train you to become a full-charge bookkeeper, but rather a well-rounded and successful builder whose skills and knowledge are enhanced profitably by his accounting system.

Section I

Sales and Accounts Receivable

Accounting Methods • Cash and Charge Sales
Managing Receivables • Bad Debt Procedures
Sales Records and Cash Budgeting
Sales Planning • Planning For Profits

Chapter 2
Accounting Methods

The way you treat sales and when you recognize income is the foundation of any accounting system you adopt. The decision as to when income should be booked as income creates problems both for the builder and his accountant because contracts are performed in anywhere from one day to several months, or even years. Solving this key problem requires that you define what you mean by income. Just exactly how much did you earn last month? There is no single answer unless you specify the accounting method you used. There are two accepted ways to record income: the percentage of completion method and the completed contract method.

PERCENTAGE OF COMPLETION METHOD

For most businesses, recording income is a simple matter. Income is generally recognized when work is performed, and recorded when statements are mailed. In other cases income is recorded when payments are received.

But contractors sometimes receive payments for work not yet performed, or only partially completed. In such a case, the portion of income above the degree of completion is said to be *unearned* in the accounting sense. It is not earned and should not be recorded as income until the work has been done.

When you receive cash for work that is only partially performed, you have to separate out the unearned portion of income and assign it to a special account. That sum will be recognized as income later when it is earned. Then it is removed from the special account. This is known as *percentage of completion* accounting because you recognize as income only that portion of money received or charges billed that is for work completed. For example, suppose you receive a $6,000 payment which represents one-half of the total contract price. You know that the job is only 40% complete. The entry would be broken down as follows:

Total payment (50% of contract amount)	$6,000
Percentage of completion ($12,000 x 40%), earned income	4,800
Balance, unearned income	$1,200

Accountants handle situations like this by "deferring" the unearned portion of the income. Since this treatment is always based on your estimate of the degree of a job's completion, you need adequate records and a consistent method for determining the status of work in progress.

The differences between *billed* or *paid* and true income can work in the opposite way as well. You will recognize this as a more common situation. For example, a job is estimated to be 40% complete. The builder has billed out or been paid less than 40%. The difference has to be taken into account or accrued. This way you recognize income before payment has been received or even billed. It is money that truly has been earned. Suppose a contract is for $12,000 and is 40% complete. The following shows an accrual of earned income:

Total contract of $12,000 - 40% complete, earned income	$4,800
Amount paid to date	4,320
Accrued income	$ 480

While your accountant should make these adjustments in preparing a financial statement, he will depend entirely on your books and records. To do a complete job, your records must have complete data. Remember also to document the procedures you use to arrive at your estimate of the degree of completion. Naturally, you have to record both sales and accounts receivable as well as progress on partially completed jobs.

Most contractors and builders use the percentage of completion method for recording income. It is the only acceptable method for builders who have fairly large jobs in progress at the end of any tax period.

Under percentage of completion accounting, your books reflect the effect of receiving payment at any time other than when that payment is earned. Occasionally you are paid in advance. Your contract may specify that you are allowed an advance to fund expenses which will be required in the early stages of the project. It is much more common for your customers to withhold a portion of each progress billing as *retainage*. This serves as guarantee against unforeseen problems with final completion of a project and is an incentive for timely completion. Retainages are normally paid upon approved completion and acceptance.

In all cases of payment above or below the estimated percentage of completion, you accrue the deficit and defer the excess income.

The advantage of the percentage of completion method is that income and its related costs, expenses, and profits are recognized and reported as the job progresses. The term *recognized* refers to the accounting treatment of an amount and is another way of saying that the amount is acknowledged as being earned in the current period though money is neither billed nor received.

This progressive recognition provides you with a series of financial statements that give a fairly accurate picture of your financial position that includes expected profits and losses. Since many jobs have a life of quite a few months, this kind of financial statement is more meaningful to you as a businessman. Your financial statements can be based on any consistent accounting method. But some notation should be included if the reporting method differs from that used for income tax purposes.

The disadvantage of the percentage of completion method of accounting is that all income, costs, expenses, and profits are recognized as the result of estimates. When you say that a job is 40% complete, this is an estimate based on your judgment. It is not an exact degree of completion. But if you have current information and use a good estimating system consistently, you probably are making good estimates of completion. With any accounting method, all income and costs will eventually be recorded. Percentage of completion accounting simply gives you a more accurate picture of what is happening while you still have time to correct it.

The estimates you use to determine the degree of completion must be based on reliable and current information. Your construction cost estimate will be a good guide to the value of work completed. If your cost estimate has been revised, this fact must be considered in your completion estimate. The most reliable indicator of the degree of a job's completion is your total of project-to-date costs and your *current estimate* of the cost of the work yet to be done. If your original estimate were used to compute the degree of completion, the accuracy of your completion estimate would vary as your actual total costs have varied. If you now estimate that a job will result in costs of $575,000 don't use the original estimated cost of $500,000 in computing the percentage of completion.

Find the current percentage of completion by adding the *total costs to date* to the *estimated costs* to complete and divide that total into the *costs to date*:

$$\frac{\text{Costs to date}}{\text{Costs to date plus estimated costs of completion}} = \text{Percentage of completion}$$

COMPLETED CONTRACT METHOD

Recognizing income by the completed contract accounting method has the advantage of exactness. There is no need to estimate the degree of completion as all income is recognized only when (and not until) the job is 100% complete. This method is almost always used for small jobs - one-time work, material sales, isolated minor contracts, and any other job of short duration. Payment for such work is usually in one installment. There is no need to estimate the degree of completion because you are paid or send a bill for the work at about the same time you know exactly what the total cost was.

The Tax Reform Act of 1986 created a new form of accounting called the *percentage of completion-capitalized cost method*. Under this rule, you are allowed to use completed contract accounting for only 60% of the total income and expenses of a job. The balance must be reported under the straight percentage of completion method.

You can use completed contract accounting only when you'll complete a job within two years from contract date, and only if your annual gross receipts are under $10 million.

Tax reform has complicated the decision to use one accounting method or the other. You're required to choose one method and stick with it unless you apply to the Internal Revenue Service for permission to make a change. But the new rules restrict your freedom to choose a more advantageous method in many circumstances.

The selection of one method over another must be based on the restrictions of current tax law. Within that framework, seek professional tax advice in the selection of the best method — and the one that you can best manage. It should always be your goal to keep it as simple as possible. But given the complexities of the tax laws, that won't always be easy.

Suppose you bid on a contract which you estimate will yield $150,000 net profit after three years of work. Under the completed contract method you recognize all income, costs, expenses, and profits upon completion of the job. During the life of that contract, you won't recognize any of the work done or payments received. Your financial statements will be prepared as though that income didn't exist. All the income and costs will be held in reserve accounts set up to delay recognition of income. This completed contract method has the obvious disadvantage of not recognizing some very real progressive profits over the three-year life of the job. Another serious problem with the completed contract method is that it does not allow you to produce periodic financial statements that reflect the status of your company. Neither does it provide you with reliable data that you could use to improve profitability during the life of a large contract. Cash control, budgeting expenses, inventory handling, and profitability problems must be spotted early to make prompt corrective action possible. Completed contract accounting provides you only with final results. When careful analysis is needed under this method you must create the data yourself; the current financial statement won't help.

THE TWO METHODS COMPARED

Assume that you have several moderate and ongoing large contracts and you are reporting income and profits under the percentage of completion method. Quite likely the profitability of the jobs varies widely. The differences are so great that each job is unique. This might be a good description of just about any builder's current situation. A builder using the percentage of completion method will usually find that changes in profit on any single job are absorbed over several monthly accounting periods. By using this method, there is less chance that a single accounting statement will reflect all of a major change in profitability.

Now, assuming the same circumstances, what would the result be using the completed contract method? The year of completion on the larger contracts will show a major change in volume. This makes monthly or even annual comparisons useless and distorts the true financial picture of the business.

We will illustrate the weakness of the completed contract method of accounting by looking at a builder with three large contracts. The degree of completion on each of these jobs will be assumed to be fairly even throughout the life of each contract. All three contracts began at the same time and were to last for thirty-six months. Assume that the total expected profit on all three contracts is $450,000.

The first year's net profits would be:

$150,000 (percentage of completion)
-0- (completed contract)

The second year's net profits would be:

$150,000 (percentage of completion)
-0- (completed contract)

The third year's net profits would be:

$150,000 (percentage of completion)
$450,000 (completed contract)

This is a simplified example, but the message should be clear: deferring income, as you must under the completed contract method, means that you lose data that would be valuable for comparisons. Furthermore, financial statements are not realistic under the completed contract method. Note that net profit (in the example above) is thrown into one tax year. This last point is especially important to remember when you have several relatively small contracts and an occasional huge one because progessive tax rates favor spreading profits evenly over several years.

The completed contract method can be useful in some circumstances. The first step in choosing the right method for you is to check your financial data for the last few years. Discuss the choices with your accountant.

Contractors find the completed contract method acceptable in some situations. Some builders adopted this method when they first began work and never bothered to change in spite of considerable growth. A small builder specializing in residential repair work can use the completed contract method because nearly all work is of short duration. But if the type of contract work he performs begins to include longer duration jobs, he may have good reason to adopt the percentage of completion method. Progressive recognition of profits from larger contracts results in a continuing flow of profits and allows for easier growth and planning.

But many builders continue to use the same accounting method though their type of work has changed. Periodically review your accounting procedures.

COMBINED ACCOUNTING METHODS

Most contractors find it necessary to be involved in more than one type of work. Most builders take on both long duration contracts and miscellaneous small jobs. Combining accounting methods in your business can cause record-keeping problems. Consult with your accountant so that the general ledger he sets up for your business gives you an easily controlled method that lets you maintain separate records for your classifications that must be treated separately. This chapter should give you all the background you need to help your accountant make the important choices.

Like many contractors, you wish to use the percentage of completion method for long duration contracts and keep your other jobs on a completed contract basis. The remainder of this chapter explains how to use one or both of these accounting methods in your business.

In the first part of this chapter, accruals and deferrals of income were discussed. Both the percentage of completion and completed contract methods require accruals and deferrals. Under percentage of completion accounting, all retainages are accrued as income. Any prepaid income (paid before it is earned) must be deferred. Under completed contract accounting, all income is deferred until a job is 100% complete. But, upon completion, it may be that a portion of the total contract has not been paid at the time your books are closed. In that case, a portion of your income would be accrued.

In the interest of (1) including both kinds of contract treatments in one set of books, and (2) still keeping these matters apart, you will need two accrual accounts and two deferral accounts: (1) earned income account (on percentage of completion accounts), (2) earned income account (on completed contract accounts), (3) deferred income account, (on percentage of completion accounts), and (4) deferred income account, (on completed contract accounts).

The earned income accounts are assets. The deferred income accounts are liabilities. The series of accounting adjustments which can flow through these accounts is complex. Remember that the purpose of all the adjustments is to give you a true picture of earned income. The complexity of the entries can be sorted out by your accountant. You have to be concerned only with having an efficient procedure to support the adjustments.

Two things happen when these adjustments are taken in your general ledger: first, billings are adjusted to reflect earned income, and, second, the accrual (earned income) and deferral accounts are adjusted to reflect the current balances of those assets and liabilities. Figure 2-1 shows the flow of adjustments for income, starting with Total Current Month's Billings

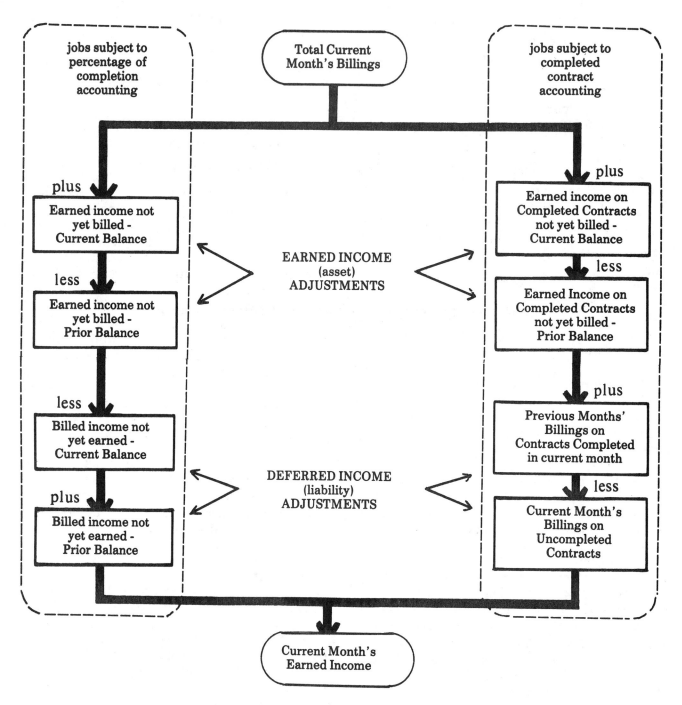

Flow of income and adjustments
Income
Figure 2-1

and arriving at Current Month's Earned Income. This schedule summarizes the entries your accountant will make and shows the effect of those entries on income. You should also know how to handle the distribution of costs and income to these accounts. For each account (Earned Income - Asset, and Deferred Income - Liability), the current balance replaces the prior balance each month. Your deposit records and check ledger support the changes and summarize the information so that you can go back to audit the account balances when necessary.

Figure 2-2 shows the flow of adjustments for the earned income (asset) account. Each month's entry consists of replacing the prior balances with the current month's balance. It may seem at first that it would be simpler to add

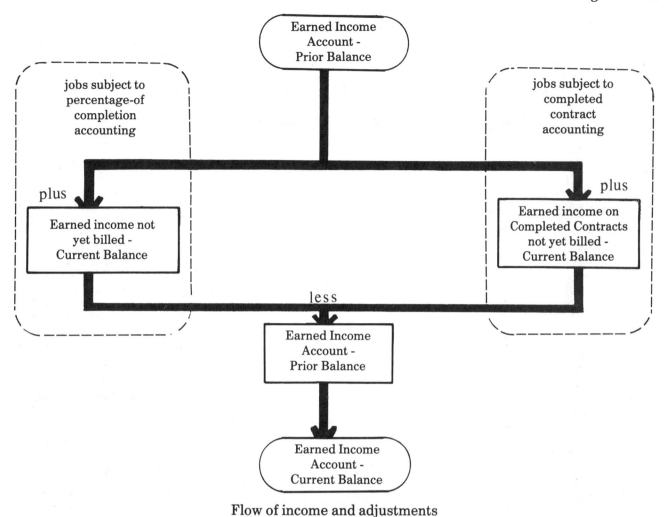

Flow of income and adjustments
earned income (asset)
Figure 2-2

the differences each month to update the balance, rather than completely replace it. But it is more reasonable and easier to compute the year-to-date totals each month. In addition, each month's worksheet computes the total account balance. Otherwise, some part of the account balance could be active for months. In those circumstances, it would be more difficult to account for the items that make up the account balance.

Figure 2-3 shows the flow of adjustments for the deferred income (liability) account. Like the asset account, the current balance replaces the prior balance. The one exception to this procedure is for the "prior month's billings on contracts completed in current month," within the completed contract accounting section. Some prior months' billings will remain in this account for several months, pending completion of a contract. The verification of the account balance is detailed in Figure 2-4 and will be explained later in this chapter. Complete your monthly worksheets using this or some similar method and you won't have difficulty verifying the detail of accrual or deferral accounts.

The processes explained above should help you understand the flow of income, as well as the offsetting effect on the accrual and deferral accounts. It is not necessary to maintain your own general ledger or become proficient in accounting theory. But you need a good working knowledge of this income procedure and its control to communicate with your accountant. A general understanding of why these accounts are required is enough. Later chapters will explain how totals are posted to the accounts described.

Figures 2-4 and 2-5 are examples of how sales are broken into categories so an account-

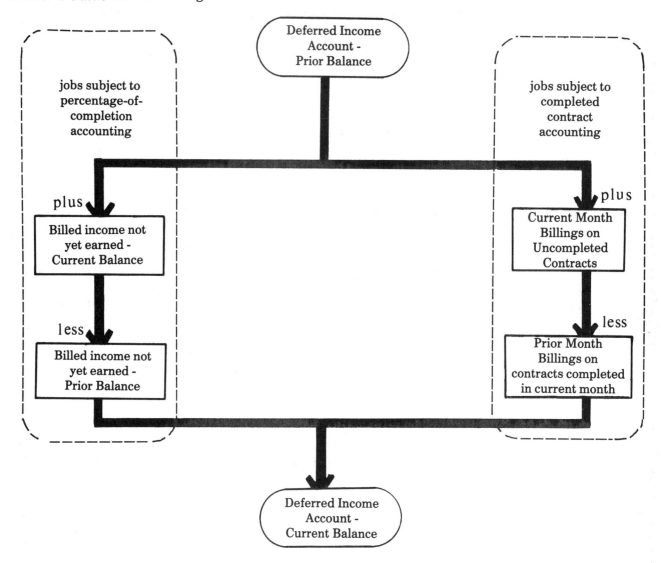

Flow of income and adjustments
deferred income (liability)
Figure 2-3

ant can make adjusting entries quickly and easily. From figures like those shown in Figure 2-5 your accountant can prepare his journal entries in a matter of seconds.

Figure 2-4 is a degree of job completion worksheet for jobs that are being accounted for under the percentage of completion method. A portion of your income from these contracts is for completed work; a portion is for uncompleted work. Since income on uncompleted work is not reported, this part must be separated out as deferred income. The income in the current month for completed contracts (plus previous months' billings for jobs completed this month) is considered earned income. As mentioned above, it is possible for a previous month's billings to remain as unearned if those contracts are not yet complete.

Note that this worksheet also adds the current month's percentage of completion billings to the completed contract billings to arrive at a control total. This total has to match "total billings" on the Sales Journal which is discussed in the next chapter.

The totals of Figure 2-4 are transferred to Figure 2-5, the Earned Income Summary. Here, figures developed under both accounting methods are summarized on one worksheet. The current month income, accruals and deferrals appear here. Figure 2-5 documents the changes in asset, liability, and income accounts — an

Accounting Methods

JOHNSON CONSTRUCTION COMPANY
WORKSHEET, DEGREE OF JOB COMPLETION
March, 31, 19___

Name	(1) Amount Previously Billed	(2) Amount Billed This Month	(3) Total Billings (1) + (2)	(4) Contracts Completed	(5) Contracts Not Completed
Black's Landscape	$ -	$1,600.00	$1,600.00	$1,600.00	$ -
M. Brown	-	482.11	482.11	-	482.11
C. Carlson	-	641.27	641.27	641.27	-
L. Carlson	-	173.05	173.05	173.05	-
Harvey Contracting	-	1,695.00	1,695.00	1,695.00	-
Hauser & Sons	-	25.00	25.00	25.00	-
J & J Paints	-	45.00	45.00	45.00	-
Jim's Contracting	-	360.00	360.00	360.00	-
L. Jones	-	74.00	74.00	74.00	-
LBN Brothers	-	355.00	355.00	355.00	-
K. Middleton	-	145.37	145.37	-	145.37
Midfield School	-	82.17	82.17	82.17	-
Mitchell & Sons	1,500.00	7.50	1,507.50	1,507.50	-
B. Ottman	-	169.34	169.34	169.34	-
L. Peterson	-	477.55	477.55	-	477.55
Regal Apartments	-	1,250.00	1,250.00	-	1,250.00
Riley Landscape	-	3,900.80	3,900.80	3,900.80	-
J. Smith, Builder	-	525.00	525.00	525.00	-
D. Thomas	-	706.08	706.08	-	706.08
Triumph Construction	3,000.00	-	3,000.00	3,000.00	-
Walton's Plastering	600.00	903.00	1,503.00	600.00	903.00
Woodman	-	425.00	425.00	425.00	-
J. Woods	-	295.80	295.80	295.80	-
Total	5,100.00	14,338.04	19,438.04		
Percentage-of-completion billings		13,819.86			
Total Billings (Control total)		28,157.90			
TOTAL INCOME EARNED IN CURRENT MONTH				15,473.93	
DEFERRED INCOME					3,964.11

Figure 2-4

important feature that supports the general ledger summary accounts.

Too often, reports that are meant to support general ledger accounts and totals do not provide enough information. You can't prove an entry unless you can trace it to its origin. Reports like the Earned Income Summary, prepared the same way each month, make tracing entries very easy.

The work accounted for under the percentage of completion method is listed by the total contract value and percentage of completion to find the earned income. This total, earned income, represents a project to date total, not just the current month's total. It is assumed that the percentage of completion is already computed when this schedule is prepared. This percentage is multiplied by the total contract amount to find the earned income. The excess of earned income over billings to date is the current (project-to-date) amount to accrue or defer.

The data on work accounted for under the completed contract method is listed in summary form, as supported by the detail on Figure 2-4. The net result, in this example $1,600.53 in income earned over amounts billed, is made up of:

JOHNSON CONSTRUCTION COMPANY
EARNED INCOME
As of March 31, 19____

Job Description	Total Contract	% of Completion	Earned Income	Billings To Date	Earned Less Billings
Percentage Of Completion:					
Anderson	$ 65,000	30%	$19,500.00	$19,500.00	$ -
Carey	50,000	45	22,500.00	24,000.00	(1,500.00)
Norwich	6,000	10	600.00	415.36	184.64
Rayne	15,000	20	3,000.00	2,880.00	120.00
Widmark	75,000	35	26,250.00	29,090.00	(2,840.00)
Windsor	280,000	80	224,000.00	219,500.00	4,500.00
Completed Contracts:					
Uncompleted contracts billed this month			-	3,964.11	(3,964.11)
100% completed contracts:					
billed this month			10,373.93	10,373.93	-
billed in prior months			5,100.00	-	5,100.00
TOTALS			$311,323.93	$309,723.40	$ 1,600.53
Summary:					
Percentage Of Completion					
Earned Income over amounts billed					$ 4,804.64
Billings over amounts Earned					(4,340.00)
Completed Contracts Method					
Deferred Income					(3,964.11)
Reversal, Prior months' deferrals					5,100.00
Total current month adjustments to income					$ 1,600.53

Figure 2-5

1) Earned income (asset): $4,804.64 (Percentage of completion)

2) Deferred income (liability): $5,100.00 (Completed contract — reversal of prior deferrals)

   $3,964.11 (Completed contract — current month's additional deferrals)

   $4,340.00 (Percentage of completion)

The percentage of completion project to date earned income asset ($4,804.64) and deferred income liability ($4,340.00) are increases to those accounts. Additional entries need to be made by your accountant to reverse completely the prior balances of the earned income asset account and the deferred income liability account.

It would also be necessary to reverse the earned income account for completed contracts if there were prior balances. There are none in this example, just as there are no instances of current month earned income on completed contracts.

As far as completed contract accounting is concerned, the $5,100.00 reversal of prior deferrals is a reversing entry, which is in this example the entire balance of that account.

We have seen how a number of asset, liability, and income accounts are put to use. But one of the problems with these accounts is that all of the totals include sales tax. Tax should not be considered a part of income. Collected sales taxes are properly treated as a liability, to be paid to the taxing agency. Since the intention here is to keep reports and worksheets as

Accounting Methods

uncomplicated as possible, the question of sales tax hasn't been raised. You can break the tax out by recomputing the sales totals without tax. But in this example, the income section of the general ledger includes an account to reverse sales taxes from income totals recorded. Below is a summary of the asset, liability, and income accounts involved in sales accounting:

Asset Accounts
 Accrued earned income accounted for by the percentage of completion method

 Accrued earned income accounted for by the completed contract method

Liability Accounts
 Deferred income accounted for by the percentage of completion method

 Deferred income accounted for by the completed contract method

 Sales taxes payable

Income Accounts
 Gross income accounted for by the percentage of completion method

 Gross income accounted for by the completed contract method

 Adjustment to gross income for sales taxes

You may want to analyze sales (earned income or billed) by type of income. Another type of analysis might be to compare by periods the types of billings — taxable sales, labor charges, resale, and so forth. The general ledger is set up to handle both accounting methods and keep totals separately and doesn't supply this detail. This type of analysis would best be based on billed totals, rather than adjusted earned income. Any analysis required can be done from the company's sales journal. The format for this is explained in the next chapter.

If your records are to be kept under one accounting method only, the procedure described above will work for you. The general techniques described here are more important than the specific treatment of sales tax, the headings of accounts, or the design of worksheets. Remember that your goal is to develop only the information you need as efficiently as possible, and at the same time provide the documentation and support that is required. Recording sales on worksheets such as shown will preserve valuable information for your own future use.

An alternative accrual method for builders is called the "cash" method. Under cash accounting entries are made and reported only when cash changes hands. This method was not discussed because it does not allow for good comparative analysis. While tax laws allow cash accounting, it is used very seldom because it does not reflect many important transactions. For instance, large cash receipts in any one month distort income, as do unusually small cash receipts. The true income for those receipts is actually earned as the job progresses (percentage of completion) or on the date that the job is fully completed (completed contract), not when payment is received. Only the simplest businesses can use cash accounting methods.

Chapter 3

Cash and Charge Sales

There are two kinds of sales you account for: cash sales and charge sales. Cash sales includes any payment received for work. It must be a payment and not a loan. It can be an advance payment and may be by check or currency. Charge sales are all sales which are made on credit and for which you will issue monthly statements of charges.

It should be mentioned at this point that a special problem of control arises when cash sales are in currency. Some customers do pay all or some part of their bills in currency. You may be tempted to use this currency to pay your own bills. The best rule to follow is: always deposit all income in your bank account. In some cases it might be very convenient to hold the cash in your office. Many builders need a fairly large office cash fund. But paying bills out of sales receipts can cause serious problems. Your bank deposits should exactly equal all cash taken in during the month. This is the best possible way to document your receipts. When cash is withheld from deposits, your cash income record is distorted.

The bank's monthly statement provides an excellent summary of sales if all sales are deposited each month. Comparing these summaries from month to month gives you the ability to prepare good cash forecasts.

If you need a petty cash fund, you should establish one by writing a check to petty cash from your general fund. This procedure and its controls are discussed in Chapter 17.

RECORDING CASH SALES

Cash sales should be listed by date so that you can create summaries for general ledger entries, fill out sales tax forms and make up other needed reports. The procedure should be simple so that your workload is minimized. Figure 3-1 shows a cash sales journal. It breaks out taxable sales and sales tax. In this example, the taxable sales are subject to a 6% tax. This tax is collected by businesses and paid to the state's taxing agency at the end of each month or quarter. For that reason, the tax you collect is a liability and not a sale. So the tax must be listed in its own column. The tax total will be entered in a different section of the general ledger than the sales (income) figures.

There is no tax on sales to other businesses that will resell to consumers. Neither are labor charges taxable. When listed as they are in Figure 3-1, totals for taxable sales, sales tax, labor, and resale are readily available in both total and detail. Notice that the last day of each week is underlined. This is done to separate the periods for bank statement reconciliation.

Cash and Charge Sales

	JOHNSON CONSTRUCTION COMPANY CASH SALES JOURNAL MARCH, 19___						
Date	Description	Invoice #	Taxable Sales	Sales Tax	Labor	Resale	Total
3- 1	J. Smith, Builder	301				200.00	200.00
3- 2	M. Brown	303	35.00	2.10	21.00		58.10
3- 3	J. Woods	304	180.00	10.80	105.00		295.80
3- 4	L. Carlson	309	92.50	5.55	75.00		173.05
3- 5	B. Ottman	310	37.11	2.23	130.00		169.34
3- 8	L. Jones	316	50.00	3.00	21.00		74.00
3- 8	J. Smith, Builder	317				150.00	150.00
3-11	C. Carlson	319	34.80	2.09	21.00		57.89
3-12	J. Smith, Builder	324				175.00	175.00
3-17	D. Thomas	327	32.16	1.93	42.00		76.09
3-19	C. Carlson	328	173.00	10.38	400.00		583.38
3-22	K. Middleton	329	42.80	2.57	100.00		145.37
3-22	Woodman Pool Svc.	332				425.00	425.00
3-23	L. Peterson	336	35.00	2.10	150.00		187.10
3-23	Midfield School	338	37.90	2.27	42.00		82.17
3-26	D. Thomas	342	280.00	16.80	300.00		596.80
3-26	M. Brown	343	301.90	18.11	104.00		424.01
3-30	D. Thomas	345	11.50	.69	21.00		33.19
3-31	L. Peterson	348	132.50	7.95	150.00		290.45
	TOTAL		$1,476.17	88.57	1,682.00	950.00	$4,196.74

Summary of bank deposits:

Date	Amount
3- 8	$ 896.29
3-15	456.89
3-22	659.47
3-29	1,860.45
4- 1	323.64
Total	$4,196.74

Figure 3-1

Bank deposits in this example are made every Monday and at the end of each month. In this way, the total receipts are equal to the total bank deposits.

Bank deposits are summarized at the bottom of Figure 3-1. This verifies that the total receipts were deposited this month. While this exact procedure is not appropriate for every builder, it is a good example of a controlled, simple method that accounts for sales and verifies deposits against receipts. An additional advantage is the arithmetic control. The detail column should add across to the right total and down to the bottom total. All worksheets should have similar controls. It doesn't take long to do this and it can save hours when trying to locate an error.

CHARGE SALES

Wouldn't it be nice to be paid in full all the time, as shown in Figure 3-1? Unfortunately, very few businesses can operate without allowing credit to most customers. Figure 3-1 is only an example of what a sales journal could look like. But it's unlikely that your business runs on a strictly cash basis. If you do any work on a credit basis, you will have to use an accounting method that records both cash and charge sales.

In the following example, Johnson Construction has between twenty and thirty charges per month, and about twenty cash sales. Figures 3-2 and 3-3, the Charge Sales Journal and Combined Sales Journal, are merely listings of

Builder's Guide To Accounting

JOHNSON CONSTRUCTION COMPANY
COMBINED SALES JOURNAL
MARCH, 19____

Date	Description	Invoice	Taxable Sales	Sales Tax	Labor	Resale	Rentals	Hauling	Freight Charges	Finance Charges	Total Sales	Cash Receipts	Discounts Allowed
3- 1	J. Smith, Bldr	301				200.00					200.00	200.00	
2	M. Brown	303	35.00	2.10	21.00						58.10	58.10	
3	J. Woods	304	180.00	10.80	105.00						295.80	295.80	
4	L. Carlson	309	92.50	5.55	75.00						173.05	173.05	
5	B. Ortman	310	37.11	2.23	130.00						169.34	169.34	42.00
8	CHARGES		5,630.00	337.80	1,890.00	1,600.00		60.00		7.50	9,525.30	6,943.00	
3- 8	L. Jones	316	50.00	3.00	21.00						74.00	74.00	
8	J. Smith, Bldr	317				150.00					150.00	150.00	
11	C. Carlson	319	34.80	2.09	21.00						57.89	57.89	
12	J. Smith, Bldr	324				175.00					175.00	175.00	
15	CHARGES		4,012.60	240.76	1,995.00	900.00				3.00	7,151.36	3,794.56	
3-17	D. Thomas	327	32.16	1.93	42.00						76.09	76.09	
19	C. Carlson	328	173.00	10.38	400.00						583.38	583.38	
22	CHARGES						180.00		5.00		185.00	3,400.00	
3-22	K. Middleton	329	42.80	2.57	100.00						145.37	145.37	
22	Woodman	332				425.00					425.00	425.00	
23	L. Peterson	336	35.00	2.10	150.00						187.10	187.10	
23	Midfield School	338	37.90	2.27	42.00						82.17	82.17	
26	D. Thomas	342	280.00	16.80	300.00						596.80	596.80	
29	M. Brown	343	301.90	18.11	104.00						424.01	424.01	
29	CHARGES		3,775.00	226.50	2,100.00	360.00		120.00			6,581.50	5,985.66	8.00
3-30	D. Thomas	345	11.50	.69	21.00						33.19	33.19	
31	CHARGES		300.00	18.00				200.00			518.00	3,200.00	
31	L. Peterson	348	132.50	7.95	150.00						290.45	290.45	
	TOTAL		15,193.77	911.63	7,667.00	3,810.00	180.00	380.00	5.00	10.50	23,961.16	23,323.22	
											4,196.74	4,196.74	
											28,157.90	27,519.96	50.00

1 CHARGE SALES
2 CASH SALES

Summary of Bank Deposits:

3- 8	7,839.29
3-15	4,251.45
3-17	4,059.47
3-29	7,846.11
4- 1	3,523.64
total	27,519.96

Figure 3-2

Cash and Charge Sales

JOHNSON CONSTRUCTION COMPANY
CHARGE SALES JOURNAL
MARCH, 19___

Date	Description	Invoice	Taxable Sales	Sales Tax	Labor	Resale	Rentals	Hauling	Freight	Finance Charges	Total Charge Sales	Received On Account	Discounts Allowed
3- 1	Riley Landscape	302	1,480.00	88.80							1,568.80		
1	Mitchell & Sons	305								7.50	7.50		
1	J. Carey	306	2,400.00	144.00	1,554.00						4,098.00		
1	Riley Landscape	paid										2,058.00	42.00
2	J. Carey	paid										2,328.00	
2	LBN Brothers	paid										245.00	
3	LBN Brothers	307						60.00			60.00		
3	void	308									-		
4	Harvey Contracting	paid										1,412.00	
5	Black's Landscape	paid										900.00	
5	Harvey Contracting	311	1,150.00	69.00	336.00						1,555.00		
5	Harvey Contracting	312	600.00	36.00							636.00		
5	Riley Landscape	313				1,600.00					1,600.00		
5	Black's Landscape												
	TOTAL WEEK		5,630.00	337.80	1,890.00	1,600.00				7.50	9,525.30	6,943.00	42.00
3- 8	Riley Landscape	314	200.00	12.00							212.00		
8	B. Anderson	315	1,900.00	114.00	1,785.00						3,799.00		
9	B. Anderson	paid										2,815.00	
9	Riley Landscape	318	400.00	24.00							424.00		
11	N. Norwich	320	312.60	18.76	84.00						415.36		
11	N. Norwich	paid										159.66	
12	Walter's Plastering	321				900.00				3.00	903.00		
12	void	322									-		
12	L. Widmark	323	1,200.00	72.00	126.00						1,398.00		
12	L. Widmark											819.90	
	TOTAL WEEK		4,012.60	240.76	1,995.00	900.00				3.00	7,151.36	3,794.56	
3-15	void	324									-		
15	Harvey Contracting	325					135.00		5.00		140.00		
15	Hiram & Hiram	paid										3,400.00	
17	J & J Paints	326					45.00				45.00		
	TOTAL WEEK						180.00		5.00		185.00	3,400.00	
3-22	Triumph Construction	paid										3,000.00	
22	C. Rayne	330	600.00	36.00	84.00						720.00		
22	LBN Brothers	331						55.00			55.00		
23	C. Rayne	paid										1,619.66	
23	void	333									-		
23	G. Windsor	334	1,375.00	82.50	1,932.00						3,389.50		
23	G. Windsor	paid										974.00	
23	Riley Landscape	335	1,000.00	60.00							1,060.00		
24	Regal Apartments	337						40.00			40.00		
25	Jim's Contracting	339	800.00	48.00	84.00						932.00		
25	Jim's Contracting	paid										392.00	8.00
25	void	340									-		
26	Hauser & Sons	341						25.00			25.00		
	TOTAL WEEK	344	3,775.00	226.50	2,100.00	360.00		120.00			6,581.50	5,985.66	8.00
3-29	Karston Landscape	paid										3,200.00	
29	Regal Apartments	346	300.00	18.00				200.00			518.00		
30	LBN Brothers	347									200.00		
	TOTAL WEEK		300.00	18.00				200.00			518.00	3,200.00	
	TOTAL MONTH		13,717.60	823.06	5,985.00	2,860.00	180.00	380.00	5.00	10.50	23,961.16	23,323.22	50.00

Figure 3-3

```
JOHNSON CONSTRUCTION COMPANY - Duplicate Deposit Slip

DEPOSIT SLIP
list checks by ABA number
CURRENCY        131.00   } L. Jones $74.00;  C. Carlson $57.89
COIN               .89
CHECKS:
  90-123        150.00     J. Smith, Builder
  3-51          175.00     J. Smith, Builder
  1-876       2,815.00     B. Anderson
  90-234        159.66     N. Norwich
  8-42          819.90     L. Widmark

                           DEPOSIT DATE:

                              March 15, 19

TOTAL         4,251.45
```

Figure 3-4

invoices. This is all that any sales journal is. A manual (hand-prepared) system like Figures 3-2 and 3-3 would be adequate for a builder with a sales volume like this.

The combined sales journal (Figure 3-2) has the same information as the cash sales journal and includes weekly sales in summary from the Charge Sales Journal. The detail of these is maintained separately in the Charge Sales Journal, Figure 3-3.

Keep the charges separate as shown so you can make use of a worksheet or pegboard method for billing. Even if you change methods some time in the future, you wouldn't need to change the whole procedure. The journals can be tailored to your business to meet your requirements.

Cash sales are, naturally, collected at the time of the sale. But for charge sales, additional controls are needed to account for payments.

Figure 3-2 is similar to Figure 3-1 but includes an additional section for payments on account. Columns have also been added for rentals, hauling, freight, finance charges, and discounts allowed. These represent a broad range of possible types of income for one builder or contractor. Every category may not be required by every builder. But some of the categories will apply to every building operation.

Finance charges are amounts added to overdue balances. In this example, one-half of one percent per month has been used as a finance charge: Past due balance times .005 = finance charge. Discounts are often given on trade accounts. They should be computed and allowed selectively. Most builders allow a 2% discount to certain types of trade accounts only, rarely to all customers.

The listing of deposits has been omitted from Figure 3-2 because a summary of charge sales deposits would seldom be a complete record of

JOHNSON CONSTRUCTION COMPANY
INCOME BY JOB
MARCH, 19___

Date	Total	Anderson	Carey	Norwich	Rayne	Widmark	Windsor	All others
3- 1 to 3- 5	10,421.59		4,098.00					6,323.59
3- 8 to 3-12	7,608.25	3,799.00		415.36		1,398.00		1,995.89
3-15 to 3-19	844.47							844.47
3-22 to 3-26	8,441.95				720.00		3,389.50	4,332.45
3-29 to 3-31	841.64							841.64
Total income by job	28,157.90	3,799.00	4,098.00	415.36	720.00	1,398.00	3,389.50	14,338.04
% Complete		30%	45%	10%	20%	35%	80%	xxx

Figure 3-5

company sales. Figure 3-3 includes a summary of bank deposits because it is the one record which contains all income and payments on account. Note that all cash sales are listed both in the total column and in the cash receipts column. This last column, cash receipts, includes all cash sales and any payments on charge sale accounts.

Figure 3-4 is the duplicate deposit slip which documents your cash receipts and the deposit total. Each item is listed by customer to back up the entry you record in that customer's accounts receivable file. This type of dual recording is not time-consuming, is easy to store, and can even replace the summary of bank deposits on the combined sales journal.

Many banks can supply you with a deposit book which is set up to handle this kind of record. Place the original deposit slip and a piece of carbon paper over a page of the record book to create a duplicate for your file. You can then write the customer information to the side of the carbon copy amounts.

The builder in the example has six contracts in progress. All of these he accounts for under the percentage of completion method. He keeps a file of correspondence, contracts, amendments, invoice copies, and other papers for each job. Payments and charges summaries are kept too, as will be explained later.

Figure 3-5 is a summary of billings for one month, including the degree of completion for each job. This kind of report lets a builder see at a glance his monthly billings and is valuable for comparing the accounts billed for several months. This report is also useful in timing the purchase of bulk materials that will be used on several jobs. Another use is in planning for the payment of subcontractors.

In Figure 3-5 the major jobs are listed by the amounts billed. "All other" summarizes billings on jobs accounted for by the completed contract method. This summary provides the same value for comparative purposes and cash budgeting. The illustration is by no means a complete record for job analysis. It does not even show earned income, merely the billings. Its usefulness is limited to cash planning and comparing monthly billings. By-the-job accounting is discussed in Chapters 6 and 27.

ACCOUNTS RECEIVABLE

The most annoying bookkeeping task for most builders is controlling their customer credit accounts. Both a sales journal and monthly billings are required for each account. A good accounts receivable system includes customer account cards which are balanced each month with the general ledger entry for the whole account. Balanced entries, back-up systems, and monthly statements can all be had

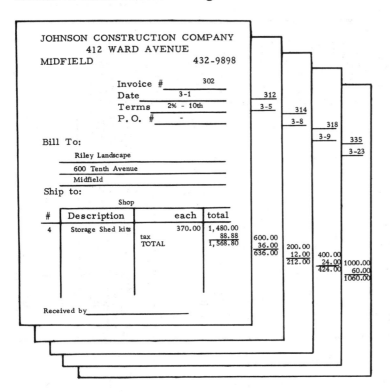

Figure 3-6

in a single entry procedure. This saves time and improves the quality of your work. The real value of a good receivables method is in the variety of information it provides—information to bill customers, control collections, make general ledger entries, and build historical files. You can have all of these features through exhaustive duplicated effort. You can also have a simple system that gives full information for all requirements by designing a procedure that fits your needs exactly. Any builder who thinks about it should want only the latter.

Figure 3-6 shows manually prepared invoices and a billing statement. Invoices should show enough detail so your customer can identify the material and services and pay according to your terms. For example, invoice 302 (indicating third month, second invoice issued) contains the information that payment is expected by the tenth, and a two percent discount will be allowed if that payment date is met. The charge is clearly identified by its description and includes the unit price.

The statement is a summary of a customer's account, including the balance forward, current invoices, payments on account, and the new balance due. In the example on the right side of Figure 3-6, Riley Landscape paid on the first of March and took a $42.00 (2%) discount.

Charge customers often accumulate balances over a period of more than one month. One very effective way to keep track of amounts due is to maintain a file of cards, one for each customer. The cards should duplicate information on the invoices (as summarized on monthly statements) and provide a summary of each account's history. Figure 3-7 is an example of a customer account card.

Customer account cards should include the name, address, telephone, and discount terms, if any. A file of cards like this is a control system for accounts receivable. At the front of the file should be a card labeled *Control*. The control card might look like Figure 3-8. The control card lists each month's charges, credits, and balance. You should be able to add up the ending balances of all cards in the accounts receivable file and come up with this summary total. The charges and credits columns are posted from the sales journal. If the control card total is correct, it will match the total of all charges, credits and the balance of individual customer cards. If the total of all customer cards does not agree with the control card total, there

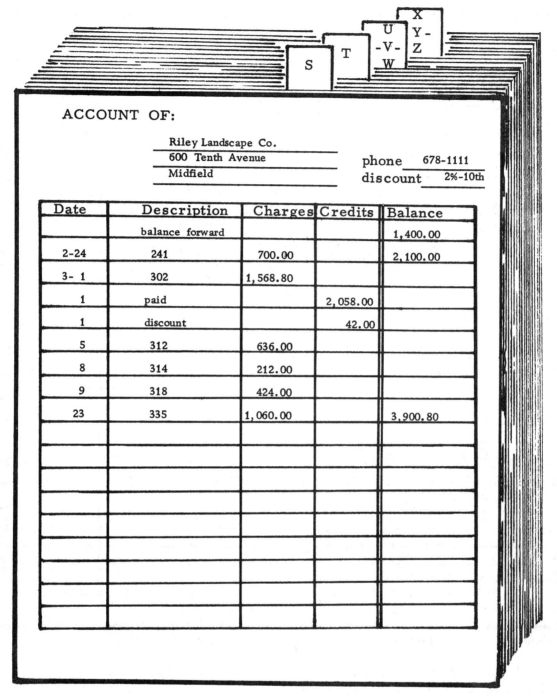

Figure 3-7

is a math error on one of the customer cards. This means that you are sending out a monthly statement which does not add up correctly. Don't let your customers lose confidence in your statements. Make every statement correct.

From the customer ledger cards you can determine which accounts are past due. While you may already know where the problems are, this will help you look at the whole picture. Doing this can give you an early warning of problems: are your overdue accounts beginning to represent a larger portion of your total receivables? The answer can be very important to you, especially in planning your future cash flow. Your aim is to maximize your available cash and minimize collection problems.

It is important for you to analyze your total receivables carefully. But retainages should be excluded from the analysis. They can not be considered as past due even though your

CONTROL

	Charges	Credits	Balance
Balance Forward			26,933.10
Month of January, 19	24,105.90		
		22,459.38	28,579.62
Month of February, 19	24,610.40		
		24,766.80	28,423.22
Month of March, 19	23,961.16		
		23,323.22	29,061.16
Month of April, 19			
Month of May, 19			
Month of June, 19			
Month of July, 19			
Month of August, 19			
Month of September, 19			
Month of October, 19			
Month of November, 19			
Month of December, 19			

Figure 3-8

billings could go back several months. Under the terms of most construction contracts you consent to the withholding of retainage. Exclude the totals for retainage. If retainages or similar withheld amounts become a major part of your receivables, set up a separate receivable file for retainage. You might then maintain customer ledger cards for retainage just as you do for accounts receivable that are presently collectible. Just as with accounts receivable, maintain a control card in front of the file. You have to keep good track of retainage to make complete cash forecasts. This means you must know when the cash will be available. You can estimate the date for receiving retainage by estimating the date of job completion and acceptance.

Control the collection of accounts receivable with an *aging list* as in Figure 3-9. Builders who are in close contact with their customers and operational facts of business may know quite well who owes them money. But an aging list can still provide needed information and a written record of slow paying accounts.

An aging list indicates what percentage of total receivables (excluding retainages) is current and what percentage of total receivables

JOHNSON CONSTRUCTION COMPANY
AGING OF ACCOUNTS RECEIVABLE
MARCH 31, 19___

NAME - ACCOUNT	0-30 days	31-60 days	Over 90	Total
Anderson, B.	3,799.00			3,799.00
Black's Landscape	1,600.00			1,600.00
Carey, J.	4,098.00			4,098.00
Harvey Contr.	1,695.00			1,695.00
Hauser & Sons	25.00			25.00
Jim's Contr.	360.00			360.00
J & J Paints	45.00			45.00
LBN Brothers	355.00			355.00
Mitchell & Sons	7.50	1,500.00		1,507.50
Norwich, N.	415.36			415.36
Rayne, C.	720.00			720.00
Regal Apts.	1,250.00			1,250.00
Riley Landscape	3,900.80			3,900.80
Triumph Const.			3,000.00	3,000.00
Walton's Plast.	903.00	600.00		1,503.00
Widmark, L.	1,398.00			1,398.00
Windsor, G.	3,389.50			3,389.50
Total	23,961.16	2,100.00	3,000.00	29,061.16

Figure 3-9

is past due. Compare aging lists for several months to see any change in the status of your receivables. This monthly search for emerging problems allows you to tighten up control of collections before the problem gets out of hand.

Refer to Figure 3-9. Triumph Construction has a balance more than three months old. However, no finance charges are being added. This could be the result of a special arrangement, or just a trade courtesy. It is sometimes the custom between businesses who have had relationships over several years to waive the normal finance charges, especially by mutual agreement. With a few exceptions, all accounts in the example are current (less than 30 days). This might not be totally realistic. Most builders have a few late-pay or problem accounts.

Because the number of monthly billings shown in Figure 3-9 is small, analysis is not difficult. The past due accounts are few and can be individually evaluated. But builders with a higher monthly volume should actively screen their accounts and analyze them on a regular basis. Because there are only a few accounts, monthly statements can be typed for each customer. Detail records can then be maintained either directly from the invoices or from the sales journal. Under this system, each entry is made three times—on statements, on account cards, and in the sales journal. While this is not a time-consuming chore, its weakness is the duplication of effort. The more times the same numbers are recorded, the more chance there is for error.

Even for a small building business like Johnson Construction, a *pegboard system* is advisable. This involves combining all three entries into one step, using either carbon or chemically treated paper to transfer information to the three records. These records are piled one on the other and held together in place by lining up a series of uniform holes. The three forms fit over a series of pegs on the pegboard. Forms

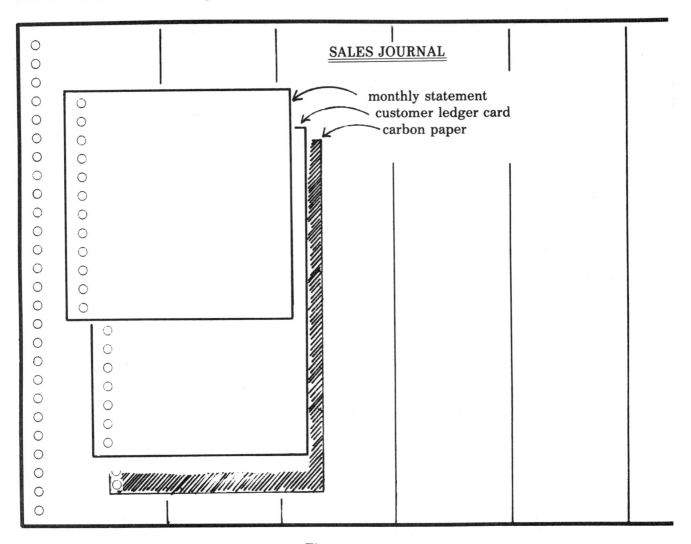

Figure 3-10

and board are available at many larger stationery stores and there are many manufacturers to choose from.

The value of a pegboard system is its control features and the effort saved. Entries are made on statements, account cards and in the sales journal all at one time. You will also see that a pegboard system makes control easier. The disadvantage of the system is that the equipment and forms needed are expensive. Forms such as your monthly statements, journals, and even checks (if you decide to use a pegboard system for your payroll or checking account) must be ordered through a company which specializes in the brand of pegboard you use. The blank forms and special printings are more expensive than a straight manual system. You may consider this disadvantage to be minor when you see the efficiency and flexibility your system provides.

The method of overlaying pegboard forms is illustrated in Figure 3-10. The bottom sheet is the sales journal, a blank form which can be set up in any manner you please. You provide your own headings. Over this is placed a sheet of carbon paper. On top of that place the customer ledger card. The top sheet is the monthly statement. This should consist of an original (which you mail to each customer with the original invoice) and a duplicate (which you keep with the duplicate invoice) for your file. The back of each statement is carbonized so that entries are transferred to customer ledger cards.

Figure 3-11 should make clear some of the control features of a pegboard system. As in the combined sales journal, you should be able to add all charge columns you assign and come up with the total of the charges. All credit columns you assign (probably only cash and discounts) must total the credits column. Verify the math

SALES JOURNAL

date	description	charges	credits	balance	previous balance

total of column, previous balance + charges − credits = balance

Figure 3-11

on individual statements by adding the figures in the previous balance column. At the bottom of each page, add the totals of the charges, credits, balance, and previous balance.

Check the math accuracy on statements by applying the following formula:

previous balance + charges − credits = current balance

If you check the balances with this formula you will always be confident that the statements you send out are correct.

Figure 3-11 shows a typical sales journal form. Most brands of forms designed for pegboard use, specifically for sales, have these headings filled in.

There may come a time when no manual system is adequate to handle your volume of transactions. When that happens, you may wish to contact a computerized billing service or buy your own small computer. The advantages of computerized bookkeeping are discussed in Appendix C.

You may want to have your accountant make a monthly general ledger entry for sales. The information for this entry comes from your combined sales journal. But since the general ledger in this example breaks down sales by the percentage of completion and those computed by the completed contract accounting methods, the totals must be taken from the earned income computation worksheets discussed in Chapter 2.

Below is a series of entries that the accountant for Johnson Construction Company would make at the end of March. No adjustments have been made for previous balances of earned income and deferred income (under the percentage of completion accounting method) on the assumption that the previous balances of those accounts were both zero.

ACCOUNTING ENTRIES FOR SALES

	Debits	Credits
Cash	$27,519.96	
Accounts receivable	23,961.16	$23,323.22

	Debits	Credits
Gross income accounted for by the percentage of completion method		13,819.86
Gross income accounted for by the completed contract method		14,338.04
Adjustment entries:		
Accrued earned income accounted for by the percentage of completion method	4,804.64	
Deferred income accounted for by the percentage of completion method		4,340.00
Deferred income accounted for by the completed contract method	5,100.00	3,964.11
Gross income accounted for by the percentage of completion method		464.64
Gross income accounted for by the completed contract method		1,135.89
Adjustment to gross income for sales taxes included	911.63	
Sales taxes payable		911.63

This summarizes the company's entire sales for one month, including the change in accounts receivable, cash receipts, sales tax liability, earned income, accrued and deferred income, and adjustments.

Together, your bookkeeping records and a flexible general ledger provide a good trail for audit verification and detailed historical information. But you can keep records that show the details of your sales and any adjustments in a few easily-prepared worksheets.

Figure 3-12 summarizes the flow of information on bookkeeping records related to sales and accounts receivable for the entire month. Note that the largest amount of paperwork occurs early in the flow of information. For example, you're likely to have more billing invoices than statements. That's because each customer's invoices will be combined on a single monthly statement. As the information flows through the system, it becomes more and more summarized. By the time final entries are made in the general ledger, they should represent a summary of a large number of detailed transactions.

This is a critical point. The detailed entries that are documented on billing invoices, checks, and represented by receipts and vouchers, might have a high volume — too high to manage efficiently. The process of recording information in the books achieves two things. First, it organizes and documents income and expenses. Second, by summarizing them in uniform journals, it makes the large volume of transactions manageable.

Cash and Charge Sales

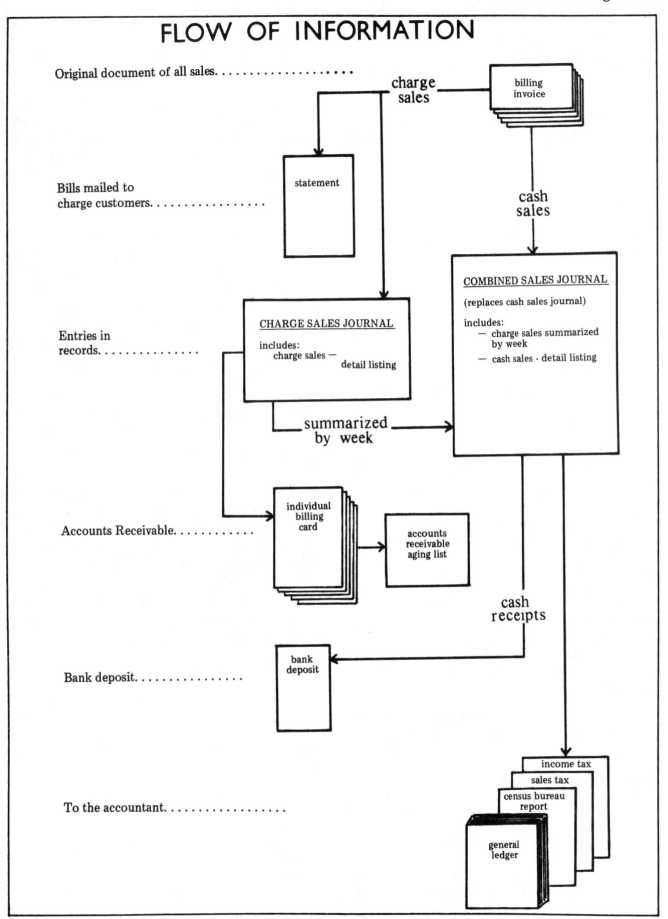

Figure 3-12

41

Chapter 4

Managing Receivables

The controls discussed in the last chapter illustrate procedures for listing and recording receivables. A good record of receivables is essential. But note that your record of receivables can provide valuable information on the payment record of your accounts receivable, important information for your business planning. Forecasting cash flow requires knowledge of probable receipts in the future. What receipts will be can be forecast from your sales and accounts receivable records.

Look beyond the functional value of carefully maintained customer accounts and accurate statements. You might want to know what cash receipts will be one year from now. You could make an estimate from your projected sales and compare that to your current sales and cash receipts. Things may be a lot different in twelve months. You will have completed many of your current projects and probably will be dealing with a new group of customers. But your best indication of the future is your past experience.

Your changing collection pattern can be forecast from the trend of your collections. Every business has trends. You acquire new business and complete old business. You are constantly moving into new sets of circumstances that never stay the same for long. But you can establish a dependable pattern for forecasting the future. This trend should be favorable in the collections area because, through experience, you learn to steer away from customers and work that result in overdue balances or bad debts. If you are not managing your receivables correctly, your trend may be negative. You may be handling work that will cause collection problems.

Trends, good or bad, are difficult to see because they hide within your changes in volume. If you experience a gradual increase in volume and cash profits over a six-month period, the slow receipt of your accounts could be obscured by a higher cash flow. You may be accumulating hard-to-collect or uncollectible accounts as volume grows. Watching the trend will alert you to major problems before they develop.

The most common way to watch the trend your accounts receivable is taking is to follow the number of days the average account is outstanding. This information for one month is useless by itself. But when several months are compared, the trend can be seen. Your goal is to keep nearly all accounts within a reasonable collection period. Regardless of your sales volume, keep your average period for collections

Managing Receivables

JOHNSON CONSTRUCTION COMPANY

Accounts Receivable Average Length Study

Month	(1) * Average Receivables	(2) Twelve Months' Charge Sales	(3) Column 1 divided by column 2	(4) Average Days of outstanding receivable (column 3 times 365)
January	$ 22,692	$ 250,611	.091	33.2
February	18,276	246,003	.074	27.0
March	20,112	280,492	.072	26.3
April	20,040	291,802	.069	25.2
May	17,486	276,014	.063	23.0
June				
July				
August				
September				
October				
November				
December				

* Excluding retainages

Figure 4-1

down. When volume increases, many builders tend to allow receivables to stay open longer. This is because a larger balance of cash is available as profits increases. The manager feels less need to collect his accounts promptly. On a smaller volume the money is needed for daily operations, so limits are enforced strictly. Naturally, this relaxation of controls can be dangerous.

An excellent way to keep tab on collections is to prepare an accounts receivable average length study, as in Figure 4-1. This indicates whether your effectiveness as a collector is stable, increasing, or decreasing. If volume is increasing and the average number of days that receivables are out is increasing, you need to tighten collection procedures. If they are decreasing or remaining about the same, your collection effectiveness is not changing. Of course, you may still wish to improve on your collection procedures. This depends on what standard you establish as acceptable for outstanding receivables.

Look at Figure 4-1. The figures in the study exclude retainage. For this analysis, retainage should be omitted from your total of accounts receivable. Retainage often represents such a significant percentage of your total receivables that it would distort your collection analysis. As mentioned in Chapter 3, retainage should be controlled separately from other accounts receivable.

The analysis in Figure 4-1 is based on an average of twelve months' receivables and charge sales. Sudden jumps in the average days of outstanding receivables can result from sudden changes in average volume (such as a large job done on credit) or from early or delayed significant payments. To minimize the effect of slow or heavy work periods, figures for twelve months are used.

Column one of Figure 4-1 is the sum of accounts receivable (excluding retainages) in each month for the last twelve months. For example, the February average is the sum of the twelve previous months from last March through this February, divided by twelve. Each month the average is updated by dropping the figure from the oldest month and adding the total for the current month:

$$\frac{\text{Sum of accounts receivable for 12 months ending with the latest month}}{12} = \text{Period of Average Receivables}$$

Column two is computed in the same way but is the sum of charge sales for the last twelve months (not divided by twelve). It represents the total charge sales for the past twelve months (excluding retainage):

$$\text{Total charge sales for 12 months ending with the latest month} = \text{Twelve Months' Charge Sales}$$

Dividing the Average Receivables by the Twelve Months' Charge Sales gives the factors listed in column three. These, multiplied by 365 (days in one year), give the Average Days of Outstanding Receivables as listed in column four.

$$\left(\frac{\text{Average Receivables}}{\text{Twelve Months' Charge Sales}}\right) \times 365 = \text{Average Days of Outstanding Receivables}$$

The management at Johnson Construction can see from this average length study that since January 1 the average length (in days) of outstanding receivable accounts has decreased from 33.2 days to 23.0 days. This is especially gratifying because the twelve-month sales are higher as of May 31 ($276,014) than they were as of January 31 ($250,611). A combination of earlier average payments, good internal controls, and fewer old accounts will result in a favorable trend like this.

The value of this kind of report is in the trend, not the numbers. You might have a longer average number of days of outstanding accounts. The actual figure depends on the type of business you handle and the policy you have established about due dates. For every builder a reasonable average could be different. Whatever average you find typical for your business, the trend of that average is something you must know.

REVERSING AN UNFAVORABLE TREND

Suppose you prepare a comparative average length study and discover that your average outstanding days are increasing. How do you reverse the trend?

Many builders are reluctant to pressure their clients for amounts due on account. If your collections are slower now than they were six months ago, try a more active approach. Call the customers and ask for a commitment to pay by a certain date. As you are probably aware, the longer you allow a balance to remain outstanding, the less collectible it becomes. Your accounts receivable aging list will reflect your collection success in the number of delinquent accounts you accumulate.

Along with staying on top of existing accounts goes prior qualification of new charge customers. Ask for credit references and make one or two phone calls before allowing credit. A responsible company or person isn't going to object to this. Your professionalism points out that you are a sensible businessman who wants to deal with those who pay their bills promptly.

Figure 4-2 is a credit application. Make it a company policy not to extend credit unless there is an application on file. The application can be completed in a few moments and is a good way to screen credit requests.

Occasionally a balance can not be collected. Perhaps the individual can not be located, or he claims to have no money. At this point it might be advisable to contact a local collection agency. If you have been active in trying to collect, chances are you won't be helped by an agency. But their fees are often based on collections, so it wouldn't hurt to give them a try.

If you have not been active in collecting the debt, a professional collection agency may be able to help. Be sure to get in touch with a reputable firm. Many states have governmental agencies that control debt collectors. It is a good idea to call the state agency involved (if your state has one) and determine that the collection company you are considering is approved or licensed.

```
                JOHNSON CONSTRUCTION COMPANY
                    APPLICATION FOR CREDIT


    This is an application for a line of credit for:
                    business_____
                    personal_____

    Name_____
    Address_____
    City_____Telephone_____

    How long in this area?_____Name of Bank____

    Please list three credit references (locally):
        Name                            Telephone
        _____
        _____
        _____

    Please describe the materials or services you would like to buy
    from Johnson Construction Company:
    _____
    _____
    _____
    _____

    Will materials be for resale?   Yes_____No_____
            (If yes, you will be asked to fill out a
             resale certificate, as required by law)

    Please estimate the total amount you would like to purchase on
    a credit arrangement?   $_____

    It is the company policy of Johnson Construction Company  to
    check credit references prior to agreeing to a credit arrange-
    ment.  This will be accomplished in as short a time as can be.
    We look forward to doing business with you.

                                        The Management
```

Figure 4-2

You or your accountant will make entries into and out of your accounts receivable account for the sales transactions described in Chapter 3, bad debts, and occasionally for accounts assigned to a collection agency.

Most builders don't have enough bad debt volume or enough collection agency assignments to justify separate accounting for these categories. Avoid distorting your analysis of accounts receivable by listing accounts assigned to a collection agency in a separate column. Exclude these figures from monthly totals. This will avoid distorting the trend of your receivables and collections.

The decision to write off an account should be made by you based on the advice of your

```
                JOHNSON CONSTRUCTION COMPANY


Dear Customer:

    Your account is now past due.  Your prompt payment
will be appreciated.

    Please forward your payment in full for $_____,
as soon as possible.

    Cordially,



JOHNSON CONSTRUCTION COMPANY
```

Figure 4-3

accountant. Document the collection efforts you make on accounts from the time they become past due. This verification process supports the decision to write off an account. In the case of accounts large enough to justify lawsuits, a log of your collection attempts could become a valuable record.

An efficient way to keep a log would be by day, rather than one page for each account. Most of your customers pay in full even if they are a few weeks late. A lined blank book is adequate for your collection log. Columns can be drawn and labeled as you please. A typical collection log might include the following:

Date
Account and amount due
Contact (phone, personal, etc.)
Spoke to:
Discussion: (payment arrangement, date promised, etc.)

Any other information you feel would be of value to you in collecting the debt should be included. Debtors who get highly personalized treatment often respond more favorably to an appeal. Let your debtor know that he is not just a number in your books. You are interested in his account and plan to stay interested until it is settled. Record in your log the promises and excuses you receive and use this information if you have to pursue the matter further. The log can contain a minimum amount of information and still be useful for those accounts that fall delinquent. Since most of your customers will eventually pay, try to avoid duplicating the records that exist on your customer account cards.

It might be helpful to design a series of collection notes as an aid in collection. Your communication should be progressively stronger as accounts become progressively late. If you send collection letters to customers, have an attorney review them to make sure you are within the law. Many states regulate collection procedures to protect the public against excessively aggressive practices. The standards vary from state to state.

The information you compile on your

Managing Receivables

JOHNSON CONSTRUCTION COMPANY

Dear Customer:

Your account is past due in the amount of $_____.

You were expected to make payment under the terms which were specified when we agreed to extend a line of credit to you.

Please call our offices and let us know when your payment will be sent in. If there is a problem, we will be glad to discuss an agreed schedule of installment payments.

Take care of this today to avoid collection actions on our part.

JOHNSON CONSTRUCTION COMPANY

Figure 4-4

accounts receivable is essential to your planning and forecasting. One of the most critical tasks you face as a builder is the cash budgeting required to prepare for major financial commitments. The fact that you bid major jobs implies that you are prepared to buy or lease the equipment and machinery that may be needed. And developing an adequate labor force requires that you keep good men busy and meet regular payrolls even when you have not collected for the work done. Insurance, a variety of taxes, licensing and permit fees, office expenses, union welfare payments, and normal business overhead are obligations that require regular cash outlays.

All of these financial obligations must be anticipated. Any growth you expect in the future will be based on your ability to generate enough working capital. Having that cash when it is needed depends on the effectiveness of your collections and your planning.

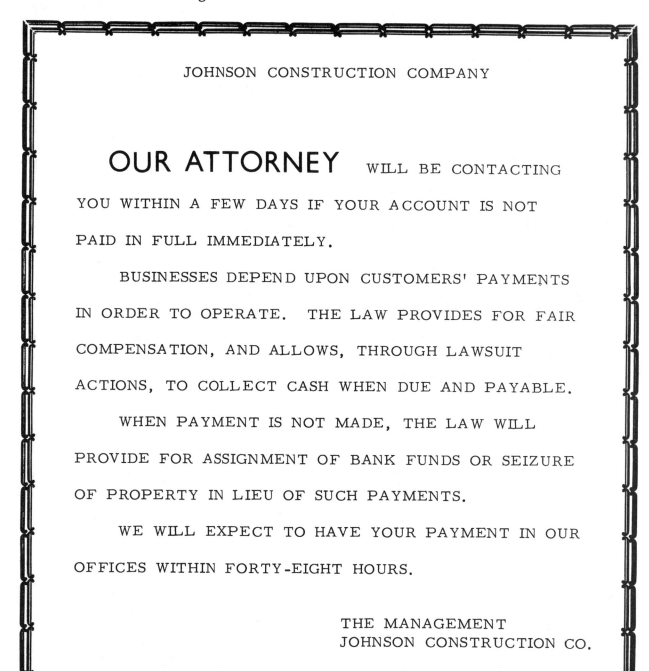

Figure 4-5

An average length of collections study as in Figure 4-1 should take you only a few minutes per month. This, along with an accounts receivable aging list, will help you control your receivables. If your customer ledger is maintained properly, if you qualify your accounts before doing work on credit, if you prepare and mail monthly statements regularly, you will very likely receive payments as agreed.

It may help to offer incentive discounts, such as "2% - 10 days, net 30." This is a standard arrangement for shops that offer discounts and seems to bring payments in quickly. The obvious disadvantage is the 2% loss. But you should also consider the cost of doing without those funds for thirty days or more. It may be worth much more than the 2% discount to collect 98% of your receivables promptly.

Chapter 5

Bad Debt Procedures

If your business has few or no bad debts you can skip through this chapter fairly quickly. Unfortunately, most builders do have fairly significant bad debts in spite of the protection most states offer under lien laws. Bad debts can be a big profit drain on your business. Handling them correctly should help you understand how to reduce your bad debt losses.

When you sell on credit, the money you are owed is income and is *booked* or recorded as an account receivable. But you know that you won't collect all of this money. From time to time you will have bad debts. Your accounting system must give you a way of recording the actual bad debts and analyzing the trend of collections so you can budget expenses and plan for the future.

When an account becomes uncollectible, the income you had recorded must be reversed. This process, called "writing off" an account, must be done in both the general ledger and on the customer's account ledger. You could write off the account on the sales journal itself, but most accountants would recommend that you make up a special journal entry rather than record the amount uncollectible as a sale in reverse. This process, adjusting by journal entry, distinguishes the write-off as a special transaction. The value of this is that you have isolated the entry as an unusual occurrence.

WHEN DOES AN ACCOUNT BECOME UNCOLLECTIBLE?

In general, you should write off a bad debt only after exhausting all efforts to collect. This means telephoning if the amount is significant until you are convinced the customer can not or will not pay. In some cases you will receive notification that the customer has filed for bankruptcy. At that point the law provides a procedure for filing a claim for payment. Further efforts at collection may violate federal law.

For tax purposes, bad debts can be recognized only when a debtor's business is abandoned and worthless, when he can not be contacted, or when he passes away and has a worthless estate. So for a true bad debt, the possibility of collection must be just about nonexistent. If a debtor dissolves his business but has net assets, his liabilities may be paid. If he moves away, you must try to contact him before you can claim that he has disappeared. And if he passes away, the debt is transferred to his estate.

JOURNAL ENTRIES FOR BAD DEBTS

If you do a substantial volume of business on credit, your bad debt losses will be fairly predictable. Your general ledger should include an allowance for losses anticipated. The accounting method used to provide for future bad debt losses is the establishment of a reserve for those losses. Regardless of how the amount of these losses is estimated, the provision for bad debts (a reduction of your total accounts receivable) is an estimate of bad debts for the month. The allowance is balanced by a similar amount in bad debt expense. The entry in your journal would look like this:

	Debit	Credit
Bad debt expense	xxx	
Reserve for bad debts		xxx

This is a *provision*, an amount which you estimate will cover future losses. When an account becomes uncollectible, the bad debt is recorded. Since you have already provided for this and other losses, and have recorded the expense, the entry to record the bad debt is:

	Debit	Credit
Reserve for bad debts	xxx	
Accounts receivable		xxx

This entry reduces the reserve for bad debts by the amount of the actual debt. It is proper to record the expense as an estimate earlier because you want to recognize the loss when you record the sale. The losses you take in the future will occur on accounts you now have as receivable.

Setting up a reserve for bad debts has several important advantages. First, your books are useful to you only if they reflect what is really happening in your business. Your profit and loss statement each month should tell you how well your company is doing. If you are not reducing monthly profits by the bad debts you expect, your books don't give a true picture of your profitability. Second, and perhaps most important, your business pays taxes on its profits at the end of each year. If your accounts receivable total is not adjusted to account for bad debts that are sure to occur, you pay federal and state income taxes on profits you will never see. For these reasons, most companies that have significant and regular bad debt write offs set up a provision for bad debts.

ESTIMATING BAD DEBTS

There are two common methods for estimating bad debts. First, you can analyze your accounts individually and base your estimate on the critical examination of each case. This is practical if you have only a limited number of accounts and are in close contact with your customers. In a situation like this, your educated guess would probably be better than any formula for predicting future losses. As an alternate, you can make an estimate based on your business records. Your record of collection experience can be put to good use here. For estimating bad debt losses, you need only the total of credit sales and bad debts for a substantial period. But remember that your estimate for bad debts should be reasonable as well. For example, a single large bad debt in the past year might distort the current estimated uncollectible accounts. Exclude unusual situations from your analysis to get a fair picture of your future. On the other hand, you might know of one or more large problem accounts currently on your books. If this is so, you might want to increase your estimate of uncollectibles for the coming year.

Because you don't know what your real losses will be, your bad debt reserve, like other reserves, is always an estimate. The estimate is more accurate if you use your business records developed over several years. Compare actual losses to sales on account for a two year period to find your typical loss ratio. If your loss experience begins to exceed this ratio, you should be aware of the reasons for the trend.

Figure 5-1 shows a bad debts study. Just as in Chapter 4, the March numbers have been extended and are used here for May to show a longer trend period. The figures in column one are actual bad debts. These amounts have become uncollectible in each month, and were charged against the reserve for bad debts.

Credit sales (column two) are broken out from total sales because only credit sales become bad debts. The total sales figure could be used, but any variation in the ratio of cash sales to credit sales would invalidate your estimate.

Column three shows the ratio of bad debts to total credit sales. A ratio expressed as "1 to 48" means that one out of every forty-eight credit sale dollars became a bad debt.

Bad Debt Procedures

JOHNSON CONSTRUCTION COMPANY
BAD DEBTS STUDY
19____

Period	(1) Actual Bad Debt Losses	(2) Total Credit Sales	(3) Ratio (1÷2)	(4) Year to Date Ratio *	(5) Month-end Accts. Rec. Total **	(6) Current entry, Reserve for Bad Debts ***
Prior year total	$ 5,180.00	$248,200	1 to 48	1 to 48		
January	600.00	23,682	1 to 39	1 to 39	$ 26,268	$ 547.00
February	-0-	23,900	-	1 to 79	26,933	561.00
March	450.00	24,106	1 to 54	1 to 68	28,580	595.00
April	522.16	24,610	1 to 47	1 to 61	28,423	418.00
May	-0-	23,961	-	1 to 76	29,061	427.00
June						
July						
August						
September						
October						
November						
December						

* Current year's total actual bad debts (sum of column one) plus current year's total credit sales (sum of col. 2)
** Excluding Retainages
*** Column 5, divided by column 4, adjusted quarterly

Figure 5-1

$$\frac{\text{Total Credit Sales}}{\text{Actual Bad Debt Losses}} = \text{Bad Debt Ratio}$$

Column four shows the year-to-date ratio. This is computed in the same way as the monthly ratio, except that year-to-date totals are used rather than only monthly totals. For the first quarter:

$$\frac{\text{Total Credit Sales in January, February and March}}{\text{Actual Bad Debt Losses in January, February and March}} = \text{Year-to-Date Bad Debt Ratio}$$

The year-to-date ratio is the key figure to watch. It is the monthly update of the year-to-date ratio that shows the trend of your bad debt experience.

The month-end balances of accounts receivable (column five) exclude retainage. The figure in column six is recorded as the reserve for bad debts and is computed by dividing the month-end receivable balance by the year-to-date bad debt ratio. In this example, the divisor is the year-to-date ratio at the end of the previous quarter (the underlined ratios). Using the ratio for the previous quarter avoids distortion that would result from a heavy or light write-off in a single month. The ratio you

JOHNSON CONSTRUCTION COMPANY
BAD DEBTS TREND
FOR THE YEAR 19____

Period	Total Credit Sales	Total Sales	Ratio	Y.T.D. Ratio	Delinquent Rec.	Total Rec.	Ratio	Y.T.D. Ratio	Bad Debts Ratio **
Prior Year	$248,220	$288,640	1/1.16	xxx	$ 3,286*	$ 23,406*	1/7.1	xxx	1 / 48
January	$ 23,682	$ 26,901	1/1.14	1/1.14	$ 3,788	$ 26,268	1/6.9	1/6.9	1 / 39
February	23,900	27,006	1/1.13	1/1.13	4,080	26,933	1/6.6	1/6.8	1 / 79
March	24,106	27,281	1/1.13	1/1.13	4,418	28,580	1/6.5	1/6.7	1 / 68
April	24,610	27,460	1/1.12	1/1.13	4,862	28,423	1/5.8	1/6.4	1 / 61
May	23,961	28,158	1/1.18	1/1.14	5,100	29,061	1/5.7	1/6.3	1 / 76
June									
July									
August									
September									
October									
November									
December									

* Average

** Bad debts ratio is from Figure 5-1, "Bad Debts Study"

Figure 5-2

develop should vary only with sharp changes in your accounts receivable balance.

Month-end balance of
<u>Accounts Receivable</u>   = Reserve for Bad Debts
Year-to-Date Bad Debt   for the current month
established as of close
of previous quarter

The bad debt study gives you a quick method for reviewing your bad debts, creates a useful record of bad debt experience and shows the trend of problem accounts. Figure 5-1 shows that the previous year's bad debts were one dollar in every forty-eight dollars of credit sales. The year-to-date bad debts as of May 31 were one dollar in every seventy-six of credit sales. If this ratio holds throughout the year, Johnson Construction has improved its collection picture.

If you want to refine your analysis of bad debts, take a few minutes each month to prepare a bad debt trend report such as in Figure 5-2. This report requires that you use your past records as a guide to evaluate future collections. The ratio between credit sales and total sales in the fourth column from the left is an important figure. The trend of these ratios shows the trend of your business. In the example, total sales remain about 1.14 to 1 over credit sales, but bad debts are decreasing as a percentage of credit sales. This is a favorable trend. If bad debts were increasing against a steady credit sales volume, the trend would be unfavorable.

The ratio of delinquent receivables to total receivables (third column from the right) shows a decrease. This is a good trend. Remember, the older an outstanding balance becomes, the more likely it is to become a bad debt.

Ratios are very useful in understanding the relation of numbers. The numbers on your general ledger and on your financial statements represent very real things—a sum of cash in the bank, lumber stacked up in your warehouse, or a loss from an uncollected account. The significance of those numbers is more apparent when expressed as a ratio. The dollar amounts don't tell the whole story. For example, assume that bad debts for six months were:

January	$20.00
February	35.00
March	55.00
April	160.00
May	215.00
June	385.00

True, bad debts are rising at an alarming rate. But if the volume of credit sales were also increasing rapidly, the ratio of bad debts might actually be decreasing:

	Bad Debts	Credit Sales	Ratio
January	$20	$900	1 / 45
February	35	1,850	1 / 53
March	55	3,000	1 / 55
April	160	9,000	1 / 56
May	215	12,500	1 / 58
June	385	23,000	1 / 60

It is unlikely that you would experience a rate of growth like that shown, but the example does show that the dollar figures themselves don't always tell you what you need to know. The ratio is the key figure to watch.

The following calculations were made to compute the two ratio columns in Table 5-2.

$$\frac{\text{Total Sales}}{\text{Total Credit Sales}} = \text{Credit Sales Ratio}$$

$$\frac{\text{Total Receivables}}{\text{Delinquent Receivables}} = \text{Delinquent Ratio}$$

Delinquent receivables are the amounts due more than 30 days. They have a predictable effect on bad debts. For that reason, Figure 5-2 includes the year-to-date ratio of bad debt losses to credit sales (from column four of Figure 5-1). Gather all of this information on one worksheet. This gives you a good picture of your collection experience.

Summaries like this bad debts trend report are valuable because they support the financial statements required by your bank or bonding company. Back up your claim that a high percentage of your accounts receivable is collectible with a report like Figure 5-2 covering a two or three year period. Lenders must rely on the loan documents you provide. Since they do not know how collectible your accounts are, a good historical summary will help you to prove your case.

Figure 5-3 is a year-to-date ratio summary. This information comes from the ratios for Figures 5-1 and 5-2, but excludes dollar amounts. The ratios in Figure 5-3 help determine what steps you should take to avoid uncollectible accounts.

Read the information in Figure 5-3 for January as follows: "There was $1.00 of charge sales for every $1.14 of total sales. In the same month, $1.00 in accounts receivable was delinquent for every $6.90 of total receivables. $1.00 in every $39.00 of credit sales was written off."

The example shows these trends:

- The portion of total sales made on credit is about the same in May as it was in January.

- The portion of delinquent receivables declined from January to May.

- The amount of total credit sales resulting in bad debts is about half in May of what it was on the average in January. $1.00 in every $39.00 was written off in January; only $1.00 in every $76.00 in May.

This is valuable information because you have to understand what you are doing right and wrong before you can do more of what you should be doing.

In the example, the trend is favorable. Delinquent receivables and bad debts are declining together, the result of good management. This is happening while the relationship between credit sales and total sales remains about the same. But an unfavorable trend would be found if:

- Delinquent receivables were increasing.

- Delinquent receivables were remaining the same while bad debts were increasing.

- Delinquent receivables and bad debts were remaining the same while credit sales were decreasing in relation to total sales.

Developing information about your re-

JOHNSON CONSTRUCTION COMPANY
YEAR-TO-DATE RATIO SUMMARY
CREDIT SALES AND BAD DEBTS
For the year____

MONTH	CREDIT SALES TO TOTAL SALES	DELINQUENT RECEIVABLES TO TOTAL RECEIVABLES	BAD DEBT LOSSES TO CREDIT SALES
January	1 to 1.14	1 to 6.9	1 to 39
February	1 to 1.13	1 to 6.8	1 to 79
March	1 to 1.13	1 to 6.7	1 to 68
April	1 to 1.13	1 to 6.4	1 to 61
May	1 to 1.14	1 to 6.3	1 to 76
June			
July			
August			
September			
October			
November			
December			

Figure 5-3

ceivables requires only a few minutes each month. When you have compiled data for several months, a comparative study of the information will be more valuable to you.

You must control your accounts receivable. To do so, you only need to know the trend of your credit sales and bad debts. A builder who has a high ratio of bad debts does not have well-defined procedures for extending credit, billing his customers, and making collections.

Don't ignore future bad debts when making your cash forecast. Your estimate of available cash can be totally wrong without a good projection of bad debt losses. Cash budgeting is explained in Chapter 14.

Two accounts were introduced in this chapter — Reserve for Bad Debts and Bad Debt Expense. You make month-end journal entries to record the provision for bad debts and the offsetting expense. When actual bad debts occur, reduce the provision and the accounts receivable total by the amount being written off. When an account becomes uncollectible, update the customer ledger cards to record the same bad debt that flows through the general ledger. If you don't do this, your customer ledger control card will not balance to the general ledger's total for accounts receivable.

The Reserve for Bad Debts account is a reduction of your current assets, and should appear in the general ledger immediately after the accounts receivable. When listed on a financial statement, the following order is standard:

Current Assets:
    Cash (one item or by type)   XXX

    Accounts Receivable   XXX

    Less: Reserve for Bad Debts   ⟨XXX⟩

Net Trade Accounts Receivable XXX

Retainage   XXX

Total Accounts Receivable   XXX

Following this partial listing would be the remaining current assets, such as the earned income account.

BAD DEBTS AND CASH SYSTEMS

The procedure is different when you keep your books on the cash accounting method. The procedures described for booking bad debts are relevant only if you're accruing income. You reverse income by journal entry because you can't collect the money.

On a cash basis, the bad debt is just as bad, but you haven't recorded as income money that you will never receive. Remember, in the cash accounting system, income is booked only when received. So the bad debt actually never makes it onto your books.

How do you keep track, then? Some firms have receivables but keep their books on the cash basis. This is perfectly legitimate. However, you still need to keep track of receivables and, of course, of bad debts. In this situation, the records you keep for accounts receivable are identical to those you keep in the accrual system. Here's the difference: The accrual entries remain in the subsidiary system and are never entered into the official accounting records.

In this example, there's a need for records beyond the accounting system. In the cash basis accounting, the books consist only of cash transactions. Matters such as receivables and bad debts take place outside of these records. Refer back to Chapter 4 for more information on recording and managing receivables.

If you do business on credit, there's no way you can eliminate all bad debts. But if you follow the procedures outlined in this chapter, you can keep bad debt losses under control and learn from your mistakes before they do serious damage.

Chapter 6

Sales Records and Cash Budgeting

Many builders make the mistake of getting into unprofitable work. You may believe that all of the work you do produces an acceptable level of income. In reality, one or more types of work may be draining income from your business. The only way to find out where you are making and losing money is by analyzing your sales records by the types of work you do.

Sales records have been discussed previously. Your sales record is more valuable if figures are kept on each job. A set of versatile forms and a few simple procedures will let you keep a record for each job nearly as easily as keeping a single sales journal.

To analyze sales, you must be able to look at your receipts for a particular type of work and compare your results there to the rest of your business. Chapters 7 and 8 explore this analysis further. This chapter is intended to familiarize you with the basic kinds of sales analysis and to show you how to budget for future income. Some profit ratios useful to builders are also explored.

Figure 6-1 is a summary of cash receipts. Notice that this is neither a summary of billings nor of earned income—it is cash receipts or payments only. This is essential for cash budgeting: planning for future available funds.

In Figure 6-1 the Johnson Construction Company cash receipts for the month of March have been broken down by job name except for a last category, *All Other*. The builder knows about how long each named contract will take to complete. From that, he can estimate cash receipts and costs at each stage of completion. The final category, *All Other*, is a more serious problem because, in this case, it represents the largest single part of receipts (68%). This category is examined further in Figure 6-2.

Figure 6-2 divides the *All Other* into specific types of income. This is the first step toward developing a realistic projection of receipts. The more detail available, the more likely that a forecast will be accurate. Figure 6-2 shows that Johnson Construction has two principal types of "other" income — repairs and improvements and material sales. The income from repairs and improvements provides a steady flow of cash each month. This regular income helps carry the payroll and inventory during slow periods. *Material sales* are resales to other contractors,

JOHNSON CONSTRUCTION COMPANY
SUMMARY OF CASH RECEIPTS
MARCH, 19____

Name	Cash Received	% of Total
Anderson	$ 2,815.00	10%
Carey	2,328.00	8
Norwich	159.66	1
Rayne	1,619.66	6
Widmark	819.90	3
Windsor	974.00	4
All others	18,803.74	68
Total Cash Receipts, March 19__	$27,519.96	100%

Figure 6-1

landscapers, roofers, and so forth. Also included in the material sales category are retail sales made to the general public from Johnson Construction's lumber yard. The last section, *other income,* is for all sales that do not fit into the other two categories. These would include income from hauling or deliveries.

Figure 6-2 makes clear the meaning of *All Other*. A more detailed analysis is possible — material sales could be broken down into retail and wholesale; repairs and improvements could also be divided. But that kind of detail is unnecessary as it provides more information than is needed to forecast future receipts.

Figure 6-3 is a detailed cash receipts summary and is more informative than Figure 6-1. It is a summary of the cash received by job or type of business. It shows where income is generated and is an excellent forecasting tool, as will be shown shortly. The 68% of *All Other* is detailed into the three principal categories shown in Figure 6-2. Notice that Figure 6-3 puts the work done by the Johnson company in March in perspective. A high percentage of cash receipts generated from one place would indicate that moving into new areas might be dangerous. Or it could mean that the builder is overly dependent on one small part of his total operation to support a large portion of his cash needs.

Unfortunately, **this report is not prepared directly from already existing records. There is no good reason to break down cash receipts by line of business, as such records are best maintained for earned income, direct costs, and expenses. So cash receipts by the job must be constructed from detailed records that are kept for other purposes.**

JOHNSON CONSTRUCTION COMPANY
CASH RECEIPTS, OTHER INCOME - BY TYPE
MARCH, 19____

Name	Total	Repairs and Improvements	Material Sales Only	Other Income
Black's Landscape	$ 900.00	$	$	$ 900.00
M. Brown	482.11	482.11		
C. Carlson	641.27	641.27		
L. Carlson	173.05	173.05		
Harvey Contracting	1,412.00		1,412.00	
Hiram & Hiram	3,400.00	3,400.00		
Jim's Contracting	392.00		392.00	
L. Jones	74.00	74.00		
Karston Landscape	3,200.00		3,200.00	
LBN Brothers	245.00			245.00
K. Middleton	145.27	145.27		
Midfield School	82.17	82.17		
B. Ottman	169.34	169.34		
L. Peterson	477.55	477.55		
Riley Landscape	2,058.00		2,058.00	
J. Smith, Builder	525.00			525.00
D. Thomas	706.08	706.08		
Triumph Construction	3,000.00	3,000.00		
Woodman	425.00		425.00	
J. Woods	295.80	295.80		
Total	$18,803.74	$ 9,646.64	$ 7,487.00	$ 1,670.00

Figure 6-2

Prepare a summary such as Figure 6-3 for yourself every few months, or when you are considering changes in your business such as going into new markets or out of old ones. It is also a valuable summary when you want to forecast future available cash. Such a study done only two or three times a year would show seasonal changes in the volume of business in the different types of work you do.

Now use the information in Figure 6-3 to project where cash will be coming from and how much will be coming in during the next few months. Figure 6-4 shows a realistic estimate of cash receipts in three, six, and nine months in the future. A projection could be done for each of the next nine months, but that much detail isn't necessary unless very careful analysis is required. Detail budgeting for cash is discussed in Chapter 14.

Figure 6-4 projects cash receipts for the remainder of the current year using the information developed for the previous figure. Four of the six bid contracts will have been completed by year-end and are assumed to be replaced by two additional projects. In previous years, Johnson Construction has started at least two jobs during the summer and autumn and between three and six jobs in late winter and early spring. Material sales are expected to decline with the average temperature. Colder weather cuts into material sales but increases the demand for repairs and improvements. With the winter slowdown in home building, repairs and improvements increases from 35% to about 55% of total cash receipts. Records at Johnson Construction support these statements. You can

JOHNSON CONSTRUCTION COMPANY
DETAIL CASH RECEIPTS SUMMARY
March, 19____

Name	Cash Received	% of Total
Anderson	$ 2,815.00	10%
Carey	2,328.00	8
Norwich	159.66	1
Rayne	1,619.66	6
Widmark	819.90	3
Windsor	974.00	4
All others:		
Repairs and Improvements	9,646.64	35
Material Sales Only	7,487.00	27
All Other Income	1,670.00	6
Total	$27,519.96	100%

Figure 6-3

see how valuable this information is in planning for each year.

This kind of spot budgeting can be done quickly with well-kept records and dependable historical data. Records such as this help you understand your business and the market where you compete. It seems that Johnson Construction has developed a good way to keep busy all year. This shifting of emphasis from one line to another helps take advantage of seasonal demand and avoids large seasonal changes in inventory and the work force.

In using historical information to project even the immediate future, be careful to take into consideration known unusual circumstances in past years or the future. Several factors can affect the accuracy of your predictions:

- Building moratoriums

- Dramatic increases or decreases in market demand

- Changes in home prices and the availability of new and used home financing

- Mortgage interest rates for buyers

JOHNSON CONSTRUCTION COMPANY
A CASH PROJECTION BY INCOME TYPE
FOR THE YEAR 19____

Sources of funds	Cash receipts projected for the month:					
	June, 19____		September, 19____		December, 19____	
	Amount	%	Amount	%	Amount	%
Anderson	$ 1,430	5	$ 3,140	10	$ -	-
Carey	2,860	10	1,256	4	-	-
Norwich	1,144	4	2,512	8	2,808	12
Rayne	1,716	6	1,570	5	1,170	5
Widmark	572	2	-	-	-	-
Windsor	-	-	-	-	-	-
Future Bid Contracts	1,430	5	3,140	10	1,170	5
Others:						
Repairs and Improvements	10,010	35	10,990	35	12,870	55
Material Sales Only	8,580	30	7,850	25	4,680	20
All Other Income	858	3	942	3	702	3
Total	$28,600	100%	$31,400	100%	$23,400	100%

Figure 6-4

- Changes in the desirability of living in the community in which you will build

- Changes in your labor force, available cash, and equipment

Another valuable analysis you can make from your sales records is a comparison of net income to sales. The correct method of figuring your direct costs and overhead expense for each job is explained in Chapter 15. For now, assume that your net income has been figured for each job. Calculate the profits by job or job type in your business, and you know where you are making and losing money. Figure 6-5 compares projects under way during a 12 month period. The projects listed were begun and ended in a variety of seasons, business environments, and over two years. The value of this summary is that you compare a variety of jobs and should begin to see emerging patterns. What markets are worthwhile and what markets are a waste of your assets?

In some cases, profitable markets should be abandoned because the risks aren't worth the rewards. Your risk includes your investment in capital and equipment, manpower time, and management. Devoting all of these resources to a type of contract that yields a small return means that more profitable markets may escape your grasp.

As shown in Figure 6-5, Johnson Construction has had a variety of returns on its efforts. A builder looking critically at this list can see that

JOHNSON CONSTRUCTION COMPANY
SUMMARY OF PROFITS AND SALES

Job Description	Net Profits	Gross Income	Ratio
Beale	$ 2,615	$55,000	1 / 21
Cauley	10,803	105,000	1 / 10
Davis	991	6,200	1 / 6
Dean	2,800	85,000	1 / 30
Farragut	1,466	35,000	1 / 24
Harrison	9,130	190,000	1 / 21
Ingrams	387	18,000	1 / 47
Nulty	4,000	90,000	1 / 23
Parker	676	65,000	1 / 96
Pearman	(2,490)	26,000	-
Sims	1,860	78,000	1 / 42
Repairs & Improvements:			
last twelve months	7,860	143,904	1 / 18
previous twelve months	4,914	78,272	1 / 16
Material Sales			
last twelve months	6,718	80,012	1 / 12
previous twelve months	6,190	76,300	1 / 12
All Other			
last twelve months	560	8,347	1 / 15
previous twelve months	844	10,410	1 / 12

Figure 6-5

the Dean, Ingrams, Parker and Sims jobs yielded the lowest profits. Were there aspects to those jobs that show a pattern? Do they represent a type of work to avoid in the future? Analysis might show that too much equipment was required on these jobs or that idle time was more than estimated. Or these jobs might have required unbudgeted costs and expenses, cutting into net profits. With a little experience and analysis you can begin to distinguish unprofitable work from more profitable jobs and emphasize the type of work that helps you most. This is especially true if other types of work are available and within your company's capacity. Look again at Figure 6-5. If the Dean project, which yielded profits of one dollar for each thirty dollars of sales, could have been replaced with work like the Cauley job, net income could have been $8,500 instead of $2,800 on the same volume of sales. It's impossible for most builders to participate in both low profit business and high profit business to maximum advantage if a large portion of available assets and manpower have to be concentrated in high-volume markets. But remember that low profit business may be more valuable than it seems on paper. Builders may purposely bid on low profit work to cover necessary operating expenses in slack seasons when more attractive work is not available. It could even be to your advantage to accept work which will bring a net loss, as will be shown in Chapter 8. Bidding on very competitive jobs could be the only way for a builder to avoid laying off a large part of his labor force. The men who contribute most to the quality of your business may find other work and be lost forever. To lose their productive labor is as much a loss as any other asset you own.

Keeping sales records is also important because it allows you to make financial statements by the job. When showing the differences between comparative periods or between one type of income and another, by-the-job records are necessary. Financial statements can be prepared in detail by the job, but only if a complete set of records is available. Another way to present financial statements is by comparative accounting methods (percentage-of-completion or completed contract). To do this, you must be able to reclassify deferred and accrued income for each method. For this you need by-the-job sales records.

To understand the significance of sales and their varying values and yields on investment, you need:

- Detail records

- Historical information

- Meaningful reports

To analyze your detail records, historical information and reports to direct your financial future, you need:

- Realistic income projections

- An understanding of your markets

- A knowledge of profit yield on various types of work

- A plan of what work you want to do

Above all, you need to be able to add the information in your books and records to your own realistic understanding of your trade. A competent builder can be assured of success by well-informed planning. True, the books, records, and reports you produce do not dictate results. But they do provide the facts you need to make informed decisions.

Chapter 7
Sales Planning

The successful builder seeks profits, not simply volume, when he takes on more work or looks for new lines of business. Volume does not necessarily produce profits that justify the building activity. You could go after all the volume you could handle if profits were not the primary consideration. The *quality* of your sales volume, not the amount, is the key to higher profits. Low-profit volume never adds to the growth of an operation. Instead, it saps income, ties up working capital and equipment, and demands time and energy that, with planning, you could devote to more profitable projects. You want to avoid low-yield jobs when considering new types of work. To do this you need a thorough knowledge of your own current and past yields and a reliable estimate of the risks and potential profits in new work you are considering.

A builder who knows his market will have a higher degree of success than one who takes his operation into new ventures without careful planning. Know at least as much about the new field as your competition. Gather facts before you commit yourself. Breaking into a new construction service, product line, or project type is often the only way to grow. But first ask questions and get answers that satisfy you. Find out if the new market is really the one for you.

Many questions you should ask before taking on a new type of work might seem obvious from a practical viewpoint. But think of their accounting and financial consequences, too.

- What is the competition like?

- What is the demand for this type of work?

- Is the market speculative and how much risk is there?

- What amount of cash will you have to tie up?

- Will special equipment and machinery have to be bought or leased?

- Is your current labor force adequate? Do you have people who are qualified for the work required?

- What portion of total sales will the new work represent? Of total gross profits? Of total net profits?

- What is the estimated return on the investment? Compare this to your current return.

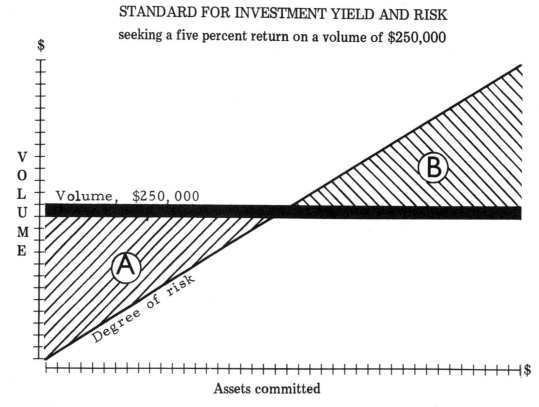

A — The risk will be justified by the possible yield

B — The possible yield is not justified by the risk

Figure 7-1

Look closely at projected gross profits and net income in the new market. If you estimate that a new line will produce large volume but a small percentage of your total profits, then your yield on the investment will be smaller than if the volume were smaller and the profit were greater. Consider your commitment of time, labor, cash, and equipment in the new work. Will you still be able to compete in your established line — the one that produces a higher yield? Or will the low yield of the new line drop your overall yield and put a drain on your operation? Say that a few customers would account for the bulk of your new volume or profits. Consider the effect of a single large bad debt on your income and cash flow. Think about the loss of a major account. Like Johnson Construction in Chapter 6, many small accounts under "Repairs and Improvements" may add up to the largest single portion of your volume and profits. In this case you could absorb a few bad debts or customer losses. But if you committed your operation to three large accounts in a new line, could you afford delayed payments on a third of your new business?

Compare your estimates on the new work to a standard or goal. This should be what you expect as a result — the yield on your investment, the net profit of the new line. The higher your risk, the higher the yield should be. The lower the risk, the lower the yield can be to offset it. If you have to commit much of your available working capital to finance equipment for the new line, your possible profits should be high enough to give good yield on the investment. You may want to take smaller risks instead in other work where you already have adequate equipment. Often this means a smaller yield.

Look at Figure 7-1. Assume that your company has a volume of $250,000 per month or per year. That volume is represented by the dark horizontal line. The horizontal line at the bottom of the figure measures the assets that must be committed to undertake the work you do. If your volume remains the same, the commitment of additional time, money and equipment means that the risk of a large loss grows greater. That does not necessarily mean

JOHNSON CONSTRUCTION COMPANY
HISTORICAL INCOME STATEMENT BY PRODUCT
based on an assumed volume of $250,000

	Total	Bid Contracts	Repairs and Improvements	Material Sales	All Other
Gross Sales	$250,000	$100,000	$90,000	$50,000	$10,000
Direct Costs	137,500	55,000	49,500	30,000	3,000
Gross Profit	112,500	45,000	40,500	20,000	7,000
Selling Expenses	56,250	24,000	18,000	9,500	4,750
Selling Profit	56,250	21,000	22,500	10,500	2,250
Fixed Overhead	40,000	16,000	14,400	8,000	1,600
Net Profit	16,250	5,000	8,100	2,500	650
Yield Ratio	1 / 15	1 / 20	1 / 11	1 / 20	1 / 15

Figure 7-2

that your business is less profitable. High risk business may be very profitable business. But you should be aware of the increased risk involved before you commit yourself and your business. If you see an opportunity to expand, but know that more risk is involved, make sure that the potential for profit outweighs the associated risk. Notice on Figure 7-1 that you can commit additional assets without increasing your risk if your volume grows proportionately to your assets. Raising the volume line decreases area B and increases area A if the assets committed remain the same. If the volume line climbs up the scale proportionately to the increase in assets committed, the size of areas A and B will remain the same.

Analyze your present business on paper before you decide on a new venture. Make up an income statement as in Figure 7-2, which assumes a volume of $250,000. This gives you yield, volume, and profit information you can use to help decide if a potential new market measures up to your yield standards. Base your product and service breakdowns and yields on five-year averages, since this absorbs large variations within a reasonable time span. Round to thousands, too, as was done in the figure. The subtotal for selling profit distinguishes the before-overhead results from the bottom line. Any method of breaking down fixed overhead will be arbitrary, but this extra piece of information provides more data with which to compare new business to your present operation.

The ratio at the end of each section shows net profits to gross sales. This is the yield ratio explained in the last chapter. Repairs and improvements shows the best yield here, but this may be seasonal. Johnson Construction could not support itself on that line alone and needs the additional work it does to offset the slow months' decreases. The bid contracts and material sales categories yield the minimum 5%, but account for 60% of the total volume.

Figure 7-3 shows the same information as Figure 7-2 but in percentages. Gross sales is 100% and all figures are expressed as a percent-

JOHNSON CONSTRUCTION COMPANY
PERCENTAGE HISTORICAL INCOME STATEMENT BY PRODUCT

	Total	Bid Contracts	Repairs and Improvements	Material Sales	All Other
Gross Sales	100.0%	100.0%	100.0%	100.0%	100.0%
Direct Costs	55.0	55.0	55.0	60.0	30.0
Gross Profit	45.0	45.0	45.0	40.0	70.0
Selling Expenses	22.5	24.0	20.0	19.0	47.5
Selling Profit	22.5	21.0	25.0	21.0	22.5
Fixed Overhead	16.0	16.0	16.0	16.0	16.0
Net Profit	6.5%	5.0%	9.0%	5.0%	6.5%

Figure 7-3

age of gross sales. Bid contracts and repairs and improvements have the highest gross profit (each 45%), but the repairs and improvements category alone has the highest selling profit and net profit. Johnson Construction could not survive by doing repairs and improvements alone. Not only is it a seasonal line, but the company's other building activities feed it business. A good deal of repair and improvement work comes from contacts in other lines.

Selling expenses for "all other" are 47.5% because of higher automotive expenses for hauling and freight in this line. Truck and auto expenses are not classified as direct costs, but as selling expenses. This distinction is helpful when you analyze the profitability of your existing work and compare this data to new markets.

From Figure 7-3, Johnson Construction determines that its lowest acceptable yield is 5%. Any new market they consider cannot have an estimated yield below this. Additional auto and truck expense would require that Johnson Construction aim for a higher selling profit to offset the higher overhead expense.

Johnson Construction has compared its five-year yield data to estimated high and low figures for the new business it considers taking on. This establishes high and low yield limits for the company, the new business included, within which it feels it could operate. You might want to add a small percent to the selling profit or the net profit to allow for errors.

Johnson Construction applies its yield data to a potential new market and sets yield limits that would be acceptable overall. See Figure 7-4. These are the company's yield assumptions including the new business. The present results column shows the minimum acceptable level. This is the current yield based on a five year average. The high expectation column shows the figures for the most desirable level that could be realistically expected including the new business. The low expectation column shows the lowest acceptable yield with the new business, assuming that risks and time commitments were also minimal.

Allow for errors in your estimate by including an error factor, a small amount added to the selling profit or the net profit. When you expand into the new work stay within the acceptable yield. You define your goals and limits when you establish high and low

Sales Planning

JOHNSON CONSTRUCTION COMPANY
SUMMARY OF MARKET ASSUMPTIONS
based on an assumed volume of $250,000

	Present Results		High Expectation		Low Expectation	
	dollar amount	%	dollar amount	%	dollar amount	%
Gross Sales	$250,000	100.0	$250,000	100.0	$250,000	100.0
Direct Costs	137,500	55.0	112,500	45.0	162,500	65.0
Gross Profit	112,500	45.0	137,500	55.0	87,500	35.0
Selling Expenses	56,250	22.5	68,750	27.5	35,000	14.0
Selling Profit	56,250	22.5	68,750	27.5	52,500	21.0
Fixed Overhead	40,000	16.0	40,000	16.0	40,000	16.0
Net Profit	$ 16,250	6.5%	$ 28,750	11.5%	$ 12,500	5.0%
Yield Ratio	1 / 15		1 / 9		1 / 20	

Figure 7-4

standards by which you can accept new business.

Chapter 8 discusses the other aspect of sales planning — deciding whether to stay in a line or get out. It can often be a good move to abandon a product or service if you are losing money. You need complete analysis by the job and by the job type for sales, direct costs, and expenses. Taking on new types of work can be a big boost or a drain. Similarly, getting out of certain types of work may have surprising results.

Chapter 8

Planning For Profits

It's wise to consider abandoning any part of your business that doesn't produce profits. Many successful builders move in and out of markets as required by circumstances. But you must know the effect of any change you consider before you make it. Your books and records will be invaluable in helping you make this decision.

A type of work or area where you build that is losing money saps profits from your total business. It drains off time, money and assets while dragging down overall performance. For example, assume that your operation is divided between three types of work. You suspect that two of these are money-makers and the third is a profit drain. Your total yield may be cut in half if the figures look like this:

Type of Work	Volume	Net Profit	Yield
A	$100,000	$8,000	8%
B	100,000	5,000	5
C	100,000	(4,000)	(4%)

The combined yield of work A and B is 6½%. Because work in the C area is losing money, it drags the total yield down to only 3%.

But look carefully at the figures before you start turning down any work. Fortunately, unlike new work you might like to get into, you have plenty of your own figures to help you make your decision. First, you need a profit and loss statement by type of work. But the net results by business type don't really tell the whole story. Avoiding low profit work doesn't necessarily cut your overhead proportionately. There are fixed overhead costs, such as rent, which do not necessarily decrease when you reduce your volume. You have to make some estimate of what costs will be reduced by abandoning certain types of work. To do this you allocate fixed costs among the types of jobs you do. The goal is to break down all costs and expenses by responsibility — in what phase of business does the specific expense occur? You can find fault with any method of assigning these expenses, because there is no totally accurate way to do it. Allocations are discussed further in Chapter 15. For now, be aware that no "general" method of assigning expenses is perfect.

Direct costs are more easily assigned because with good records you can identify the related area of income for each direct cost item. Figure 8-1 is a profit and loss statement for Graham Construction Company for one month. This builder splits his business into commercial and residential lines. Commercial work includes all bid contracts. Graham is a specialist in office building remodeling and generally has several

Planning For Profits

GRAHAM CONSTRUCTION COMPANY
PROFIT AND LOSS STATEMENT
FOR THE YEAR ENDED DECEMBER 31, 19____

	Total	Commercial	Residential
Gross Income	$348,216	$184,728	$163,488
	100%	53%	47%
Direct Costs	193,401	77,599	115,802
Gross Profit	$154,815	$107,129	$ 47,686
	100%	69%	31%
Operating Expenses:			
Office Salaries	$ 16,400		
Rent	9,200		
Depreciation	9,115		
Utilities	7,118		
Operating Supply	3,691		
Office Supply	1,284		
Insurance	8,411		
Automotive	7,700		
Union Welfare	31,416		
Taxes and Licenses	14,800		
All Other	8,280		
Total Expenses	$117,415	$ 62,230	$ 55,185
		53%	47%
Net Profit	$ 37,400	$ 44,899	$ (7,499)
yield	10.7%	24.3%	(4.6%)

Figure 8-1

large jobs going. The steady volume makes this type of work desirable. Graham's management knows months in advance what a good portion of its income and expenses will be. And because the line of business involves a few large contracts, the related costs and expenses are more controllable than in most other lines.

Residential work on the Graham statement includes all material sales, repairs and improvements, freight, hauling, and other work sold directly to homeowners. Graham Construction obviously prefers commercial construction. Figure 8-1 shows that Graham might in fact be better off without the residential line of business. You can see that net profits for commercial work are $44,899, a 24.3% yield. Abandoning the residential business would save a large part of the bookkeeping volume, management time, and daily telephone calls.

GRAHAM CONSTRUCTION COMPANY
PROFIT AND LOSS ANALYSIS BY LINE OF BUSINESS

	Total	Less: Residential	Commercial Only
Gross Income	$348,216	$163,488	$184,728
Direct Costs	193,401	115,802	77,599
	100%	60%	40%
Gross Profit	$154,815	$ 47,686	$107,129
Operating Expenses:			
Office Salaries	$ 16,400	$ 8,200	$ 8,200
Rent	9,200	-	9,200
Depreciation	9,115	-	9,115
Utilities	7,118	-	7,118
Operating Supply	3,691	2,215	1,476
Office Supply	1,284	321	963
Insurance	8,411	2,103	6,308
Automotive	7,700	1,925	5,775
Union Welfare	31,416	18,850	12,566
Taxes and Licenses	14,800	7,400	7,400
All Other	8,280	-	8,280
Total Expenses	$117,415	$ 41,014	$ 76,401
Net Profit	$ 37,400		$ 30,728
yield	10.7%		16.6%

Figure 8-2

The Graham Company does not have a good method for allocating expenses. All expenses are assigned on the same basis as gross sales. Graham should break out selling expenses separately as Johnson Construction does. This is a more accurate way of allocating expenses.

Graham realizes the shortcomings of its by-line statement. For example, if operating expenses had been split on the basis of gross profit, the results would have been very different. But the Graham Company decided to analyze the effect of getting out of residential work by preparing the statement shown in Figure 8-2.

Graham used conservative assumptions in preparing this summary. Abandoning almost half of the operation's volume seems to be a drastic move. The following estimates were made:

- Office salaries (clerical help), taxes and licenses would be cut in half without the residential line.

SUMMARY OF SUCCESSFUL VOLUME				
Description	A	B	C	D
Sales	$50,000	$100,000	$150,000	$200,000
Direct Costs	27,500	55,000	82,500	110,000
Gross Profit	22,500	45,000	67,500	90,000
Selling Expenses	13,750	27,500	41,250	55,000
Fixed Overhead	6,250	11,250	15,000	15,000
Total Expenses	20,000	38,750	56,250	70,000
Net Profit	$2,500	$6,250	$11,250	$20,000
yield	5.00%	6.25%	7.50%	10.00%

Figure 8-3

- 60% of operating supplies and union welfare expenses would be cut.
- 25% of office supplies, insurance, and automotive expenses would be cut.

These estimates are based on the builder's realistic appraisal of expenses in his operation. He also relied on his familiarity with his business.

The problem Graham would face in leaving the residential market is apparent. Figure 8-2 shows that the 24.3% yield on commercial work in a consolidated operation falls to a 16.6% yield if only commercial work is taken on. This shows two things. First, the basis used for allocating operating expenses is not as accurate as it could be. A better division would have been based on gross profit. But this is still only a broad estimate. Second, abandoning residential work would not increase profits. The residential work is not a money maker. But it does help carry the business overhead.

This does not mean that Graham can't afford to leave the residential market. But the move must be the result of careful planning. If Graham really does not want to be in that business, he must either seek out new markets to replace the residential business or settle for a reduced profit.

Remember that the present combined yield is 10.7% and the commercial line is estimated to yield 16.6%. Chapter 7 pointed out that a higher yield on volume is always more desirable. Graham Construction might want to consider phasing out of residential work as it replaces lost volume with more commercial work. A realistic goal would be to hold profits of $37,400 (as in Figure 8-1) on a reduced volume. When the profits climb to that level the company could consider abandoning residential work completely. The result would be a smaller volume of business with no cut in profits.

At a profit ratio of 16.6%, Graham Construction would have to increase its gross to

SUMMARY OF UNSUCCESSFUL VOLUME				
Description	E	F	G	H
Sales	$ 50,000	$100,000	$150,000	$200,000
Direct Costs	27,500	60,000	97,500	140,000
Gross Profit	22,500	40,000	52,500	60,000
Selling Expenses	13,750	24,250	31,500	38,000
Fixed Overhead	6,250	11,250	15,000	15,000
Total Expenses	20,000	35,500	46,500	53,000
Net Profit	$ 2,500	$ 4,500	$ 6,000	$ 7,000
yield	5.00%	4.50%	4.00%	3.50%

Figure 8-4

about $225,000 to reach $37,000 in monthly profits. This is about 22% more than the current volume. Actually, Graham's fixed overhead would not vary, so an increase of somewhat less than this could bring the same results. Graham must consider whether it can increase volume 22% in a competitive market. There might not be enough work available at attractive margins in their market area.

If Graham did decide that it could go after a bigger share of the commercial market, it should watch costs and expenses carefully during the transition period. You can calculate the required volume by starting with the monthly profit assumption:

Net Profit Goal	$37,400
Fixed Overhead	35,000 *
Selling Expenses	75,000 **
Gross Profit Goal	147,400 (69% of sales)
Sales Volume Goal	$213,623 (increase of 16%)

* Rent, depreciation, utilities, and all other

** Calculated based on Figure 8-1:

Total operating expenses	$117,400
Less fixed overhead	35,000
	82,400
69% (percentage of commercial to the total)	56,856
Plus estimated increase, 22%	18,128
Total	74,984

A last consideration is that fixed overhead can increase or decrease with significant differences in volume. Going into new markets may require more rental space. More equipment and machinery would increase depreciation expenses. Some other expenses would probably increase as well. A decrease in volume might allow you to cut some fixed expenses.

Figure 8-3 shows favorable volume increases assuming direct costs at 55% of gross receipts.

Planning For Profits

COMPARATIVE YIELDS
SUCCESSFUL AND UNSUCCESSFUL VOLUME

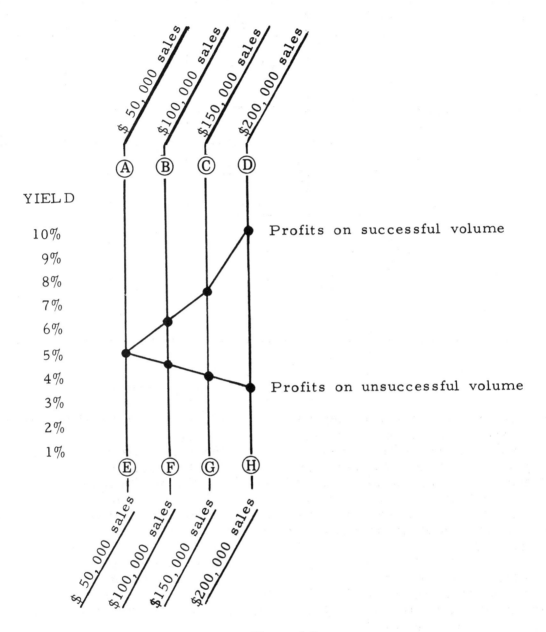

Figure 8-5

Profits vary with volume and expense levels. Figure 8-4 shows unfavorable volume increases and illustrates what can happen when direct costs are not controlled. Column E shows direct costs calculated at 55% of sales. Column H has direct costs at 70% of sales. This is the kind of trap that Graham Construction must avoid if they decide to increase their commercial business. The degree of success depends on the control of costs and expenses.

Figure 8-5 compares the yields for the favorable and unfavorable volume increases in Figures 8-3 and 8-4.

It is important to understand the relationship between volume and profit. You should recognize this relationship when you decide to concentrate your efforts in one market and abandon another. The margin of profit will increase with the volume if costs and expenses

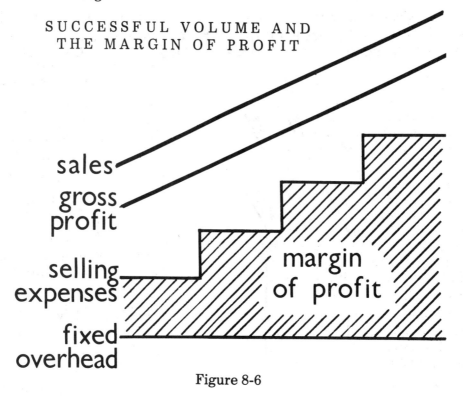

Figure 8-6

are a declining percentage of gross receipts. Without that advantage, the margin of profit will remain the same with increased volume, or may even decrease.

Controlling expenses means both keeping increases to a minimum and knowing what will happen when new markets are acquired or old ones are abandoned. With a smaller volume, Graham Construction would have difficulty maintaining its profit margin because certain fairly inflexible expenses suddenly represent a much higher percentage of total sales. As volume increases, holding expenses to a lower percentage is easier because those same fixed expenses are slow to rise at first.

Figure 8-6 illustrates the relationship between volume and profit. A constant overhead expense (the horizontal line) represents a smaller percentage of sales with higher volume and a larger percentage of sales with smaller volume. Success in increasing the margin of profit rests on control of fixed expenses.

Selling expenses usually rise in steps and are related to the volume of business. Selling expenses tend to follow sales volume fairly exactly, but are controllable even more than fixed overhead. Supplies, insurance, automotive expense, and office salaries should be budgeted and controlled. As volume and cash profits increase, some builders tend to relax their control over selling expenses. But maintaining the standard you have set for yourself requires that these be kept under control.

Management Guidelines

- There is no precise way to allocate your expenses among the various sales categories of your business. Even the most controlled allocation method involves some judgment. Decide whether the basis you use for allocation is accurate by comparing the yields you develop for each part of your business.

- Don't be too critical of seemingly low-yield markets until you calculate the effect of doing without that volume. Your true profit might be higher than you think.

- Like getting into a new market, getting out of a market requires careful research, planning, and analysis. Unprofitable markets should be abandoned, but only after determining that they are truly unprofitable, likely to remain so, and are a drag on your business.

- Decide how you will replace lost volume. The growth of your business depends on the availability of work and the competition for it. Know beforehand that you will be able to take over enough of a market to make up for lost volume.

- Map out a plan for replacing abandoned volume. Include a volume goal. Begin with the profit you want and figure back to the volume you need. Then phase out of the old, low-yield market and into the new volume you have budgeted. All through this process, carefully control your direct costs, selling expenses, and fixed overhead.

- Keep in mind the relationship between successful volume and the margin of profit. Control selling expense and direct costs as increases in volume occur.

- Check your yield from time to time. This is the best indicator of the effectiveness of your controls. Profit margins that remain unchanged or that decrease while volumes increase are the result of uncontrolled changes in costs and expenses.

- Be realistic. Don't play with the numbers just to make your plans work out to expectations. The purpose of profit planning is to come up with a realistic estimate, not to justify a foregone conclusion.

- Be willing to change your plan if your previous profit forecast doesn't work out. Remember, your market environment is constantly changing. Your planning needs to change in order to keep up with the market.

- Keep your original objectives in mind. There's no point in pursuing new markets (or keeping old ones) you really don't want. *You* are in charge and should remain there. When you find yourself doing things you don't want to do, ask yourself why you started your own business.

- Be cautious about the idea of "acceptable" losses. You have probably heard of this idea — that it's okay to lose money initially because you'll make it up with future profits. That may be true. But it's better to start making profits right away, if possible. And few of us can afford two or three years of losing money.

Section II

Costs and Expenses

Check Writing and Recording

Accounting For Materials • Payroll Accounting

Overhead Expenses • Equipment Records

Cash Budgeting • Cost and Expense Records

Accounting For Costs and Expenses

Petty Cash Funds

Balancing The Checking Account

Accounting For Estimates

Chapter 9

Check Writing and Recording

No amount of control over costs and expenses is effective if you don't follow good check writing and recording procedures. The information in your check record must be complete and informative. Your cost analysis is only as good as the source document, your check record. Obviously, you must have an adequate check writing system for your business. You also have to decide how many accounts to maintain. Multiple checking accounts can save time, but they can also become a headache. This chapter is intended to help you decide on the type of check writing system which is appropriate for your business. Your checkbook or check record reflects your costs and tells a lot about your business. Your check record can also be used to calculate your future cash flow and profits. Make sure that you are getting all the information you need from your check record.

This chapter explains the controls and procedures you should use in writing and recording checks, including the check book itself, keeping a check book in balance, the check register (the summary accounting document that classifies your payments in a logical manner), and the handling of accounts payable in the check-writing process. Other problems covered include the proper control of multiple checking accounts in one business, and the handling of checks when more than one business is operated from the same office.

MAINTAINING THE CHECK BOOK

Keeping a complete and accurate record of checks is neither time-consuming nor difficult. But if it is neglected, useful information is lost or must be reconstructed. Your financial reports, statements, and forecasts will be of minimum value if you don't know where your money is spent. The small amount of time required to create and maintain a well-organized, controlled check recording system will pay off immediately if you have been negligent in recording checks in the past. First write out complete names and descriptions for each check and carry a balance of payments by account. This will:

- support summary entries recorded elsewhere;

- be an aid to balancing your bank account;

- give you detail needed for adjustments to accrue or defer costs and expenses;

- provide records for analysis and forecasting.

You need a summary of checks to update your general ledger each month. This task cannot be completed until all checks have been listed and your account reconciled with your bank statement. Balancing your bank account requires that you account for checks cleared as well as checks outstanding. Rather than going through your check stubs, prepare a check listing that gives you all the banking information you need. This is explained later in this chapter.

Chapter 2 explained how income is deferred or accrued under both the percentage of completion and completed contract accounting methods. Similarly, the related costs and expenses are adjusted. The amount of detail needed for adjusting sales is minor when compared to the adjustments you have to make on the money you pay out. Adjusting costs is more complex because there are many categories of costs associated with each job. Good, detailed check stubs on the payments make for more accurate financial statements and forecasts and simplify handling of accrued and deferred expenses.

The check stub and the cancelled check are called source documents. Payment summaries found elsewhere in your books (general ledger accounts or line-by-line summaries on financial statements) can all be traced back to checks. So the check stub or copy must have clear, concise information.

Figure 9-1 shows check stubs in a check book. There are many styles of check books. If you use a pegboard system, your checks don't look like Figure 9-1. With the board method, there is no check stub at all. Pegboard checks will be discussed later.

Here is the basic information you must include on the stub when you write a check, regardless of the style of your check system:

- The check number

- The date

- Who was paid

- What the payment was for or account number

- The amount of the check

Checks include a pre-printed number on the check and the stub. If your check stubs do not include a pre-printed number, be sure to write the check number on each stub.

Once you have written and mailed a check, the stub becomes the explanation of what the check was used for and the amount of the check. The reason for payment on the stub could be a coded number, a written description, or an invoice number. You are probably familiar with your own vendors and don't need to write a complete description. A regular material supplier's check could be written without a full description. As far as you are concerned, the company name tells the whole story. But what if someone who is completely unfamiliar with your business needs to analyze your check book? Your records are like a detailed history of your business, and there are occasions when others must look at your check book. Your accountant may hire a new assistant. Or you may need a statement for a loan application audited by an outside accounting firm. You could have an insurance or Internal Revenue Service audit. Make sure you document each check so no further explanation is required. Complete each check stub on the assumption that the next person to see it won't know anything at all about your operation.

Figure 9-1 shows a series of lines between the stub and the check itself. Not all check books are exactly like this, but all have some area for keeping a running balance. Use this area to bring a balance forward from the previous page and subtract the amount on the check. Record deposits and increase your balance in this column so you know what the current cash balance is.

Keep your account as up to date as possible. All adjustments should be added or subtracted as soon as you find out about them. These include correcting math errors and adjusting for bank service charges and returned checks.

Avoid math errors in bringing forward the balance by checking each page backwards for accuracy. From the bottom, *add* all checks and *subtract* all deposits. The total should agree with the amount carried forward at the top of each page. This should save you time when you balance your bank account each month and give you more confidence in the balance you're carrying. Make it a habit to check each page before you go on to the next.

Here are the types of adjustments you can expect to make when you balance your account.

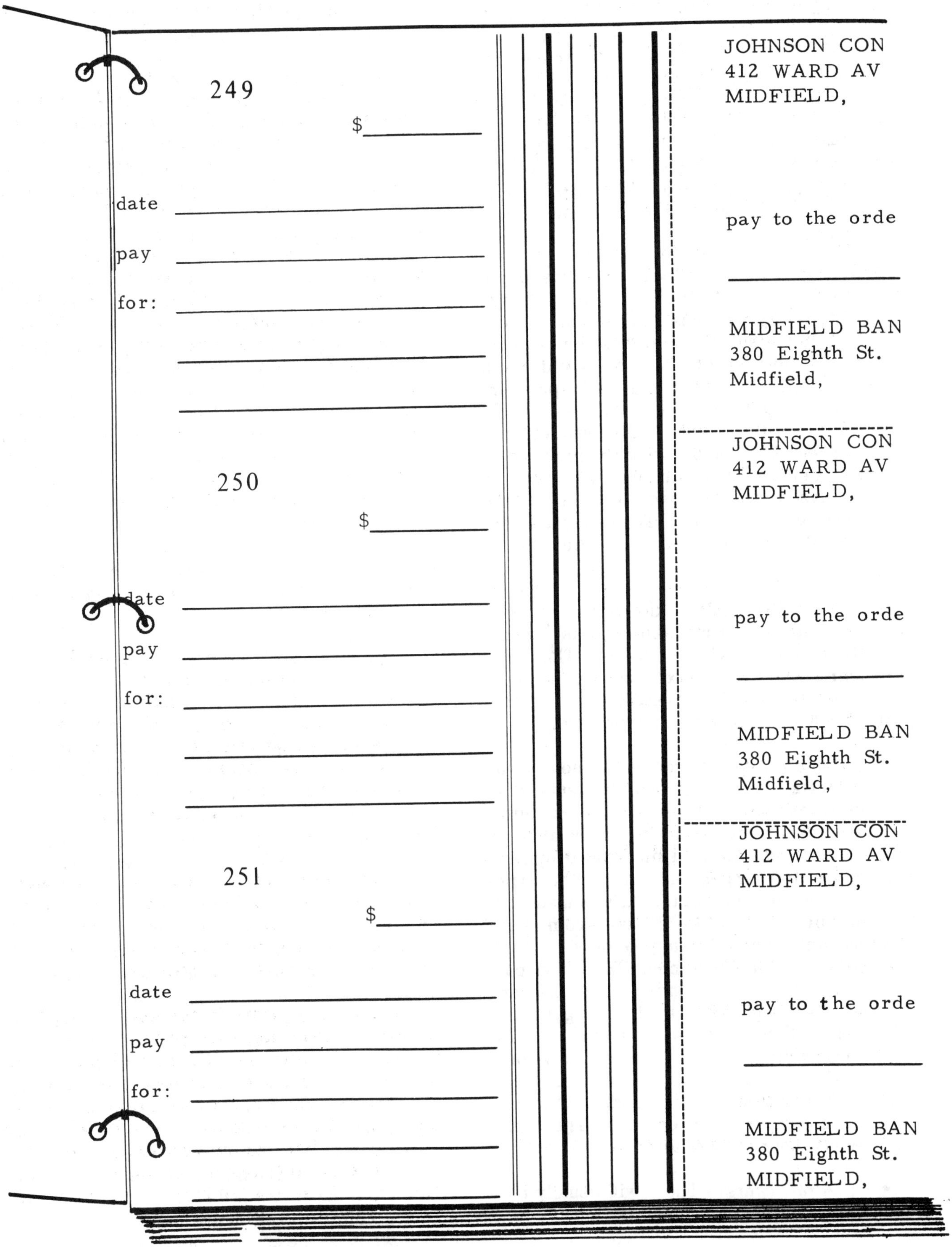
249
$
date
pay
for:
JOHNSON CON
412 WARD AV
MIDFIELD,
pay to the orde
MIDFIELD BAN
380 Eighth St.
Midfield,
250
$
date
pay
for:
JOHNSON CON
412 WARD AV
MIDFIELD,
pay to the orde
MIDFIELD BAN
380 Eighth St.
Midfield,
251
$
date
pay
for:
JOHNSON CON
412 WARD AV
MIDFIELD,
pay to the orde
MIDFIELD BAN
380 Eighth St.
MIDFIELD,

Figure 9-1

- *Bank service charges* Banks usually charge a monthly fee on checking accounts. It may be called a service charge, handling fee, or activity charge. Some banks charge a set monthly fee; others base their charge on the number of checks, deposits and your average balance. You may not know the amount of the charge until your bank statement is received. The fee reduces your cash balance.

- *Returned checks* Checks you have deposited in your account are not always honored by your customer's bank. There may be insufficient funds, the account may have been closed, there could be an unacceptable signature, or a stop-payment order may have been put on the check. Whatever the reason, use a form letter or phone call to find out what the problem is. And don't forget to increase your accounts receivable by the check amount. In the meantime, your cash balance must be reduced by the amount of the checks returned.

- *Special charges* Banks normally charge for printing checks by taking funds directly from your checking account. This is a legitimate business expense and should be recorded in your general ledger. Special charges reduce your cash balance.

- *Automatic payments* For convenience, you can arrange to have loan payments automatically deducted from your account. You must reduce your account balance for such payments and make appropriate entries to your general ledger. Other types of automatic payments, such as insurance payments, can be taken from your account. These often appear on your statements as debit memos. Since the payments are sent to outside parties, the bank makes the payment by honoring a draft just as though you had written a check. For example, an insurance company could send your bank an authorized draft. Your bank then transfers the funds to the insurance company in your name. These also reduce your cash balance.

- *Overdraft charges* There will usually be a charge on overdrawn accounts.

- *Error adjustments* Your bank notifies you about errors they find. For example, if you have a math error on a deposit, the bank will send you either a debit memo (advising you to decrease your balance) or a credit memo (advising you to increase your balance). If you receive a notice similar to Figure 9-2A, *increase* your balance. You are being given credit for more funds. If you receive a notice like Figure 9-2B, *decrease* your balance. You are being charged for the amount shown. Banking terminology can be confusing. Charges, debits, and debit memos are all ways of reducing your balance. Credits and credit memos increase your balance.

- *Non-bank adjustments* When you balance your bank account, you may find other adjustments that are required. Bank errors occur occasionally. For these, do not adjust your balance. Instead, call the bank and point out their error. More likely you will find your own math or recording errors. Be sure to adjust your balance for these.

- *Voiding checks* A check can be voided at the time it is written or after it is written. Void a check while it is being written if you make a mistake while completing the check. Simply carry the same balance forward to the next stub and record "void" on the stub of the voided check. But if the check is voided later—for example, as a duplicate payment returned to you—record an *increase* in your balance to reverse the check. Remember to record the reversal either in a special journal or on your check register so that the adjustment can be recorded in your general ledger.

- *Stop payments* Occasionally you write and mail a check and discover later that the payment was made in error. Occasionally a check will be misplaced or lost after it is written. It would be foolish to simply write another check. Both payments might be cashed. Instead, stop your bank from making payment on the check. Most banks prefer that you go to the bank in person to stop payment. Some require you to do so. Stopping payment involves filling out a form that requests the check number, amount, date paid, payee, and reason for stopping payment. The bank searches its accumulated checks in your account to see whether the check in question has cleared. If it has not, they put your stop order into effect and refuse to honor the check if it

Check Writing and Recording

```
                    MIDFIELD BANK
                     main office

    account number            date           credit number
   ☐☐☐☐-☐☐                   ☐☐☐☐         │C│4│4│7│6│2│7│

your account has been CREDITED for:

date          adjustment explanation          amount
3-20          deposit error                   $ 18.00


M   ┌                                    ┐   $ 18.00
A
I      Johnson Construction Co.
L      412 Ward Avenue
       Midfield                               ___JN___
T                                             By
O   └                                    ┘
```

Figure 9-2A

```
                    MIDFIELD BANK
                     main office

    account number            date           debit number
   ☐☐☐☐-☐☐                   ☐☐☐☐         │D│8│9│8│3│3│8│

your account has been CHARGED for:

date          adjustment explanation          amount
3-16          Andrew Menke - check returned
                   (refer to maker)           $ 62.19


M   ┌                                    ┐
A      Johnson Construction Co.
I      412 Ward Avenue                       $ 62.19
L      Midfield
T                                             ___JN___
O   └                                    ┘   By
```

Figure 9-2B

81

Builder's Guide To Accounting

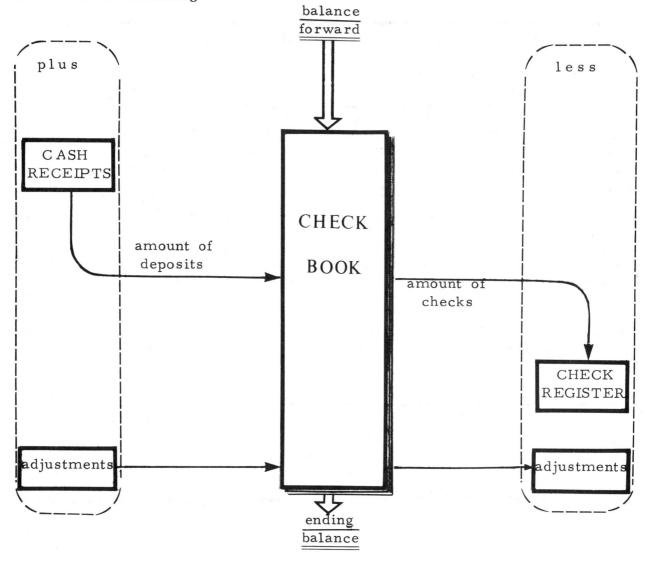

Figure 9-3

does arrive. Most banks charge a fee to put a stop payment order on a check. This fee reduces your cash balance.

A summary of all adjustments (including the ones you find yourself when balancing your bank account) is presented in Chapter 18. This summary should serve as a reminder list to insure that the proper general ledger accounts show the changes made. From this summary you can prepare a monthly journal of bank related adjustments. The journal will also help you balance your cash account in the general ledger.

All changes to your cash account balance should be fully documented. Most of this documentation comes from your bank statements. Your bank account will stay in balance once you have established a good procedure. Figure 9-3 summarizes the flow of entries and adjustments in your check book.

KEEPING A CHECK REGISTER

Your check register shows at a glance what categories of expense you have and where your money is being spent. This document is the basis for the largest monthly entry in the general ledger. It lists in one place the accounting for all checks written, showing who received payments and in what amounts.

Most businesses need to maintain a monthly check register, just as they need monthly sales journals and other important records. The time required to complete a check register is minor. The register should be prepared at least monthly. You might find it more convenient to work on it weekly, or even daily. You can't figure your monthly profit without a completed check register. If your monthly financial statement is prepared by a computer service bureau, you probably also receive a computer prepared check register.

Monthly summary financial statements are a minimum requirement for most builders. Don't go longer than thirty days without preparing your check register. Most builders write a lot of checks. Bringing a 60 day old check register up to date can be a big job. If you write only fifty checks each month, it might take half an hour or forty-five minutes to do this job. But how long would it take you to go through six hundred checks once a year?

The check register is prepared directly from your check book. From check stubs, each payment is listed in the order made. The information on each stub is crucial at this point, as the completeness of these source documents affects the quality of the check register.

When you design your check register, consider the nature of payments in your typical month. Select the important account categories you want to detail out and place them at the top as shown in Figure 9-4.

Checks shown in Figure 9-4 were listed in the register directly from check stubs, both in total and in distribution. The date, payee, check number, amount, and description are needed for each check. Each check takes up one line. The description on your check stub must be detailed enough to make distribution to the right account easy.

In Figure 9-4 all payroll checks are listed by net amounts only. Although payroll checks are written from this account, the detail is included elsewhere. Only the total appears here. There are several reasons for this:

- Listing only the weekly payroll total means that just four payroll lines are required rather than forty or more.

- Fewer vertical columns are required because only the payroll net amount is listed.

- A special analysis should be prepared for labor expenses each week, and the check register is not the proper place for this. (See Chapters 11 and 16.)

- Summarizing the payroll simplifies the check register. Controls for the payroll are handled in a separate register.

Your check register provides the basic general ledger input and is a neat summary for your own use. You can hire an accountant to prepare your check register, but anyone in your office should be able to do the job if you include enough information on your check stubs.

One major function of a check register is that it identifies math errors. The total column is checked by adding distribution totals across the page. See Figure 9-4.

Johnson Construction Company chose to break down its check information into Materials, Net Payroll (with a summary for each week at the end of the register), Operating Supplies, and Automotive Expenses. This last category includes gas, oil, and repairs. The final section, Other Spendings, lists all additional expenses and payments. Note that the Expenses column includes all payments that affect profit and loss. The general ledger includes all balance sheet related payments. You might prefer to use an appropriate code number from the chart of accounts instead of a written description in the All Other column.

The categories you choose for detail listing are those you think have the largest number of checks each month. This depends on the nature of your business. You might find that you need more vertical columns in your check register. Most forms allow for much more detail than shown in Figure 9-4.

WRITING CHECKS TO "CASH"

For a variety of reasons, builders sometimes write checks payable to "cash." Try to avoid this practice. Business deductions for checks made payable to "cash" are questionable and more difficult to prove than checks drawn to a specific payee.

To reimburse your petty cash fund, show the distribution in detail. The amount should be supported by a complete set of documents. To write a check to yourself as a draw against salary (for individuals and partnerships), make out the check in your name and describe it on the stub

Builder's Guide To Accounting

JOHNSON CONSTRUCTION COMPANY
CHECK REGISTER
MARCH, 19____

Date	Paid To:	Check #	TOTAL	Material	Net Payroll (Note 1)	Operating Supplies	Automotive Expenses	OTHER SPENDINGS Expenses	OTHER SPENDINGS General Ledger	Description
3- 1	Downey Management Company	482	$ 600.00					$ 600.00		Rent
2	Tri-County Lumber	483	2,486.40	$ 2,486.40						
2	Readi-Bag Company	484	300.00			300.00				
3	Kilroy Oil Co.	485	482.60				482.60			
3	Midfield Bank	486	600.00						600.00	Note No. 2
3	Midfield Bank	487	400.00						400.00	Note No. 3
4	Dearfield Building Supply	488	2,010.86	2,010.86						
5	Midfield Repair Service	489	260.00				260.00			
5	PAYROLL	490-501	1,525.82		1,525.82					
3- 8	John's Lumber	502	1,899.00	1,899.00						
9	Local 46 Trust Fund	503	425.95					425.95		Union Welfare
9	Downtown Cleaners	504	260.00	260.00						
11	North County Lumber	505	3,311.86	3,311.86						
12	Midfield Bank	506	1,880.02						411.51	F.I.C.A. - Feb.
									1,057.00	Federal Tax - Feb.
12	Department of Taxation	507	201.80						68.00	Disability Insurance
									133.80	State Tax - Feb.
12	PAYROLL	508-519	1,432.38		1,432.38					
12	State Light & Power	520	48.56					48.56		Utilities
3-15	Central Telephone Co.	521	160.12					160.12		Telephone
17	Matthews Tire Center	522	447.09				447.09			
19	PAYROLL	523-534	1,506.38		1,506.38					
3-22	Gallagher Oil Company	535	193.00				193.00			
23	Hunt's Supply Company	536	450.00			450.00				
23	Spanner Lumber Yard	537	1,140.08	1,140.08						
26	PAYROLL	538-549	1,483.00		1,483.00					
3-29	Heady & Krauss	550	200.00					200.00		
31	Midfield Bank	551	1,100.00						1,100.00	Accounting
										Note No. 1
31	State Board of Sales Tax	552	812.92						812.92	Sales Tax - Feb.
31	Sycamore Building Supply	553	1,612.60	1,612.60						
31	v o i d	554	-0-							
31	Rivas Stationery	555	29.91					29.91		Office Supplies
31	Nightwatch Patrol Service	556	25.00					25.00		Security Service
31	H. R. Goodwin, Broker	557	660.00					660.00		Insurance
	TOTAL		$27,945.35	$12,460.80	$5,947.58	$1,010.00	$1,382.69	$2,561.05	$4,583.23	

Note 1 - Summary of Payroll	gross payroll	disability insurance	F.I.C.A.	Federal tax	State tax	net payroll
March 5	$1,960.00	$ 19.60	$ 118.58	$ 259.30	$ 36.70	$1,525.82
March 12	1,840.00	18.40	111.32	243.40	34.50	1,432.38
March 19	1,935.00	19.35	117.07	256.00	36.20	1,506.38
March 26	1,905.00	19.05	115.25	252.10	35.60	1,483.00
Total	$7,640.00	$ 76.40	$ 462.22	$1,010.80	$143.00	$5,947.58

Figure 9-4

Check Writing and Recording

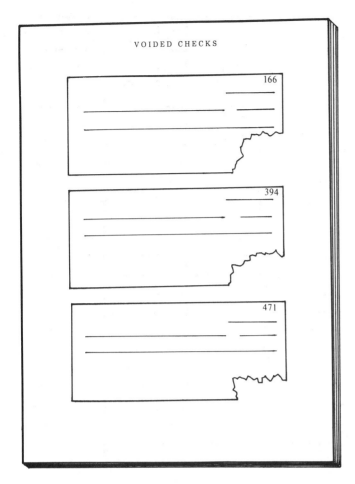

Figure 9-5

- Tear off the corner of the check where your signature would appear.

- Stamp or write "void" on the face of the check.

- File the voided check.

If you accumulate voided checks in a file for several years, you have a folder that is fat at the bottom and thin on the top. This is a bulky and inconvenient way to keep documents. Besides, they would be difficult to keep in numerical order. Figure 9-5 shows one way to keep voided checks neatly. Set up a looseleaf binder or booklet of blank pages. Tape voided checks (as many as will fit on one page) to the pages in check number order. The file stays flat and you have a numerical reference for all voided checks.

You will need to refer to your voided check when you require audited financial statements (for example, in applying for some kind of loans), insurance audits, and stockholder audits (corporations).

PEGBOARD CHECK REGISTERS

Check registers kept on a pegboard differ from manual types in both the style of the checks and the procedure for maintaining the register. A pegboard system saves a lot of time when the volume of work is substantial.

In a pegboard system there is no check stub. Writing the checks creates the check register directly. This eliminates transferring information from stubs to the check register. The obvious advantage is that maintaining the check register and writing checks takes only slightly more time than writing the check alone. The total payment must be distributed to one or more columns. The total for each page must still be added. But the actual process of copying the date, check number, and payee is eliminated.

The disadvantage of pegboard check systems is that the design of the check is very limited. Important information must be transferred through the check from a carbon strip on the back of the check to the check register. All important information must appear above this carbonized line. You have to order pegboard checks from a company that manufactures checks for the pegboard system you use. The checks must be written by hand; they can't be typed.

as a "draw." There is always a better description than "cash."

HANDLING VOIDED CHECKS

Check 554 in Figure 9-4 was voided. Either an error was made in recording the check or the cancellation of a liability did away with the need for a payment. This happens commonly in every business. Even though there is no amount to be recorded, the voided check is accounted for. Check 554 is listed to explain what happened to that check. It is shown with a -0- total. This makes it clear that all checks are listed in sequence and are accounted for.

Many builders throw away voided checks. But voided checks are source documents just like supporting invoices, cancelled checks and statements. They should be kept as part of a complete and self-explanatory check control system.

Follow these steps in voiding checks and maintaining records:

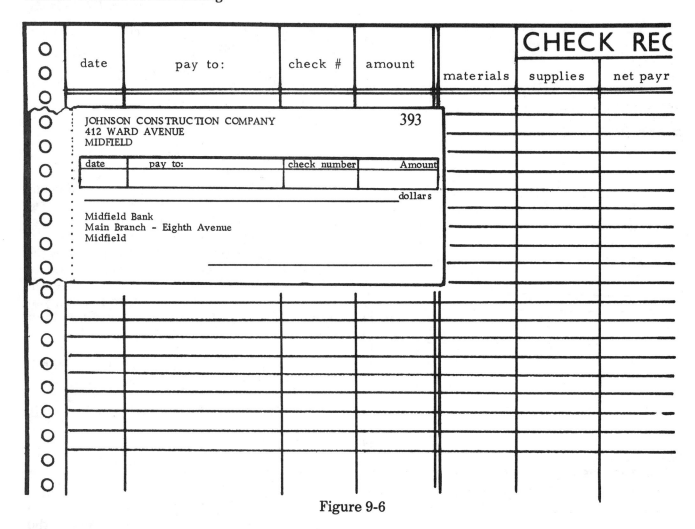

Figure 9-6

With pegboard checks you complete your check register as each check is written. This advantage should outweigh the limitations on check choice and design if you write a large volume of checks. A pegboard check should take less time to complete than it takes to write a manual check and fill in the check stub.

Figure 9-6 shows a sample check register and pegboard style check. Filling in the date, payee, check number, and amount on the check creates an entry in the check register with the same information. The check total is then distributed to the correct column or columns at the right of the form.

When a page is complete, add down the columns to get totals. You can check your math by adding distribution columns across to prove the total column. A balance forward is carried to the top of the next page to accumulate totals for each month.

On any check register system you need a column for other categories of spending. As in the previous examples, there are usually two columns labeled Other Expenses and General Ledger. To make your classification job go more quickly, include a summary of these distribution columns at the bottom of the check register.

Figure 9-7 shows how this can be done. The various amounts in the Other and General Ledger columns have been summarized. This identification would normally be by account code, which saves time because it eliminates writing out lengthy descriptions for each category. Add subtotals for these summaries to find the total. It is better to catch math errors before you begin posting totals to the general ledger. At this late date, problems in addition can waste a lot of your time.

Using pegboard style checks may require some adjustments in your routine. Pegboard checks are usually printed so that they overlap one another and you must fold them back to write each one. Otherwise, the next check would

Check Writing and Recording

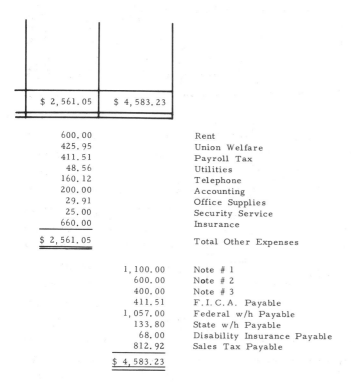

Figure 9-7

flap over the one you want to write. But once you are accustomed to it, a pegboard check system is easier to control. Many professional building organizations use pegboard checks exclusively.

PAYING ACCOUNTS PAYABLE

An accurate financial statement requires reliable information, prepared consistently from month to month. Under the *accrual* method of accounting, this means that you must make adjustments to your general ledger balance for accounts payable.

The accrual method of accounting requires adjustments for events that have not happened yet—such as recording income already earned or bills already payable. The accumulated balance of all bills you owe is your true accounts payable. If you know a month in advance that you will owe money to someone, it is an account payable if the services or goods have been received. Future liabilities for goods or services should not be included.

The other principal accounting method is the *cash* system; you recognize only cash transactions and do not allow for money due or owing. Under the cash accounting method, accounts receivable and accounts payable are not recorded. While this method is acceptable for income tax reporting (when used every year), it does not reflect the true financial picture in a building business because the money owed to you and the money you owe has significant affect on your profit and loss statement and balance sheet.

Some builders record accounts payable only yearly. They carry that balance as the payable liability all the following year. This method gives an accurate picture of the financial condition of the business only if the monthly changes in accounts payable are minor. The accounting entry made in the last month of the year would show:

	Debit	Credit
Various cost and expense accounts	xxx	
Accounts Payable (liability account)		xxx

At the same time, an entry must be made to reverse the previous year's accounts payable accrual:

	Debit	Credit
Accounts Payable	xxx	
Various cost and expense accounts		xxx

Since the same balance of accounts payable will be carried all year, the offsetting modifications to the various cost and expense accounts must also be carried all the following year. Any significant changes in the true balance of accounts payable is not reflected in the monthly financial statements. To this extent, that financial statement does not reflect the facts.

One way to solve this problem is to compute the true accounts payable each month. Make a monthly journal entry to record accounts payable and reverse the previous accounts payable total. But this requires keeping your accounts for the month open until all current billings are received (for example, until the 10th of the following month). Most builders find it inconvenient to keep the books open that long.

The most efficient way to keep track of accounts payable on a monthly basis is to maintain two check register—one for billings pertaining to the current month and one for billings pertaining to the previous month. This method allows you to pay bills received in the current

87

Builder's Guide To Accounting

ACCOUNTS PAYABLE
DOUBLE REGISTER METHOD

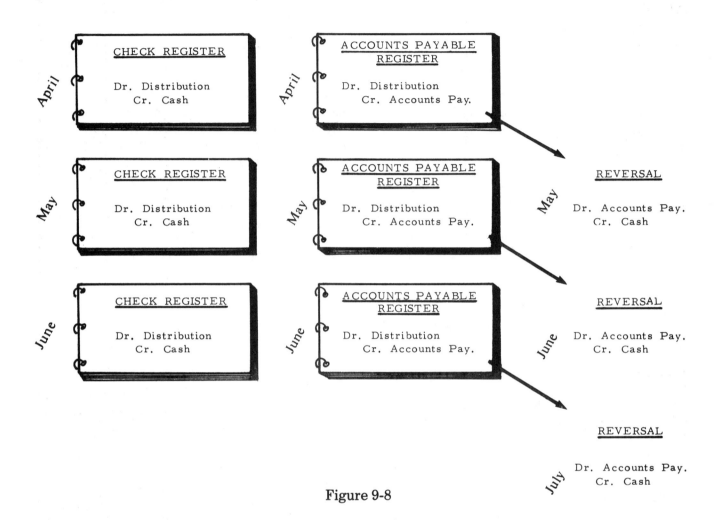

Figure 9-8

month for amounts due in the prior month and record these as payments for services in the prior month. Figure 9-8 illustrates the procedure. *Cr.* stands for credit and *Dr.* stands for debit. In the month of April, a check register is maintained for April payments. Beginning on May 1, a check register is started for May payments on May payables. But an accounts payable register is also opened for May payments on April payables.

Once the register for April is closed, the balance of accounts payable is reversed as shown at the right in Figure 9-8. The net effect of this is that the balance of accounts payable is fixed at the end of each month and the total is reversed each month following the closing of the books. This method is ideal for builders who have a large volume of checks. It works best if

you use two different check number series, one for the check register and one for the accounts payable register. Without this precaution, you lose the control inherent in writing checks for each month in numerical sequence.

Figure 9-9 should help you better understand the result of using a dual register system. This summary form shows the effect of the various transactions on the accounts involved. The cash account entry shows all cash payments. The accounts payable account is reversed to zero each month. But as of the close of any period, the accounts payable total should show the actual amount owed.

As mentioned above, one of the problems of this method is that you must use checks with two number series. This complicates balancing your

	Cash Account		Accounts Payable		Distribution	
	Debit	Credit	Debit	Credit	Debit	Credit
APRIL:						
Check Register		xxx			xxx	
Accounts Payable				xxx	xxx	
MAY:						
Reversal		xxx	xxx			
Check Register		xxx			xxx	
Accounts Payable				xxx	xxx	
JUNE:						
Reversal		xxx	xxx			
Check Register		xxx			xxx	
Accounts Payable				xxx	xxx	

Figure 9-9

bank account because outstanding checks also fall into two number series. There is a good way to simplify recording accounts payable and it might also help you balance your bank account. This involves using two checking accounts and alternating their use from month to month. This is illustrated in Figure 9-10. Beginning on the first day of April, checks are written from bank account B. This is continued through the month. On the first day of May that account is used exclusively for writing checks against the accounts payable register. From the last check for May's check register to the first check in the May accounts payable register, the number sequence would be unbroken.

Beginning on June 1, bank account B is again used for paying the current month's bills. Account A is then used for the May accounts payable register.

This method reduces the number of outstanding checks for each account in alternate months. Each account is inactive for almost a month at a time. The list of outstanding checks is smaller every second month because no checks are written for two or three weeks prior to the end of each alternate month. This is an important consideration for builders who write a large volume of checks. It could save a lot of time when you write out your lists of uncleared checks. Of course, it is a good idea to balance your bank account every month. With this system your bank accounts are balanced only every other month.

Maintaining two checking accounts requires some planning of bank deposits. You must be careful to make deposits to the proper account to cover expected checks, whether they are for current payments or for the previous month's accounts payable. If both accounts are in one bank, you can usually request a transfer between accounts over the phone. Otherwise, you can write a check from one account to the other. The problems involved in using a dual account system are usually more than offset by the advantages.

Another use of multiple checking accounts is in setting aside funds for special purposes. For example, some builders have three accounts:

- *General Account* For the bulk of costs and expenses. This account should receive all bank deposits.

- *Payroll Account* For payroll only. Deposit

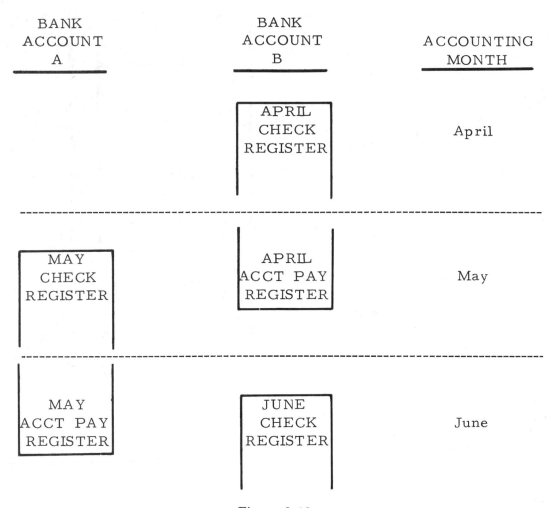

Figure 9-10

in this account only enough money to cover the payroll and the estimated service charges. Maintain a minimum balance in this account.

- *Tax Account* Like the payroll account, deposits are made to this account from the general account only as required. The amount deposited is the estimated total taxes due. These might include payroll taxes, sales tax, highway use tax, property, public utilities, and any other taxes.

Whether multiple accounts are used for different kinds of payments, or just to stagger activity in the check register and payroll register, maintain the controls required. The more accounts you have, the more bank accounts you have to keep balanced, the more printed checks you have to buy and the more service charges you have to pay.

You simplify your use of multiple accounts if all your deposits are made into one account. Other accounts can then be fed transfers of funds as needed. This is an ideal procedure but is not always practical. If you need to deposit receipts into two or more accounts, you might set up a *cash clearing* account in your general ledger. This device is useful because it does away with the problem of balancing actual cash receipts to booked bank deposits. Cash clearing serves to *wash* your receipted bank deposits, which are then distributed into several actual bank accounts. All deposits are recorded as increases to Cash Clearing, and all checks are recorded as decreases. Each month's remaining balance is distributed to the various accounts. Any balance remaining in Cash Clearing should represent the balance of unbooked adjustments to cash accounts. This procedure can help you balance your total cash each month. Even if the cash clearing account never has a zero balance, the account balance at month end should be identified and cleared in the following month's process of bank balancing.

Check Writing and Recording

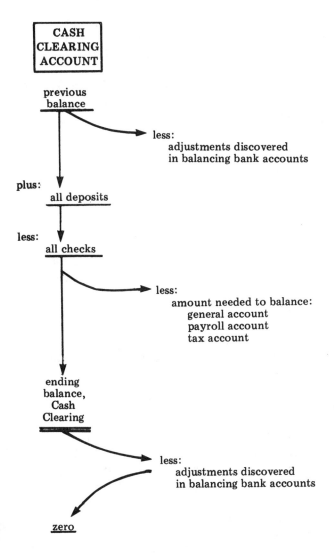

Figure 9-11

Figure 9-11 shows the flow of deposits, checks, and balances through a cash clearing account. This account can make your job easier, since you do not have to balance a variety of cash accounts each month.

MULTIPLE BUSINESSES

Some builders run more than one distinct business from the same office. This can cause a lot of bookkeeping problems, such as the allocation of shared overhead and inventories. In specific, some bills paid by one business are billed in part to the other business. This requires some established control method for inter-company dealings.

Accounting for multiple businesses under single ownership can be difficult, especially when it comes to the check book. For example, suppose you decide that one-third of your rent should be paid by the second business. Should you split the bill and write two checks? Or should one business pay the bill and be reimbursed by the other? Either way, problems arise. It is nearly impossible to allocate overhead expenses precisely between two businesses. The basis on which this is done must vary with each type of expense. Even with the best and most complete analysis, allocation is still an estimate. And money flowing back and forth between the two businesses will obscure the true profit picture of both.

But the two business shop is common. The builder with two businesses can only handle the accounting in the best way available to him. The best accounting system for a builder who owns both businesses is to avoid transfers between the two operations except for periodic adjustments. This involves setting up a balance sheet "net worth" control account on the books of both companies. The net worth account of a dual-operation shop would look like this:

Paid-in Capital	xxx
Retained Earnings	xxx
Johnson Freight Co., Control	<u>xxx</u>
Total Net Worth	<u>xxx</u>

This example assumes that Johnson Construction had a second operation, Johnson Freight Company. This freight company's net worth section would include a similar control account, "Johnson Construction Company, Control." The balance at any time would be identical in both accounts. One would be a positive number. The other would be negative, depending on which company had accrued greater liabilities.

This control should be cleared to zero occasionally by issuing a check from one company to the other. This kind of transfer should be kept to a minimum to make it clear that the two businesses are separate.

As a matter of practice, the two operations should be kept strictly separate. Many dual-business operations have far more than their share of problems. Keeping the bookkeeping as simple as possible will reduce the problems somewhat.

If Johnson Construction makes a payment, one third of which is due from Johnson Freight

Company, the entry would be:

	Debit	Credit
Distribution (two-thirds)	xxx	
Johnson Freight Co., Control	xxx	
Cash		xxx

Johnson Freight Company would make this entry in its books:

	Debit	Credit
Distribution (one-third)	xxx	
Johnson Const. Co., Control		xxx

The two control accounts would have the same entry. The Johnson Freight Company entry, picked up from an allocated charge, should be supported by a voucher to be issued like a bill from Johnson Construction. This is the right way to document the assignment of a bill from one company to another.

Good records require this minimum amount of paper flow. But there is no advantage to paying a bill for another company to avoid writing out a separate check. The payment still must be documented. In this example, some supporting document is required for two sets of books — the charging company and the charged company.

Dual operations cause more paper flow, not less. If you need to operate more than one business, keep the books for each business separate and share expenses by writing multiple checks.

OTHER CHECK-WRITING METHODS

Computerized check writing can be useful to builders who write a large number of checks. This is most effective when payments are made regularly to the same vendors or subcontractors.

Computerized check writing requires that you create a coded vendor list. A common system is to assign a five digit number to every payee your business deals with. Any time a payment must be made to someone not on the list, the name and address must be entered before a check can be issued. The vendor list should remain alphabetical, so the numbers assigned originally must be spaced wide enough to allow for new additions in alphabetical order.

Each time a payment is required, the following information is fed into the computer:

- Vendor number
- Date of the check
- Amount to be paid
- Account to be charged (code number)

This information is then applied to issue a check automatically. The vendor list is searched until the vendor number is found. Coded information under that number produces the vendor's name and full address, which is printed on a blank check. Then the information you have input allows the computer to print the date, check amount, and code on the check.

The check is usually printed with a carbon duplicate for your files. A completely automated check system can produce a distribution run that lists your payments by account code, date or in any order you want them. Most programs produce a daily summary so that you can check for errors before mailing out checks. In addition, the monthly distribution run shows a summary of each general ledger account charged for the entire month. If your whole general ledger is on a computer system, it is automatically updated. Check stubs and check registers are eliminated completely.

A computerized system is not necessarily an error free system. If you make a mistake in inputting information to produce checks, that mistake will be carried through on the entire system. Computers are fast but not independently intelligent. They won't correct your mistakes, no matter how obvious they may be.

Even an automated procedure has its disadvantages. Computers are ideal for handling large volumes of routine numeric data. Most application programs don't handle exceptions very well. And most builders deal in exceptions. When something unusual comes up, everything tends to stop until the new problem is mastered. Errors can be particularly annoying. If an error is discovered in a check, it cannot be corrected until the next run. It is a waste of time to run an entire program to correct one mistake. So errors are usually held over a day.

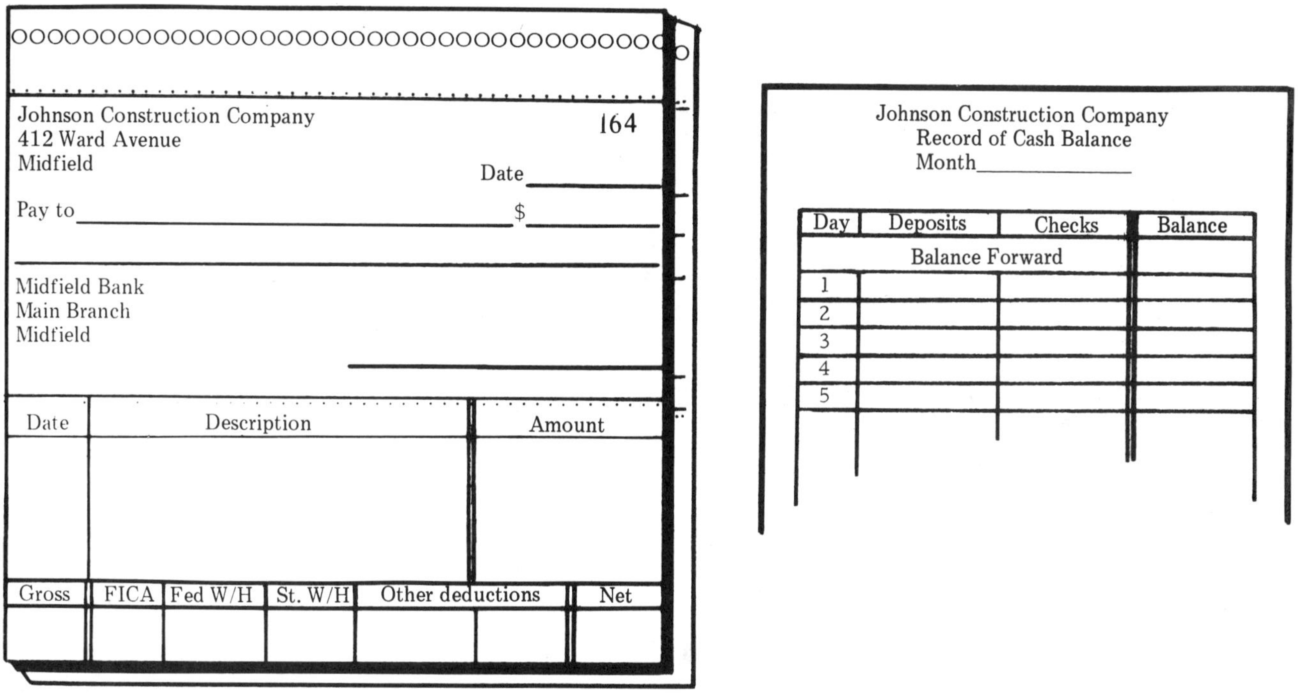

Figure 9-12

On a manual system, you merely cross a line through a check to be voided. On a computer system you would have to reissue the check and put through a special instructions to reverse the error.

Computers are useful and efficient tools, but only for a builder with tremendous volume. When the time saved in running checks automatically is greater than the time spent handling exceptions, computerized processing may be appropriate. The volume most builders have will not justify even limited computer use.

Another check writing system uses a two-part or three-part check. The original is issued for payment and the copies are for the company files. One copy is attached to the invoice or statement it pays and is filed alphabetically. The other copy is used to produce the check register. This system requires that you produce a check register (manually or with automatic equipment) and maintain a separate running bank balance. On a pegboard system, a running cash balance can be kept off to one side so that there is no need for additional paperwork. But with voucher checks, a manual check register must be prepared. Of course, this is routine if you have a service bureau or your own computer to prepare your monthly financial report.

A voucher check is shown on Figure 9-12. Also shown is a record of cash balance form. All checks written on a particular day are added together and entered under the "checks" column. Deposits are entered under the "deposits" column. Any bank adjustments can be added to a day's total for either checks or deposits. The advantage of voucher checks is that they can be typed and do look better than hand written checks. But the value of nice-looking checks is small when you consider the amount of work involved. And typing checks invites errors.

The check writing system you use should be the most efficient you can find. Don't adopt a procedure that gives you less than you need or more than you want. For most builders, a pegboard works well enough. It is flexible but allows room for growth. Fortunately, the volume of checks in most operations does not grow as much as receivables. So any workable method will have a long useful life.

Chapter 10
Accounting For Materials

Direct costs are costs directly related to production on each job you have. As sales increase or decrease, so do direct costs and gross profits. The largest direct costs to a builder are subcontracts, materials and labor. Materials include everything purchased that will be installed on your jobs. Count as direct material cost both material used from your yard as inventory and material delivered directly to jobsites. Freight charges on that material are direct costs because they are part of the cost of getting the material into the structure. Labor is probably your largest direct cost. Manhours are used like materials for specific tasks on each project. The number of manhours varies with each project and is directly related to the production of income. Without labor you can't maintain sales volume.

Direct costs should be distinguished from indirect costs. Indirect costs are the expenses of taking a particular job that don't involve any work on the job itself. These may include repairs to adjoining property, sewer and water connection fees, building permits, jobsite toilets and the like. Overhead is yet another category of expenses. Overhead is the cost of the items you need to conduct business that cannot be charged against a particular job. These include office expenses, small tools, printing, postage and the like. Overhead remains fairly steady regardless of the work in progress. Direct costs and indirect costs change as your volume changes. This chapter examines direct material costs and shows how to use ratios to analyze the effectiveness of material handling. Direct labor costs and payroll records are the topic in Chapter 11. See Chapter 15 for designing your own records by job, and Chapter 16 for by-the-job accounting methods. Estimating labor is discussed in Chapter 19.

MATERIALS

The more professional the building operation, the more emphasis materials handling and control will usually receive. About 50% of your direct costs will be for materials. It has been estimated that about 8 to 10% of labor costs on residential construction is required just to move materials from the curb line to where they are installed. But that is just the beginning. Early deliveries mean overstocked yards and onsite vandalism, arson, theft, and weather-related losses. Chronic overstocking increases your costs for insurance, handling, property taxes and distribution, and can mean losses from obsolescence. Materials arriving late cause expensive delays, interrupt your job schedule, and create costly idle time. Subcontractors are delayed. Without a good materials handling program, even the best builders lose profits.

Agree on delivery terms with your suppliers. This is as important as large discounts, lower prices, and quality materials in cutting costs. Suppliers who are dependable in meeting your delivery specifications will save you more money than discount suppliers who can't deliver.

A well-organized builder needs to know exactly what he wants, and when. Develop a firm onsite schedule. Stick to this schedule and the job will move along smoothly and profitably. Idle time is cut and deliveries are there when you need them. Know as far in advance as possible what will be needed at each site on a day-to-day basis.

Volume buying is a big part of a good purchase plan. A good supply of materials in shop inventory gives you a contingency reserve you can use when deliveries are short. Volume purchasing also brings discount terms, which can make the difference between high and low profit. But discounts should be sought only if the purchase can be justified. Be sure you can use the discount material in the reasonably near future. Otherwise, your yard becomes loaded with dead inventory. This extra handling means higher operating expenses and an unnecessary commitment of your capital.

You can have a smooth flow of materials through the yard if your operation is diversified. For example, Johnson Construction does both bid contract work and repairs and improvements. It can order and use a steady supply of inventory, switching from one type of job to another during slack periods or to meet varying demand. But Johnson Construction can still overstock if they make large materials purchases that they can't use on upcoming jobs.

Your mix of jobs should keep your inventory moving steadily. But you need control over materials entering your yard from suppliers. Written purchase orders are the most simple control device. Specify to your suppliers, wholesalers, and shippers exactly what materials you want, when you want them, and where you need them delivered. You avoid misunderstandings and late or early deliveries. And you manage your purchases yourself. Your field superintendents or foremen can be authorized to make purchases on your purchase order form. This may be your only way of controlling job costs before you are committed. Control of the form itself should help you cut down on unauthorized purchases.

With a purchase order system, you often receive a confirmation from the supplier including terms and conditions. This lets you compare different suppliers' terms to find the best place to buy specific materials.

Refer to Figure 10-1. A purchase order form should include the following information:

- Billing and delivery addresses
- A preprinted number
- Order and delivery dates
- Shipping instructions, including the shipper and terms of delivery
- Quantity, description, unit price, and total price
- Authorized signature

You need at least three copies of the purchase order. Distribute copies as follows:

- Original — To the supplier.
- First copy — For the office file. Keep the file in numerical order.
- Accounting copy — To be matched to the invoice when received. Keep these in alphabetical order until bills are paid. Check the prices on the invoice against purchase order (agreed) prices.
- Fourth copy — For receiving purposes if necessary. You may want a document in hand when the delivery arrives.

A well-documented purchase system provides you with a ready reference, good records, and better communications with suppliers. Purchase orders are the basis of a smoothly run delivery schedule as they coordinate field activities over a wide geographical area.

Summarize your purchase orders on a purchase journal so you can analyze total company material purchases at a glance for a given period. Organize your summary in purchase order number order. See Figure 10-2. A properly designed form provides both inventory summaries and breakdowns by job.

```
                JOHNSON CONSTRUCTION COMPANY
                       PURCHASE ORDER

                                                  No.  P.O.89376

Date_____       Job Number _____

Billing Address:     412 Ward Avenue, Midfield

Delivery Address: _____

Shipping Instructions: _____
_____
_____
_____
```

quantity	description	price ea.	total
	Total		

```
Date confirmation                 Signature_____
requested_____
                                  Title _____
Confirmed by_____
```

Figure 10-1

Accounting For Materials

	JOHNSON CONSTRUCTION COMPANY
	PURCHASE JOURNAL
	Month_____ Year____

Date	Purchase Order Number	Total Purchase	Inventory	Job # ____	Job # ____	Job # ____	Job # ____	Job # ____	Job # ____	Job # ____
TOTAL										

Dr purchases / Cr acct pay

Figure 10-2

This is a good way to compare budgeted spendings to actual costs. The first distribution column, "Inventory," tells you by date what purchases have been made toward your yard inventory by referring to the specific purchase order on the same line. You control materials purchased and delivered directly to each jobsite using the purchase date and purchase order number reference as a distribution guide. You know what you need for each job in the immediate future. This form helps route deliveries where they are needed and keep the inventory levels adequate for your work output. It helps keep down unnecessary material purchases that cause timing problems and cash shortages. This journal also assures that your jobs are allocated the right amount of materials from the suppliers on the right dates.

Some builders use the purchase journal as a source document for accounting entries. The total purchases are treated as accounts payable, and the entire purchase journal is used as a supplementary accounts payable ledger. But using the accounts payable and check register method discussed in Chapter 9 eliminates the need for a purchase journal as an accounting document. The two procedures are not compatible. The real usefulness of a purchase journal is in better control of your inventory.

If your purchasing system requires that you use a purchase journal as an accounting document, here is the entry you must make each month to record purchases.

Debit: Purchases (Direct Cost)

Credit: Accounts Payable (Liability)

As purchases are billed and paid for, an entry must reverse the one above:

Debit: Accounts Payable (Reversal)

Credit: Cash (Paid out)

Unless all purchases are recorded on the purchase journal, this procedure can cause trouble. Be careful to code all material payments

```
          JOHNSON CONSTRUCTION COMPANY
               PURCHASE REQUISITION

Date _____        ISSUED   ☐
Job # _____        RETURNED ☐

| Number | Description | Price Ea. | Total |
|--------|-------------|-----------|-------|
|        |             |           |       |
|        |             |           |       |
|        |             |           |       |
|        |             |           |       |
|        |             |           |       |
|        |             |           |       |
|        |             |           |       |
|        |             |           |       |
|        |             |           |       |
|        |             |           |       |
|        |             |           |       |
|        |             |           |       |
|        |             | TOTAL     |       |

Signature _____
```

Figure 10-3

correctly. Some payments will be coded to Materials Purchased (those not recorded on the journal) and some to Accounts Payable (those recorded on the journal). Otherwise, both the liability for accounts payable and the materials purchased accounts will be overstated. This is an example of a troublesome accounting procedure. It creates more problems than it solves.

INVENTORY

Controlling your inventory requires a purchasing plan that is in line with a strict inventory level policy. How much do you need to keep on hand, and of what materials? Lack of planning and control can leave you with too much material at any given time. This ties up working capital that could be better used. Keeping too little material on hand is expensive as well. You cannot tell exactly what piece you need tomorrow or next week. Having a reasonable amount of surplus materials on hand saves time and brings in more short-term contracts.

Establish minimum and maximum levels of shop inventory, and then stay within those limits. To do this you need to know what you have on hand. This means keeping track of your inventory on a perpetual basis.

Purchase orders summarized on the purchase journal tell you what has gone into inventory. You also need a systematic way of knowing what comes out. If you carry more than a few thousand dollars worth of inventory in your yard, every piece of material drawn from your yard should be documented with a requisition. This cuts your losses and controls your stock.

Figure 10-3 is a purchase requisition. The information needed to complete this form includes the date, job number, material description, number of items, price each, and total amount. The amount should be equal in value to the amount at which items are carried in the inventory. Valuation is discussed in the next section of this chapter.

At the top of the purchase requisition are two boxes, labeled "Issued" and "Returned." When materials are taken from inventory, check the issued box. When materials are returned to inventory, check the returned box. To keep a running account of inventory, carefully divide your requisitions into issues and returns and record them by date on an inventory form. See Figure 10-4.

You can keep a running inventory on a daily, weekly, or monthly basis. Beginning with a physically counted balance, subtract issues from inventory and add returns and orders received. You need to take a physical count from time to time to determine whether your requisition procedure is working and to give you a beginning balance from which to begin your inventories. Try to take a count more often than once a year. If your level of inventory changes greatly from month to month, look for low levels as a good time to make a count. A good way to check inventory is to count different categories at different times, rather than taking a whole count all at once. Categorize your materials and count each category its own month. This requires carefully marked purchase orders and requisitions with the category code numbers.

VALUING INVENTORIES

Inventories are usually valued on some type of cost basis, since the methods used to report

JOHNSON CONSTRUCTION COMPANY
INVENTORY

Date	Orders Received	Issued from Inventory	Returned to Inventory	Balance	Physical Count

Figure 10-4

sales, profits, and yields are all based on actual costs. Job estimates are based on costs as well, so it is logical to maintain your inventories the same way. There are three cost-basis valuation methods worth considering. The right one for you depends primarily on the size of your average inventory.

Specific cost Materials are labeled with their exact purchased price or cost. As they are used, this cost is read off the labels and assigned to the cost summary of the appropriate job. This method requires the individual labeling of each piece of material in the inventory.

First in-first out The first materials purchased are assumed to be the first ones used. The amount charged to jobs as materials are used is the earliest price paid for the materials on hand.

First in-last out The latest materials purchased are the first ones used. This is the easiest way to maintain a perpetual inventory because you don't have to label stock or look up prices for materials you purchased long ago. You know the current market value of material as it is used.

There are four major categories of inventory. They should be distinguished from each other, since you may have to keep separate records unless all your material falls into the first category.

Raw materials This includes all direct purchases for use on various jobs.

Supplies This could include items like nails, bags or boxes.

Work in process You need this category if you prefabricate in your shop.

Finished goods Products that are ready to sell in a prepared state fall into this category. They are different from raw materials because they are put through a process of completion.

Some builders who do shop work keep inventories of both shop materials and jobsite materials. This means they need a complete requisition system for each site, unless they take a physical count regularly. You can avoid this problem if you consider goods delivered to construction sites as out of inventory and thus committed to the jobs.

ANALYZING MATERIALS AND INVENTORY

There are two principal ratios used to judge whether the level of your inventory is too high or too low for the volume of work you do. One is the cost of goods sold to average inventories. The other is sales to inventories. The sales to inventories ratio is not a precise indication of effectiveness because inventories are based on the cost of materials and sales are based on a computed cost plus profit. But you can use this ratio if sales are stable from month to month. A far more useful ratio is the cost of goods sold to average inventories. The cost of goods sold represents all direct costs, adjusted for the change in inventory levels. This is shown below:

Inventory at beginning of period	XXX
Labor	XXX
Materials purchased	XXX
All other direct costs	XXX
Total direct costs	XXX
Direct costs plus beginning inventory	XXX
Less: Inventory at end of period	XXX
Cost of Goods Sold	XXX

Find the second part of the ratio, average inventories, by adding the balances of the physically counted and perpetual inventories from each month and dividing by 12. Shown below are physical inventories based on the calendar year, but partnerships and corporations can base their physical inventories on the fiscal year.

January	(perpetual)	xxx
February	(perpetual)	xxx
March	(physical, quarterly)	xxx

• • •

December	(physical, yearly)	xxx
Total balances		xxx / 12

Since there are four possible types of inventories, there are four variations on the ratio.

Raw Materials:
$$\frac{\text{Cost of raw materials used}}{\text{Average inventories of raw materials}} = \text{ratio}$$

Supplies:
$$\frac{\text{Cost of supplies used}}{\text{Average inventories of supplies}} = \text{ratio}$$

Work in Process:
$$\frac{\text{Cost of goods completed}}{\text{Average inventories of work in process}} = \text{ratio}$$

Finished Goods:
$$\frac{\text{Cost of goods sold}}{\text{Average inventories of finished goods}} = \text{ratio}$$

Compute these ratios regularly. Monitor trends as they progress and take action to correct the situation. A low ratio can mean that you are keeping your average inventory too high. A high ratio can mean that you are maintaining too small an inventory and may result in lost sales. The correct level depends on your operation and its volume.

Good inventory control does the following:

- Minimizes investments of capital inventory.

- Reduces the need for storage space.

- Reduces exposure of materials to theft or arson.

- Economizes on taxes, insurance, and other costs of maintaining an inventory.

- Allows you to take advantage of discounts, reductions, and special prices through volume purchasing.

- Provides needed materials fast.

- Helps you avoid construction delays.

- Minimizes obsolescence losses.

- Helps you know the amount of material on hand and reduces the cost and time of physical counts.

THE NON-SYSTEM SYSTEM

Whether you account for materials by hand or with a computer, the important thing is to install the procedures you need for your particular business. A system that does more than you need is as ineffective as one that gives you less than you need.

Many builders can get by without any significant method for accounting for materials. They simply work from job to job without a material tracking system. You don't need a system if you don't keep inventory on hand. Many contractors in different specializations order materials for specific jobs and have them delivered directly to the site — or they pick them up as needed and take them to the site themselves.

Without inventory, there's no back order problem. There's also no need to count inventory, insure it, pay for storage space, or deliver it to job sites.

You may still have to analyze material costs as they relate to total job costs and profits. Some builders depend on supplier discounts to make their material margins instead of marking up materials substantially. This is fine as long as you can get an acceptable discount, which means you pay by the deadline and order a regular volume of materials. If you don't qualify, you need to add on some markup when you bill your customers. Analyzing your costs versus profits will tell the story.

Here's an example. Say you've noticed that you make a higher percentage of profit on smaller jobs, or jobs requiring more labor than materials. This is an indication that you're not giving material costs due consideration. Since you have to invest capital in materials during the job, you should mark up your costs to compensate yourself. How much? That depends on the local custom. If you go too low, you're cutting into potential profits. If you go too high, you'll lose business to more competitive builders. Your records will help you make the call.

Chapter 11
Payroll Accounting

Labor is the largest and most variable cost in building. Because of the nature of labor costs and the legal and practical needs of payroll bookkeeping, you must keep complete records. This chapter covers the record-keeping requirements and procedures related to the company payroll. Regardless of whether your operation employs one or a hundred employees, you should have these records:

- Detailed earnings records for all employees

- Summaries of payroll data for reports you file with the federal and state governments, state employment agencies, and health and compensation insurance companies

- Payroll accounting information for your general ledger

- Good payroll records for preparing company budgets and in-house labor cost reports

- Labor cost data broken down by specific types of job or job phase to use for future contract estimates

Figure 11-1 summarizes the flow of information and paperwork in payroll bookkeeping.

You issue payroll checks at regular intervals, probably weekly. So you generate a large number of payroll checks every month. Your check register can become cluttered and you can lose control of it if you list payroll checks in full detail along with your routine payments. Payroll checks require five or more columns, so the detailed listing of your other payments suffers. This means that breaking down routine payments for analysis is harder. For this and the following reasons, list payroll checks in a separate check register.

- A check register should always list checks in strict numerical order. The style of most payroll checks is different than regular checks, so you need a separate number series.

- Payroll record-keeping must include accurate earnings records for all employees. Controlling two functions on one check register is an invitation to error.

- A separate register for payroll allows special cost analyses. A standard check register cannot be effective as a control re-

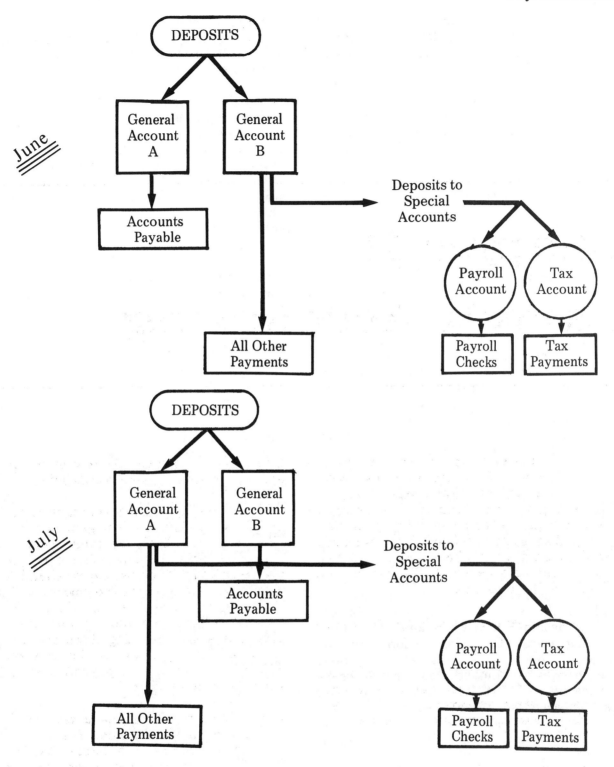

Figure 11-1

cord and a payroll summary. There isn't enough room for a practical format.

There are four common ways of keeping payroll separate from other payments.

1) You can pay a bank or an independent data service center to prepare the payroll and provide backup reports.

2) You can use the general account for both

payroll and other payments, but you have a special style of check with its own numbering system for payroll. Enter payroll on the check register in summary form.

3) You can issue payroll checks from the one general account and from the same check series. Checks must be designed to handle any kind of payment.

4) You can set up a separate payroll account with its own check style used only for payroll.

Most major banks can prepare company payrolls as well as current and year-to-date records, general ledger, a check register entries summary, and all other employee records at modest cost. These services are also available from non-bank data service centers. Many payroll service systems also prepare quarterly payroll tax returns. They can automatically pay deposits when due and can provide employee tax reports once a year. Banks can transfer funds out of your general account to cover payroll and tax payments if you request it, and let you know how much has been used. The cost of an outside payroll service may be justified by the convenience and time it saves.

The third procedure above has disadvantages. In Chapter 9, Johnson Construction used a single check register and weekly payroll summaries. Payroll was written on the same account as other checks. But it was prepared and analyzed on a separate register where all necessary details were listed. This procedure causes confusion since the checks are issued in a single number series for all payments. Checks must be designed to pay any kind of account, including payroll. Or a separate form must be prepared to detail gross checks and deductions for each employee. But either method makes for duplication of effort. Itemizing each payroll check must be done three times:

- To write the check stub or summary of earnings

- To complete the payroll register

- To complete the employee's earnings record

A check designed for both payroll and regular payments is more expensive to print and is not practical in most building operations.

A much more practical way to handle payroll and still maintain a single account is to use a special check style with its own number system for payroll. Use this with a full pegboard system and you eliminate duplication of effort.

Finally, you can establish a separate payroll account. This isolates the payroll procedure from the rest of your checking. You don't have to make payroll fit a check register that was never intended for it. No other kind of account is as easy to keep in balance because most checks are given out in person and cashed promptly. This means you have a smaller listing of outstanding checks.

Budgeting your cash is easier with a separate payroll account, because you set aside a lump sum each pay period in that account for payroll only. Make deposits to payroll by writing a check before each payday from the general account.

You can budget for payroll taxes in one of two ways if you keep a separate account for payroll. First, you can deposit just enough in payroll to cover that week's checks. You keep a minimum balance in the account at all times. Thus payroll becomes a clearing account. At the same time, you write a check from the general account to cover payroll taxes and deposit this amount to a tax account. This third account is practical for builders, since they must pay a variety of taxes that come due all at once. You set aside funds each week for this future payment. Each deposit pays for payroll taxes for a single pay period and helps pay all other taxes as well. The second way to budget for payroll taxes is to use the payroll account both to pay employees and accumulate payroll taxes from week to week. There is no tax account to balance each month. The payroll account builds up until taxes are payable and includes both withheld taxes and your share of employment as well as payroll taxes. The funds for payroll checks and for taxes are available for either kind of payment. Each week's deposits should cover the following:

- Gross payroll (amount paid before deductions).

- Employer's share of F.I.C.A. — an amount which matches the total amount withheld from paychecks.

- Employer's liability for federal unemployment tax, state unemployment tax, and

Payroll Accounting

Project Number	hours worked								Foreman's Approval	Work Description
	Mon	Tues	Wed	Thu	Fri	Sat	Sun	Total		
Total										

TIME CARD — JOHNSON CONSTRUCTION COMPANY
Name _____ Employee Number _____ Week Ending _____

Figure 11-2

any other state or local taxes for which you are liable.

Some builders include a provision in this weekly deposit for union welfare payments and workmen's compensation insurance.

Figure 11-1 shows how multiple accounts are used and how funds are transferred to the payroll and tax accounts. Deposits to these two special accounts are always made from the check register, not the accounts payable register. Checks are written alternately from general accounts A and B in Figure 11-1.

Chapter 10 covered the advantage of using multiple general accounts in alternating months. The accounts payable/check register method allows a monthly balancing of the bank balance.

TIME CARDS

You pay employees in two main ways — by salary and by wage. When you pay a salary you pay the same amount each week or month regardless of the number of hours the employee puts in. Partners in partnerships and business owners in corporations generally pay themselves salaries. Partners can also take draws, which are not salaries but profits withdrawn from the business. You pay wages by the hour. The hourly rate you pay your crews is often determined by union contract. Some clerical workers may be paid a salary, but most can be paid by the hour.

Many states require that time cards be kept as a basic record to prove how many hours a wage-earning employee has worked. They are important documents in unemployment benefit disputes. Time cards also give you a basis for complete cost analyses. They can be printed with the names of specific jobs or contract phases so you can easily add up the hours spent on each task.

Figure 11-2 shows a time card used by Johnson Construction Company. Each wage-earning employee turns in an approved time card each week. Hours are listed by day and are broken down by project. The foreman approves the work description section and the weekly total hours for each man in his area. This doesn't take much time, and provides information for making all the cost analysis required.

Johnson Construction's timecard format makes each foreman responsible for his own weekly investment of labor. Field reporting can be made still more uniform. Johnson could develop a comprehensive list of job types performed in his operation. He could code each timecard by type as well as by project where he thinks he needs a greater cost breakdown. One job type might be shop time, which could

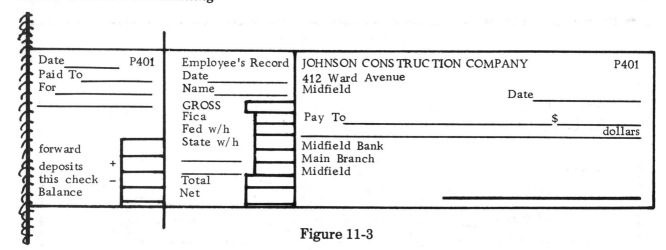

Figure 11-3

include work done on shop fabricated components, inside repairs, and inventory-related tasks. Idle time might be another valuable job type on time cards. Idle time is expensive. You pay for delayed deliveries, lost plans, bad weather and the time your men spend standing around. A labor cost analysis should show you how to coordinate your crews more effectively and cut down on this unproductive time. Make complete breakdowns of time card time to eliminate any unaccounted for hours. Include explanations of non-worked days such as paid holidays, vacations, and paid sick leave.

The foreman or field superintendent should deliver the completed time cards to the office several days before the payroll is due. Check them over and put them in a logical order. Arrange the time cards alphabetically if the employee earnings records are kept this way. This saves time when you start writing checks. Prepare each payroll check by computing from the timecard the number of hours the employee worked times the rate per hour. You figure the employee's net pay by deducting taxes from this gross amount. The timecard is thus a very important source document, both for tax purposes and for your own labor records.

PAYROLL CHECKS

Payroll checks come in many styles. Checks with stubs or deduction summaries create duplication of effort. But pegboard checks are designed for one-step processing.

The least economical of the stubbed checks is the double-stub type like the one shown in Figure 11-3. This check is designed for businesses that require one check for all uses. The middle stub, the employee's record, is not used at all when paying non-payroll bills. You must do the following to complete payroll with this type of check:

- Fill out the employee's earnings record. Do this to determine that the employee has not reached the maximum limits on any of the limited deductions.

- Complete the check stub itself, carrying forward the balance from the previous stubs.

- Fill in the employee's record stub. The employee detaches this stub from his check and keeps it for his records.

- Write the check itself.

- Write the check register entry.

A pegboard system only requires you to write the check and update the cash balance.

Another type of payroll method uses a regular check, as shown in Chapter 9. This procedure has as many steps as the double-stub method and requires that an employee record called an *earnings statement* be completed as a separate step. See Figure 11-4. Similar forms are available in duplicate sets in small booklets or single sheets. No matter how efficiently these forms can be set up, no steps can be eliminated. Figure 11-4 is a little more detailed than a double-stubbed check. It breaks down different rates of pay and shows the period covered by the check. In many states, showing the period covered is a legal requirement. Pegboard systems usually include adequate space to include such details.

STATEMENT OF EARNINGS

Employee_____ Paid from_____
Job Classification_____ to_____

Hours: Rate: Total:
    Regular
    Overtime
    Other
    GROSS

Deductions:
    F.I.C.A.
    Federal withholding
    State withholding
    Disability Insurance
    _____
    _____
    TOTAL DEDUCTIONS

NET CHECK   (Check Number_____)

Figure 11-4

Date	Name	Hours	Rate	Gross	FICA	Fed.	State	Other	Total	Net

Johnson Construction Company   P 819
412 Ward Avenue
Midfield

                                        Date _____

Pay To _____

_____ dollars

Midfield Bank
Main Branch
Midfield

Figure 11-5

Builder's Guide To Accounting

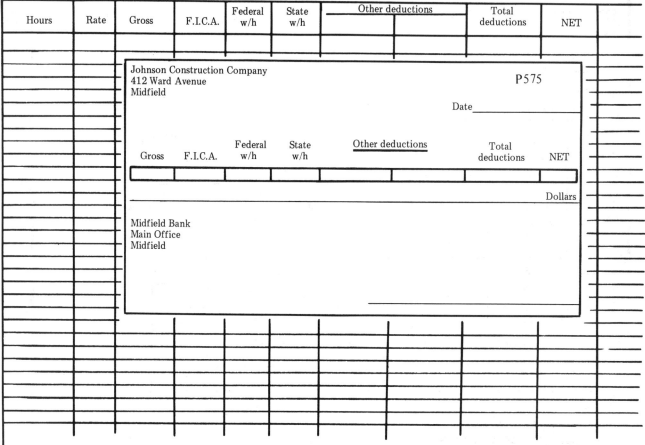

Figure 11-6

Some builders make out the statement of earnings in duplicate, giving the original to the employee and keeping the copy. This means that they keep no payroll register or earnings record once a month. They hand over copies of the statement to an accountant, who makes entries in these two records himself. This means duplication of effort plus a higher accounting bill.

Many builders type each check they issue. There is a greater chance of error in any multi-step process and typographical errors will compound payroll balancing problems.

A pegboard system for payroll accounting makes the job of recording accounting information easier by far. You can choose from several check designs. Some use a carbon strip in the middle of the check. Their main fault is that you must fill out a separate stub or earnings record. The most practical pegboard check has a removable carbon strip at the top of each check. This strip transfers all the necessary payroll accounting information to the records beneath. You fill out the strip first, then you complete the check, which is an uncarboned original. Before you issue the check you must note the amount on it to catch any transcribing errors from the strip.

Figure 11-5 shows a top-strip style pegboard check. Figure 11-6 shows a center-strip style, less practical but also less work than a strictly manual system.

EARNINGS RECORD

You must keep detailed and accurate earnings records for each employee to comply with the legal and tax requirements of federal, state, and local government agencies. You need these records to determine the point at which total earnings reach maximum levels for certain types of taxes. Above these maximum limits you pay no tax, as discussed later in the chapter. Tax reports must be sent to governments and to all employees each year. Unemployment depart-

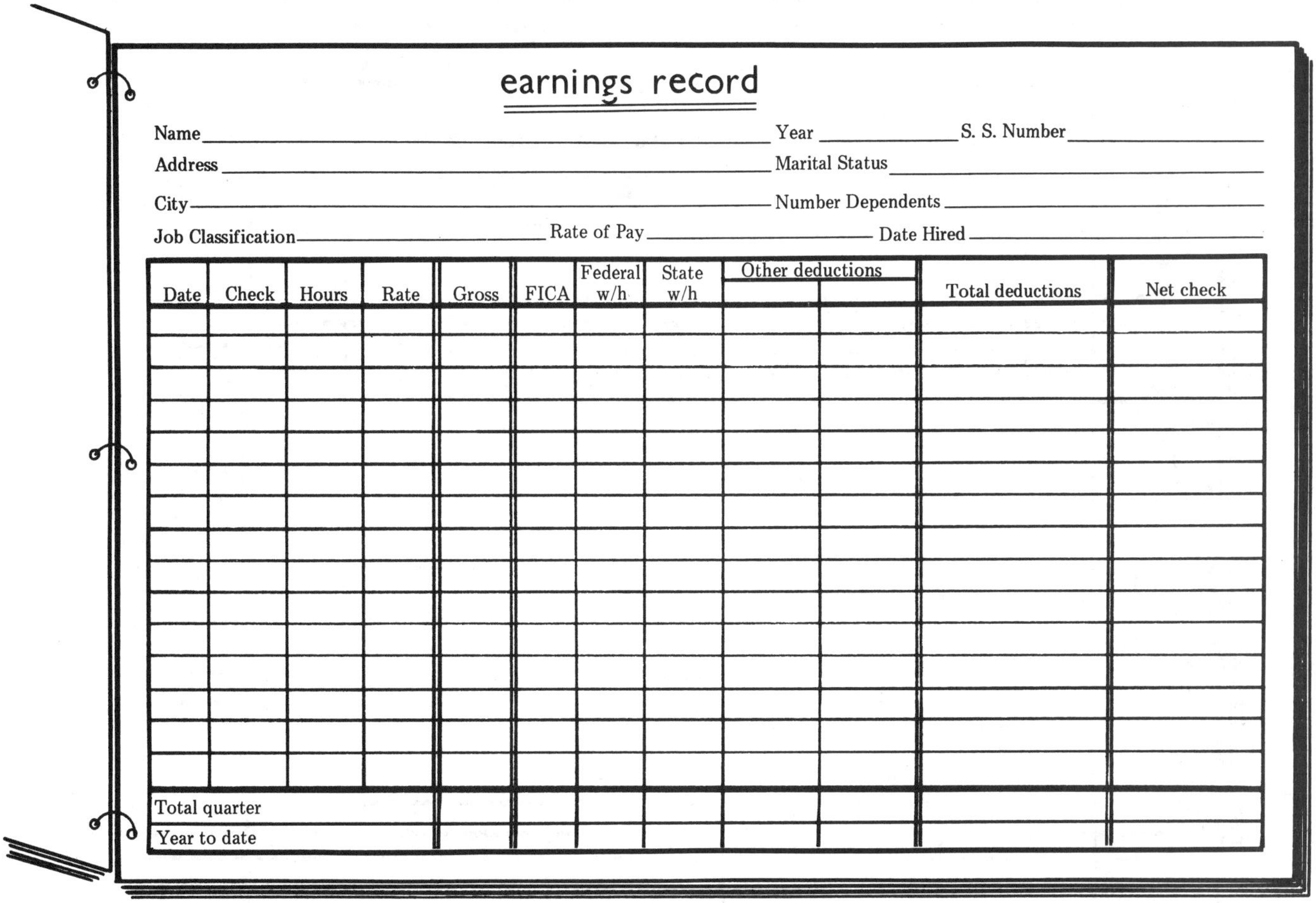

earnings record

Name ____ Year ____ S. S. Number ____
Address ____ Marital Status ____
City ____ Number Dependents ____
Job Classification ____ Rate of Pay ____ Date Hired ____

Date	Check	Hours	Rate	Gross	FICA	Federal w/h	State w/h	Other deductions		Total deductions	Net check
Total quarter											
Year to date											

Figure 11-7

ments want individual records or information from these tax reports to verify claims for unemployment insurance. Union reports and worker's compensation reports are based on earnings records.

Figure 11-7 shows the information on an earnings record. Forms are available that can handle this kind of requirement for an entire year on one page. Or a form can be designed to your own specifications.

Payroll is reported for taxes on a yearly basis. Keep earnings records for each employee for each year. Also keep quarterly subtotals, as payroll taxes are payable and reportable at the close of each quarter.

PAYROLL REGISTER

The payroll register is the same as the check register, except that it includes the detail required for payroll checks. The columns on the payroll record are the same as those on the earnings record, but they can be designed to allow more room on the right side for cost analysis. Weekly cost analysis is discussed in Chapter 15.

A pegboard system makes filling in the payroll register as easy as making out the check. See Figure 11-8. Place the payroll register on the pegboard and put a piece of carbon paper over it (unless you are using a carbonless treated-paper transfer system). Place the employee's earnings record over this. Do not put the earnings record on the pegs but push it into position. Thus you don't have to remove checks each time you place another employee's earnings card on the board. Attach the checks on the pegs directly over the earnings record. Make sure the earnings record is lined up so that the carbon strip on the back of the check will come through on the first unused line. As you make out the check, the amounts are transferred to the earnings record. These amounts are transferred in turn through the

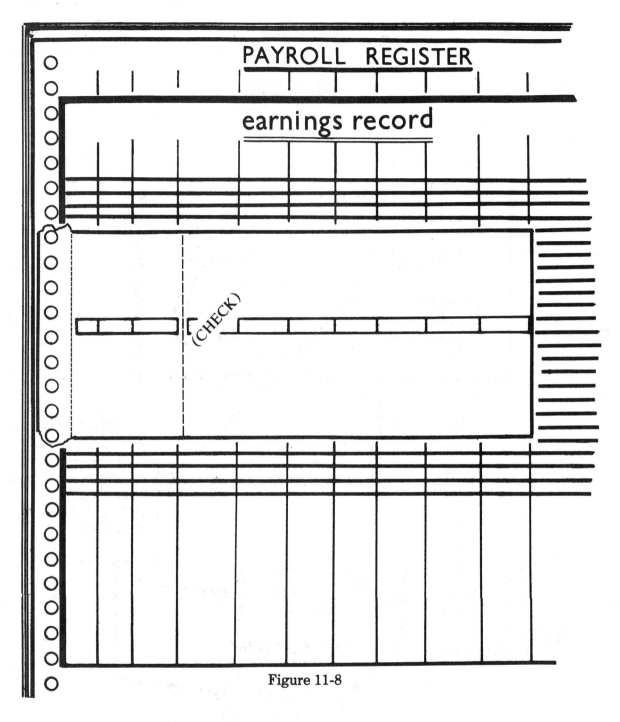

Figure 11-8

carbon paper and onto the payroll register. The flow of payroll data and paperwork is summarized in Figure 11-9.

ACCOUNTING FOR TAX LIABILITIES

Your tax liabilities arise in two ways: from taxes withheld from payroll checks, and from employer taxes. You pay taxes on behalf of your employees to various taxing agencies, withholding these amounts from employee checks. You also pay employer payroll taxes based on your gross earnings. This amount is your own liability and is not withheld from payroll.

Both taxes are payable at certain intervals to federal and state governments. Familiarize yourself with the regulations and deadlines for these payments. You can't afford to ignore tax liabilities if you want to stay in business.

Although there are many ways to account for tax liabilities, the easiest method is to have two separate general ledger tax accounts. These should be called *Payroll Taxes Withheld* and

Payroll Accounting

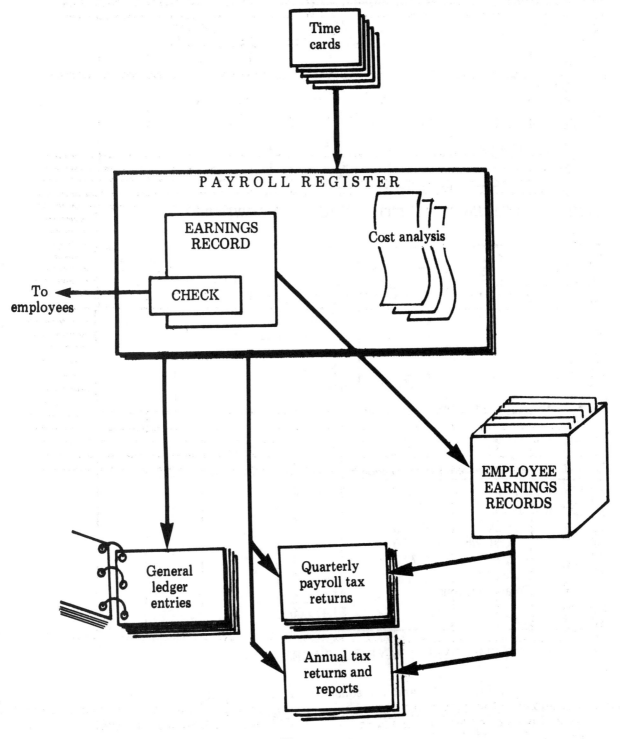

Figure 11-9

Employer Payroll Taxes Payable. The first is a clearing account, through which taxes you withhold from employees are deposited and then paid. The second account is for employer taxes. Funds accumulate in the employer tax account until minimum deposit amounts are reached.

Figure 11-10 shows the flow of funds through clearing accounts such as *Payroll Taxes Withheld*. On a gross payroll of $5,000.00, withholding taxes are $1,496.00 for the current month. This amount includes the tax liability on behalf of the employees, so it is deposited in the clearing account in the current month.

111

Gross Payroll		$5,000.00
Deductions:		
F.I.C.A.	$382.50	
Federal w/h	891.00	
State w/h	197.50	
State disability	50.00	
Total deductions		$1,521.00
Net		**$3,479.00**

LIABILITY (CLEARING ACCOUNT) - PAYROLL TAXES WITHHELD

Detail - accounts by type					Summary		
F.I.C.A.	Fed. w/h	State w/h	Dis. Ins.		DEBIT	CREDIT	BALANCE
(242.00)	(560.00)	(158.00)	(40.00)	Balance forward			(1,000.00)
(382.50)	(891.00)	(197.50)	(50.00)	Current Month		1,521.00	(2,521.00)
				Prior month paid:			
		158.00	40.00	State	198.00		
242.00	560.00			Federal	802.00		
(382.50)	(891.00)	(197.50)	(50.00)	Balance forward			(1,521.00)

Figure 11-10

Gross payroll		$5,000.00
Deductions:		
F.I.C.A.	$382.50	
Federal w/h	891.00	
State w/h	197.50	
State disability	50.00	
Total deductions		**$1,521.00**
Net		**$3,479.00**

LIABILITY, EMPLOYER'S PAYROLL TAXES PAYABLE

A) F.I.C.A. - An amount equal to amount withheld ---------------- $382.50

B) State unemployment tax - state's prevailing rate times gross payroll.
    Example:
        3.6% x $5,000.00 -------------------------------------- 180.00

C) Federal unemployment tax - Federal prevailing rate times gross payroll.
    Example:
        0.08% x $5,000.00 ------------------------------------- 40.00

    Total Employer's Payroll Taxes --------------------- $602.50

Figure 11-11

Payroll Accounting

	PAYROLL TAXES		TAX LIABILITY	
	Debit	Credit	Debit	Credit
Balance forward	$1,545.00			$442.00
Current month	602.50			602.50
Prior months paid:				
State (The offsetting entry is to "cash.")			$144.00	
Federal (The offsetting entry is to "cash.")			249.50	
Balance forward	$2,147.50			$651.00*

*The "Tax Liability" balance forward consists of:
  F.I.C.A. payable, current month $382.50
  State unemployment insurance payable, current month   180.00
  Federal unemployment insurance payable, year-to-date   88.50
  Balance forward $651.00

(Note: Federal unemployment insurance liabilities
are deposited when the amount is over $100.00)

Figure 11-12

The current month line shows this deposit. The balance forward equals the amount of the previous month's taxes now due. Withdraw this amount to make the payment, leaving a final balance of the tax liability on this month's $5,000 payroll. This current liability rests in the clearing account until it is due and next month's liability is added to it. Then, in its turn, withdraw it to make this month's payment. With the clearing account, you always have the liability for the current month, a month before it is due.

You compute your liability for employer taxes as in Figure 11-11. Check the latest tax schedules for up to date tax rates. On the $5,000.00 payroll, the employer's share of taxes totals $577.50. Employer taxes are due only on payroll reaching certain maximums. Figure 11-11 assumes that all employees are under these maximums when the current payroll is paid. These maximums, or limits, are discussed later in the chapter.

Figure 11-12 summarizes the entries you make to record employer payroll taxes. The total amount of these taxes is an expense to the employer and accumulates in the payroll taxes expense account. Also shown is the tax liability. Break down this liability into the same categories as withheld taxes.

Payments to taxing agencies clear the employer's liability account. But the liability account doesn't clear every month, since a liability is included for federal unemployment insurance. This tax is payable only when the total liability exceeds $100.00. If the liability is less than that at the end of each quarter, no entry is made to clear the account.

You must make payroll tax deposits by certain deadlines or you may be fined penalties and interest. Deadlines for state taxes vary from state to state. Pay the federal taxes as follows:

1) Due dates for liabilities of $3,000 or more for payroll taxes are broken down into eight sectors each month. If your liability for F.I.C.A. withheld, the employer's share of F.I.C.A., and federal income tax withheld,

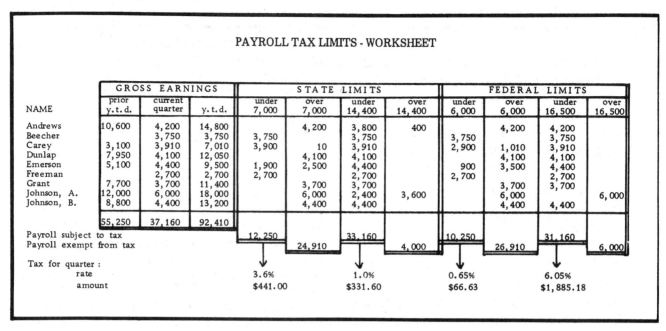

Figure 11-13

totals $3,000 or more, pay within three banking days after the pay date, on the 3rd, 7th, 11th, 15th, 19th, 22nd, 25th, or last day of the month.

2) If your month's total tax liability is between $500 and $3,000, pay these taxes by the 15th of the following month. This is true for the first and second month of each quarter. When your liability at the end of the third month is less than $3,000, pay it with the quarterly filing of Form 941.

3) You don't have to prepay amounts under $500. Let them accumulate from one month to another until they exceed $500, then pay by the 15th of the following month or with the quarterly filing of Form 941. For example, assume that the first month of a quarter shows a total liability of $320, and the second month shows a total liability of $390. The deposit, due on the 15th of the third month, is:

First month	$320
Second month	390
Total deposit	$710

4) Pay federal unemployment insurance on the end of the month following each calendar quarter (ending March 31, June 30 or September 30) only when the accumulated liability is greater than $100 to that point. The final payment is due with the annual F.U.T.A. tax return.

In most cases, you make your deposits of payroll taxes directly to your bank. The Internal Revenue Service provides each employer with a set of data processing cards for this purpose. Your check is made payable to the bank and must accompany this card. Make sure you use the card for the correct quarter and for the tax you are paying. Taxes you can pay at the bank include federal unemployment tax, payroll taxes, income tax (for corporations), excise tax, and any other taxes payable by deposit. Use a Form 501 for monthly or quarterly payroll tax deposits. Deposit federal unemployment taxes with a Form 508.

PAYROLL TAX LIMITS

All payroll taxes except income tax are subject to limits. These taxes are not payable on wages and salaries that are above certain amounts. These amount limits change from time to time and vary in different states. Your calculations of the taxes you owe must take these limits into account.

Figure 11-13 is a format you can use each quarter to complete your payroll tax forms. Use it to calculate how much tax you owe considering the under and over limit. The taxable amounts for each type of tax are listed on the forms. The worksheet in the figure uses the following hypothetical under and over limit assumptions. Check your tax form for the latest limits.

- State unemployment taxes (employer's tax) are payable on all wages under $7,000.00.

- State disability insurance (employee tax which is withheld) is payable by the employee on all wages under $14,400.00.

- Federal unemployment taxes (employer's tax) are payable on all wages under $6,000.00.

- F.I.C.A. (employer's and employee's taxes) is payable on all wages above the current year's limit. This illustration uses a limit of $16,500. Current limits are higher.

Each set of under and over columns balances to the employees' total earnings for the current quarter. Compute the taxes you owe using your payroll records and the most recent under and over limits in both tax categories.

1) *Subtract* the prior year-to-date gross earnings for each employee from the limit amount in each tax category. In the state limit column, this is $7,000 and $14,400 for state unemployment taxes and disability insurance, as mentioned above. In the federal limit column, the limit amounts $6,000 and $16,500 are for federal unemployment taxes and F.I.C.A., also mentioned above.

2) If you get zero or a negative number, the entire current quarter's gross amount for that employee is over the limit and is *not* taxable.

3) If you get a positive number, some part of that employee's current quarter gross earnings is under the limit and *is* taxable.

4) If you get a positive number that is greater than the current quarter's total, the entire amount of the employee's earnings is under. If the positive number is smaller than his current quarter's gross, put down as over only the amount that brings the prior earnings up to the limit.

As an example of 4) above, assume the following about an employee for whom you are computing your tax on a $7,000 limit.

Prior earnings:	$6,000.00
Current quarter:	1,500.00

The taxable amount under the limit is $1,000.00 ($7,000 less $6,000 = $1,000). The non-taxable amount over the limit is $500.00 ($1,500 less $1,000). To prove the accuracy of the computation, add $6,000 (previous) to $1,000 (current amount under). This equals $7,000, the amount of the limit.

PAYROLL FORMS

The federal government publishes several forms for payroll taxes. Be familiar with each of them and know how they apply to you. Form W-4 is the Employee's Withholding Allowance Certificate. This form must be completed by each employee. It documents the amount of tax to be withheld from his gross pay, and it is the employer's only source for this information. Even employees who will not be liable for any taxes should fill one out, as it supplies information you need to complete year-end reports required by law. Form W-4 includes the following information: the full name of the employee, his address and city, his Social Security number, his marital status, and his number of exemptions.

Tax tables used to compute withholdings are arranged by marital status, with columns for the number of exemptions. If an employee is married and claims three exemptions, his weekly payroll would be subject to the federal tax listed on the weekly table under Married, exemption column 3.

You are required to have a W-4 for each employee. Keep a file of all W-4's. Any time there is a change in marital status, number of exemptions, or address make sure the employee completes a new W-4.

File a Form W-2, Wage and Tax Statement, every January for each employee. Use the W-2's to report the employee's wages and taxes in summary for the preceding year. Distribute the six copies of the form as follows. Also see Figure 11-14.

Copy A - To the Social Security Administration. Summarized along with the other copy A's on Form W-4, Transmittal of Income and Tax Statements.

Copy B - To the employee to file with his federal income tax return.

Copy C - To the employee for his records.

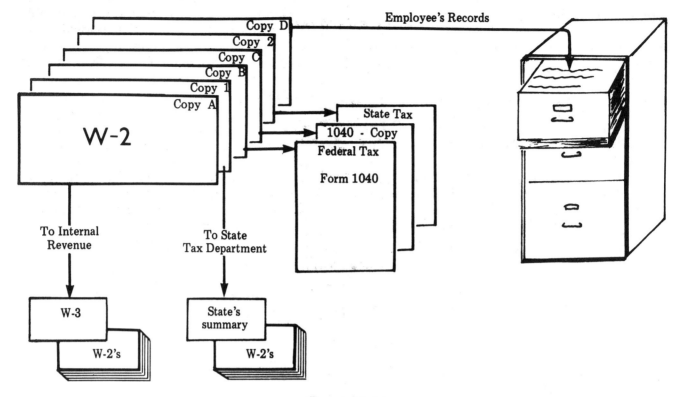

Figure 11-14

Copy D - Kept by the employer for his records.

Copy 1 - To the state taxing agency. Summarized on the state version of a Form W-3.

Copy 2 - To the employee to file with his state tax return.

Use quarterly and annual forms to report wages and pay taxes. Form 941, Employer's Quarterly Federal Tax Return, summarizes taxable wages and provides for payment of any income taxes or F.I.C.A. due. This form also requires you to list your tax liabilities and deposits. Liabilities you must list on Form 941 include F.I.C.A. tax withheld, F.I.C.A. tax (employer's share), and federal income taxes withheld.

Form 940 is an annual report used to pay federal unemployment taxes. This form summarizes the yearly taxable wages and computes the maximum taxable amount compared to the maximum for state unemployment taxes. A federal credit is allowed if the state rate is higher than the federal rate. Otherwise, the difference between state and federal rate is payable. This effectively makes unemployment taxes uniform throughout the country.

Summarized below are the federal tax forms most builders need in their payroll accounting.

Form W-2 - Filed annually, with copies sent to the Social Security Administration, state taxing agencies, and the employee.

Form W-3 - Filed with W-2's as a summary form.

Form W-4 - Completed by each employee to indicate taxable status and address. It helps the employer comply with legal requirements.

Form 501 - Used to deposit federal F.I.C.A. and withholding taxes with local banks.

Form 508 - Used to deposit federal unemployment taxes with local banks.

Form 940 - Used once each year to report and pay federal unemployment taxes.

Form 941 - Used once each quarter to report and pay F.I.C.A. and withholding taxes.

There are also information returns you must file each year. These are not actually payroll forms, but they have similar formats and are completed at the same time as the annual payroll information returns. Form 1099 is mailed to individuals receiving certain types of income from builders. Some types of payments to include on your 1099 are dividends or interest over $10.00 in one year, and rents, commissions, and other types of miscellaneous income over $600 you pay someone other than for trade purposes. Form 1099 must be summarized on a Form 1096.

A point to keep in mind regarding payroll accounting: The specific rates change from one state or county to another, and the federal (and state) rates are likely to change each and every year. With indexing of insurance and tax rates for payroll withholding, it would be unusual for any of the percentages or numbers to remain stable from one year to the next.

Payroll accounting requires a large number of records. You need to keep records adequate for preparation and payment of payroll itself, often on a weekly basis; for identification of payroll taxes due; for preparation of quarterly payroll tax forms; and for preparation of year-end information and tax returns. In addition, payroll records may be used for other purposes, including accounting entries, workers' compensation insurance billings, census reports, and union benefit calculations and payments.

Chapter 12
Overhead Expenses

Previous chapters examined direct costs in a building operation. Direct costs vary with sales. Overhead expenses, on the other hand, tend to go on independent of sales volume. Every builder has overhead expense. This chapter will explain in detail what overhead expense categories should be included in the general ledger, and will distinguish between the broad divisions of variable and fixed expenses.

Another matter of great concern to builders is covered in this chapter: overhead and related expenses. Budgets serve as a measure against which you operate. Without budgets, your company doesn't have a financial road map. The importance of general expense budgets is often ignored. You may feel that general expenses are inevitable. This chapter is intended to show you how to control what may appear to be uncontrollable. In truth, most expenses can be controlled within a reasonable budget. The control you exercise is a measure of your effectiveness as the manager of your business.

VARIABLE AND FIXED OVERHEAD

The discussion of sales analysis earlier in this book included a distinction between selling expense (a variable overhead expense) and fixed overhead. Selling expenses (commissions, advertising, etc.) usually vary with sales. Your fixed overhead (rent, phone, etc.) are fairly constant whether sales are up or down. The difference between direct costs and overhead expenses is that direct costs are identified with a project. Variable overhead expenses are not associated with any single job, though they vary with sales volume. For example, there is a difference between labor (a direct cost) and union welfare expense (a variable expense). The labor that goes into a project is a necessary part of the production process. It generates income. The union welfare paid on that labor is once removed from the generation of income. It does not produce sales. Yet it is a necessary expense of doing business. It is variable because the expense varies according to the cost of producing that income with labor. Overhead expenses such as office rent are fixed. No matter what the volume and direct cost a business experiences, rent remains unchanged over the term of the lease.

Some expenses can be both variable and fixed, depending on their definition. For example, payroll taxes on direct labor are variable expenses; payroll taxes on office salaries are fixed.

The following is a list of expenses incurred by most builders. Some expense titles appear more than once as they are both fixed and variable. Your general ledger could be set up in the same order as this list. The categories should be observed precisely in coding expenses. Otherwise, a true comparison of expense categories from month to month will be impossible.

VARIABLE EXPENSES

Payroll Taxes This should include all taxes computed as the employer's share based on direct labor, as opposed to payroll taxes on office salaries. This expense is variable overhead because it is directly influenced by changes in the amount of construction labor required.

Taxes, Other Variable taxes should be kept in detail because they must be listed individually on income tax returns. These taxes include federal excise tax (computed on fuel and varying by the amount of use), licenses (varying by the amount of sales volume in most areas), permits and fees, public utility fees (based on a business truck use), state fuel tax (varying by consumption), and state income tax (varying by the amount of profit).

Automotive, Repairs and Maintenance This includes all repairs, replacement parts, and other maintenance to company vehicles.

Automotive, Gas and Oil This includes gas and oil used by highway vehicles.

Operating Supplies This includes wrapping materials and any other incidental items used out of the office but required for doing business rather than for working on a project.

Small Tools Any purchase of tools should be included here unless high enough in value to constitute treatment as an asset. Treatment is usually determined by the purchase price. Some equipment costs little and isn't worth depreciating and is thus categorized as "small tools." This expense can vary with the volume of sales.

Equipment Rental This category should vary according to the special needs of certain projects. Equipment rented or used exclusively on jobsites is a direct cost.

Union Welfare All pension and trust fund payments by an employer are variable because direct labor cost is the basis for measuring the amount due.

Insurance Certain types of insurance are variable because they relate to specific jobs or labor functions. Some types of insurance vary with the volume of sales and should also be considered as variable.

Collection Expenses Fees paid to agencies and other costs of collection are totaled under this category. These are variable expenses because collection costs usually change with changes in volume.

Bad Debts Like collection expenses, the amount of bad debts is likely to be affected by volume.

Travel and Entertainment Travel and entertainment expenses are limited to those costs that are necessary to the generation of income. Thus they are considered to vary with income. Builders operating over a wide area have meal and lodging expenses as well as car rental costs. These are all travel expenses. Entertainment expense should be documented as to its business purpose.

Miscellaneous Any other variable operating expense.

FIXED EXPENSES

Salaries and Wages This includes payroll for officers, estimators and clerical employees. The amount is fixed because these expenses are not necessarily related to sales.

Payroll Taxes All payroll taxes on income to company officers and office personnel are fixed. They do not vary with sales or direct costs.

Taxes, Other This includes all non-variable taxes such as property taxes (on real estate or equipment and inventories not directly related to income) and vehicle registration fees (a fixed expense of doing business).

Office Supplies This amount includes all consumable supplies used in your office. The cost of purchasing a pegboard accounting or check writing system would be an office expense.

Postage This account should not vary with minor changes in volume. A significant increase in postage expenses would occur, of course, if you took on a lot of new work in a short period.

Printing This includes all letterhead, envelopes and business cards. Printing expense should be controllable and is defined as fixed for that reason.

Insurance This account should include all insurance on fixed cost items. For example, the cost of insurance on premises, the contents, and health insurance for officers would be fixed expenses.

Telephone This is another controllable expense which may vary with changes in sales volume. However, it is defined as fixed because, usually, most phone calls are local and local calls are made at no cost after paying a fixed monthly charge.

Utilities This expense does not vary with volume and is generally a cost associated with operating office and warehouse areas.

Dues and Subscriptions This includes all memberships, dues, subscriptions, books and periodicals. It has no direct relationship to the volume of sales.

Accounting and Legal These fees do not vary substantially with income. All accounting and legal fees are administrative costs rather than selling expenses. An increase in these costs would not necessarily be due to an increase in sales.

Consultants' Fees This includes all one-time outside consulting fees. There are a variety of types of consultants. Any fee that relates entirely to a job or several jobs is a variable cost and should be included with variable costs under Direct Costs, Other. Consulting fees of a general nature belong here.

Rent This includes rent and storage expenses for the general office, shop or warehouse, and any leased buildings or land.

Depreciation This includes all depreciation on trucks and autos, construction equipment and office equipment.

Amortization This category is for writing off leasehold improvements, organizational expenses, and any other capitalized asset being amortized. The expense does not relate directly to sales or to the cost of sales.

Advertising and Promotion This is usually a limited expense for builders. Any advertising or promotion of your business in general that is not sales related should fall in this category. For example, help wanted classified advertising would fall in this category.

Security Most builders need to protect their inventories and jobsites. The cost of protecting these investments belongs here.

Interest This includes any interest paid on bank loans, notes and other obligations.

Building Maintenance This includes all expense of maintaining an office and shop, such as repairing a leaking roof, doing plumbing repairs or painting.

Miscellaneous Include here any expense that does not bear a relationship to the volume of sales or to the direct cost of those sales.

Define expenses as variable or fixed according to that expense's influence on sales volume. Variable expenses, while not a direct part of the production process, vary broadly with sales. Fixed expenses are not related to sales and direct costs, and do not vary because of changes in the volume and gross profit. Because of these differences, variable expenses are controllable only as far as sales and direct costs are controllable. Fixed expenses, though, are directly controllable by management.

Figure 12-1 is a summary of variable and fixed expenses.

BUDGETING EXPENSES

Since expenses are fairly predictable within ranges of business volume, they can be budgeted. The budget is your guide to making the work you and your company do pay off in income and profits. A realistic budget establishes the profit and is the expense standard you plan to observe. The budget is, of course, only a tool. It is the people in your operation who control expenses. So think of your company as your guide to profits.

The budgeting process has three parts:

1) Preparing the budget

2) Analyzing the results

3) Taking action to control expenses

Overhead Expenses

```
VARIABLE EXPENSES

      Payroll Taxes
      Taxes - Other:
           Federal Excise Tax
           Licenses
           Permits and Fees
           Public Utilities Fees
           State Fuel Tax
           State Income Tax
      Automotive, Repairs and Maintenance
      Automotive, Gas and Oil
      Operating Supplies
      Small Tools
      Equipment Rental
      Union Welfare
      Insurance
      Collection Expenses
      Bad Debts
      Travel and Entertainment
      Miscellaneous Variable Expenses

FIXED EXPENSES

      Salaries and Wages
      Payroll Taxes
      Taxes - Other:
           Federal Use Tax
           Property Taxes
           Vehicle Registration
      Office Supplies
      Postage
      Printing
      Insurance
      Telephone
      Utilities
      Dues and Subscriptions
      Accounting and Legal
      Consultants' Fees
      Rent
      Depreciation
      Amortization
      Advertising and Promotion
      Security
      Interest
      Building Maintenance
      Miscellaneous Fixed Expenses
```

Figure 12-1

The plan must be realistic or the entire budget is useless. The budget is based on past results and your estimate of reasonable expectations. A budget can be fairly accurate if it allows for price changes and anticipated changes in business volume. The budget you develop can become a real guide to savings and profits in your business if your budget is both realistic and possible to live with, your analysis is frequent and probing enough to spot problems early, and you take the action required to correct problems.

The goal of a budget is not to forecast as accurately as possible what expenses will be. Rather, the budget should establish the ideal level of expense and income in a well-controlled operation. Comparing your budget to actual performance will reveal areas where your budget was unrealistic or your company could improve its performance in the future. Analyze over-budget costs to identify areas where money is wasted.

Budget analysis requires accurate books and records and detailed cost information. Because monthly bills are not always received at the same time each month and most expenses don't occur each month, leave some leeway for month-to-month variations. Monitor expenses over several months and you should begin to find areas where savings are possible.

Controlling expenses requires a definite plan. Review each category of expense and look for ways to keep the level of expense down. You probably already have a variety of policies which are aimed at reducing costs. Examples of this would be limiting employee raises to a percentage of their pay, with annual reviews; requiring specific approval for certain types of general expenses; careful checking of invoices; and maintaining a log of long-distance telephone calls. Keep a list of the methods and techniques that are effective in minimizing the expenses in your operation. This should include monthly analysis of detailed entries to certain accounts.

Figure 12-2 shows an expense analysis form. The components of the office supplies account have been split into the various categories. You see immediately where the cost savings have occurred and where costs have exceeded the budget.

Give special attention to troublesome cost areas. Correct cost overruns before more time goes by and expenses mount. In the analysis of office supplies in Figure 12-2, the builder or his office manager might want to see all invoices for supplies before they are paid. Make sure that the goods being purchased are really needed and used. This authority may have been delegated to the bookkeeper, who might not be as concerned about budgets as you are. You might decide that no purchase of office supplies above $50.00 could be made without your approval.

It is obvious from Figure 12-2 that a detailed analysis of All Other is needed. Most of the excessive cost falls in this category. This is a common problem in analyzing books and records. There is not always enough detail on summary reports to be fully revealing.

There are two methods for preparing budgets. Use one method for variable expenses

JOHNSON CONSTRUCTION COMPANY					
Expense Analysis - Office Supplies					
September, 19___					
Description	Total	Papers and Worksheets	Rubber Stamps	Pencils, Pens, etc.	All Other
Midfield Stationery	$ 37.80	$ 24.75	$ 13.05		
Jim's Office Supply	$ 76.50	$ 26.50			$ 50.00
Jim's Office Supply	$ 42.00	$ 42.00			
Petty Cash	$ 8.44			$ 8.44	
Total Month	$164.74	$ 93.25	$ 13.05	$ 8.44	$ 50.00
Previous Year-to-Date	806.00	$418.05	$ 76.50	$ 72.60	$238.85
Year-to-Date total	$970.74	$511.30	$ 89.55	$ 81.04	$288.85
Year-to-date budget	$600.00	$450.00	$ 50.00	$ 75.00	$ 25.00
Variance	($370.74)	($61.30)	($ 39.55)	($ 6.04)	($263.85)

Figure 12-2

and the other for fixed overhead. Budgets for variable expenses should be prepared as a percentage of income within broad ranges. As your volume changes, your variable expenses will change, but the percentage each expense category is of sales should remain within certain limits. The problem with this approach is that no dollar amounts are included in the budget. There is a tendency to skip over significant variances, blaming them on the level of sales rather than on internal control of expenses.

Another approach for variables would be to set budgeted amounts for expenses based on assumptions about the level of sales or labor. This method requires several budgets. Any variance from the budget for the sales volume closest to actual volume would be analyzed to find the reason for the discrepancy.

A monthly budget for fixed expenses should be prepared one year in advance. Most of the expense categories can be estimated fairly accurately. Set optimistic but also realistic goals. The budget becomes a guideline against which performance is measured each month.

GENERAL LEDGER ACCOUNTS

Efficiency and accuracy are essential in maintaining general ledger accounts. The more accounts a company has, the more time is required to post and control the ledger. Yet enough detail is needed to provide a meaningful breakdown of expenses by category. Two major sections are essential for the general ledger: variable expenses and fixed expenses. Variable expenses can also be called selling expenses. So your financial statement reflects a selling profit:

Gross sales minus direct costs = gross profit

Gross profit minus selling expense (variable costs) = selling profit

Selling profit minus overhead (fixed costs) = net profit

Overhead Expenses

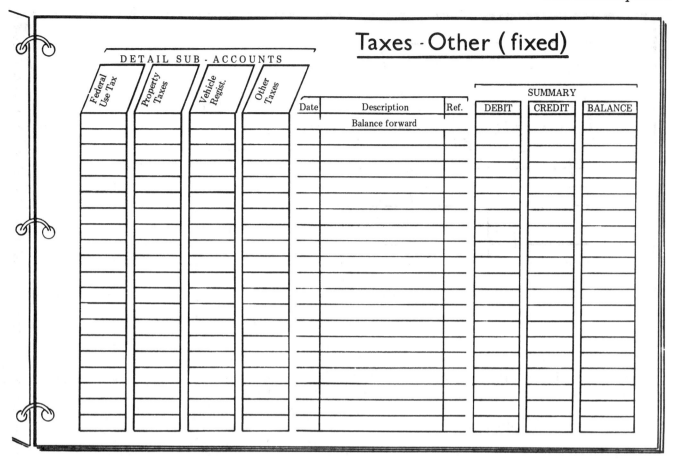

Figure 12-3

Within variable and fixed cost categories each account and sub-account should be kept individually. This does not necessarily mean that each account should have a page of its own. Your general ledger would be too large for effective control. There are two types of account pages recommended for expense classifications. The first is illustrated in Figure 12-3. This is a page for one account with sub-accounts listed off to the left. Accounts like insurance and taxes should be kept as shown in Figure 12-3.

The second method is shown in Figure 12-4. This is ideal for most expense categories. Only a few pages like this are enough to handle all the variable expense accounts most builders would have. In these examples, all accounts are listed individually but controlled in summary. This saves posting time and works as a control feature. Each line is balanced on that page itself so that no errors slip through. This is important because it is much more difficult to find and correct errors later.

The principal difference between Figures 12-3 and 12-4 is in the detail provided. The first is a page for one account, with divisions of sub-accounts. Figure 12-4 includes many accounts. Most expense categories do not require the detailed sub-accounts shown in Figure 12-3. You can break down some accounts into sub-accounts and leave others intact. Remember, the general ledger loses its effectiveness when it includes too much detail. Be sure that the breakdowns are truly needed. The ledger is not the proper place to perform detailed analysis. But it must be your tool for analyzing the summary of business transactions.

Below is a summary of the arrangement that is most suitable for listing the accounts shown in Figure 12-1:

Variable Expenses:
    Payroll Taxes (detail sub-accounts included)

    Taxes, Other (detail sub-accounts included)

    Automotive (both accounts listed as sub-accounts)

    Operating Supplies, Small Tools, Equipment

Builder's Guide To Accounting

[Table headers across top: Operating supplies, Small tools, Equipment rental, Union welfare, Collection, Bad debts, Travel, Miscellaneous | Date | Description | Ref. | Debit | Credit | Balance — with "Balance forward" row]

GENERAL EXPENSES (VARIABLE)

Figure 12-4

Rental, Union Welfare, Collection, Bad Debts, Travel, and Miscellaneous Expenses

Insurance (detail sub-accounts included)

Fixed Expenses:
Salaries and Wages, Office Supplies, Postage, Printing, Telephone, Utilities, Dues and Subscriptions, Accounting and Legal, and Consultants' Fees

Payroll Taxes and Taxes, Other (detail included for Taxes, Other only)

Insurance (detail sub-accounts included)

Rent, Depreciation, Amortization, Advertising and Promotion, Security, Interest, Building Maintenance, and Miscellaneous Expenses

You should be able to handle all of your accounts on a few pages. The collection recommended above has five pages of variable expense accounts and four pages of fixed overhead expense accounts.

RATIOS

Budgeting is only one way of analyzing overhead costs. You sharpen your understanding of your overhead by comparing the overhead costs you have to other figures. Since overhead expenses are divided into two broad categories, any combined ratio analysis will not be as meaningful as an analysis of individual items. The two ratios that are most significant to a builder in analyzing his overhead are: (a) variable overhead expenses to sales, and (b) fixed overhead expenses to sales.

The comparison of expenses is made to *sales* because there is less chance for distortion than there would be by comparing expense categories to total costs. If you compare variable expenses to direct costs, or fixed expenses to direct costs, any deviation in normal costs (such as losses in inventory or idle labor time) would distort the results.

Figure 12-5 is a worksheet used to determine the trend of variable expenses to gross sales. For this comparison, gross sales are assumed to be the earned income within a period of time. Variable expenses are taken as the adjusted accrual basis variable expenses. Compare monthly and previous twelve month variable expenses to sales for the month and twelve previous months. The twelve month average should be updated each month. As of May 31, the comparisons would be: 1) May variable expenses to May earned income. 2) Variable expenses for twelve months ending April 30 to earned income for twelve months ending April 30.

A separate worksheet can be maintained to chart the twelve-month averages for an entire year. See Figure 12-6.

Overhead Expenses

JOHNSON CONSTRUCTION COMPANY
ANALYSIS OF VARIABLE EXPENSE TREND
For the year 19____

	Current Month			Twelve Month Average		
	Expense	Income	Ratio	Expense	Income	Ratio
January	7,481	27,560	1/3.7	6,281	29,612	1/4.7
February	7,077	29,401	1/4.2	6,407	26,401	1/4.1
March	7,411	26,811	1/3.6	6,499	26,880	1/4.1
April	8,496	31,004	1/3.6	7,762	28,416	1/3.7
May	8,219	27,960	1/3.4	8,105	27,206	1/3.4
June						
July						
August						
September						
October						
November						
December						

Figure 12-5

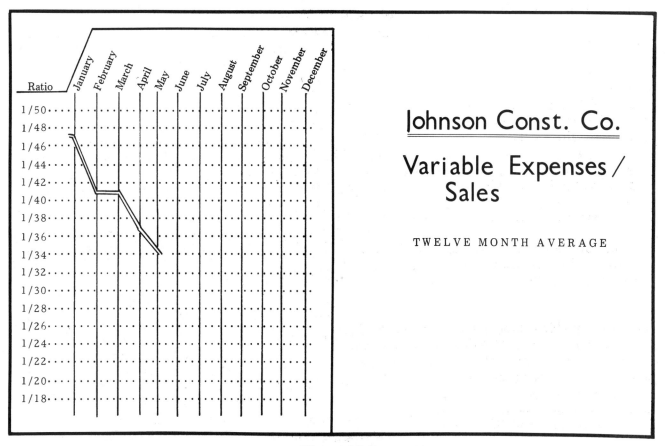

Figure 12-6

Find the reasons for any unfavorable trend in the ratio. The reports by themselves are useless unless you can reverse the trend. There could be many reasons for relative increases in variable expenses. Price increases could be part of the cause, but your company's charges might not be

JOHNSON CONSTRUCTION COMPANY
FIXED EXPENSES - BUDGET ANALYSIS
For the year 19____

		Jan	Feb	Mar	Apr	May	Jun	Jul	Aug	Sep	Oct	Nov	Dec
Postage	Budget	14	18	18	13	15	15	18	16	15	18	21	11
	Actual	17	14	17	17	12							
	Variance	(3)	4	1	(4)	3							
Printing	Budget	85	44	45	120	55	-	35	50	-	100	35	18
	Actual	-	25	72	-	210							
	Variance	85	19	(27)	120	(155)							
Insurance	Budget	145	145	145	145	145	165	165	165	165	165	165	165
	Actual	158	158	158	163	163							
	Variance	(13)	(13)	(13)	(18)	(18)							
Telephone	Budget	38	38	38	38	41	41	41	41	45	45	45	45
	Actual	28	32	29	37	36							
	Variance	10	6	9	1	5							
Utilities	Budget	57	57	57	53	44	44	35	35	35	38	41	44
	Actual	57	59	61	55	41							
	Variance	-	(2)	(4)	(2)	3							

Figure 12-7

keeping pace with your costs. There might be other causes as well. For example, equipment expense might be higher now than five months ago. This could indicate a lack of organization in assignments or the purchase of unnecessary equipment. Each item of expense should be important in your analysis. Compare expenses in different periods. Where there is a major difference in expenses, you should know what is causing that difference.

A similar analysis can be performed for fixed expenses, but the result would have less significance than the analysis of variable expenses. Since variables change more than fixed expenses in relation to sales, you measure your success in percentage trends. The control of fixed expenses depends entirely on your effectiveness as a manager. A comparison of fixed expenses to sales should show only a steady ratio if volume remained constant, an increase of the ratio with increased sales, or a narrowing of the ratio with decreased sales.

The best way to control fixed overhead is to make careful monthly comparisons of budget dollar forecasts to the actual results. Identify trends as they develop to keep down the rise in expenses. Comparison of actual expenses to the budget is illustrated in Figure 12-7. In this example, budget expenses are listed month by month. If you budget expenses for the year only, rather than for each month, it is easy to blame unfavorable trends on timing differences. Instead, examine the cause of current overspending. Document the reasons for variances from the budget. This is a helpful way to remind yourself of the danger of uncontrolled spending. Next year when you're trying to control expenses and maintain an expense budget, think how helpful it would be to have a monthly summary of this year's expenses as reference along with the remedies you found for overspending.

RECORDING EXPENSES

Expenses are recurring in most cases. That is, they occur month after month in generally set patterns. For example, rent is predictable and payable at about the same time every month. Office supplies, if paid on account, may be payable the same day of each month. Sometimes expenses are recorded earlier than their payment. This is an accrual. At the end of the accounting month, expenses incurred but not paid are accrued by recording a liability. Thus, the profit and loss statement for that month would show the expense in the month that the liability exists. The entry to accrue an expense would be:

	Debit	Credit
Expense:	XXX	
Accounts Payable:		XXX

When the account is paid, the entry would be:

	Debit	Credit
Accounts Payable:	XXX	
Cash:		XXX

Accounts Payable for this expense item is returned to zero. The effect is the same as though a cash payment had been made in the first place.

This kind of entry should be made at the close of the first month in which the expense is known and the value has been received. You will have liabilities like this every month because current liabilities are not always billed and paid in the month value is received.

Another way to record expenses in a period other than when paid is by setting up a prepaid asset. An example of a prepaid asset would be an annual payment for insurance. The one year policy is paid by a single payment at the beginning of the policy year. The balance must be deferred on your books until it is earned. For example, a July payment of $120 would be set up as a prepaid asset and deferred through each month to the following June. This is illustrated in Figure 12-8, which also shows the treatment for a single payment covering 36 months.

The entry to record a prepayment such as insurance is:

	Debit	Credit
Prepaid Assets:	XXX	
Cash:		XXX

Overhead Expenses

The entry to record each month's charges is:

	Debit	Credit
Insurance:	XXX	
Prepaid Assets:		XXX

This kind of accounting means that your books carry the proper amount of expense in each month. The examples given involve only small amounts, but prepaid insurance plus other types of expense (office supplies, printing, and so forth) can add up to a significant amount. Set up payments for large expenses which apply over a defined period and recognize a portion each month.

Another way to treat expenses is to capitalize them as deferred assets. Prepaid and deferred assets are often thought of as the same thing. But there is a difference. Prepaid assets defer recognition of an expense until value is received for that expense. A deferred asset might be a one-time charge. An example of this might be organizational expense.

The cost of organizing a business can be high. The cost should not be charged against the first year of operations. Instead, you should set up in an asset account, Organizational Expenses, and write off the cost over a period of years. This process of spreading costs over a period of years is called amortization. It is handled like depreciation. A monthly amount is charged to some expense category and the asset is reduced by that amount each month.

Another distinction between the two types of assets is that prepaid assets could be considered current assets while deferred charges usually are not current. That is, they are not likely to go to zero in the next twelve months. To avoid confusion, both accounts are normally included in the balance sheet classification as other assets.

SUMMARY

The control of overhead expenses is easier if you distinguish between variable expenses and fixed overhead expenses.

Variable expenses are those which will be affected by changes in the volume of sales and by direct costs. The distinction between direct costs and variable expenses is in how they relate

PREPAID INSURANCE SCHEDULE

		12-month prepayment	36-month prepayment
year 1 -	Jul	$ 10 ($120 paid)	$ 8 ($288 paid)
	Aug	10	8
	Sep	10	8
	Oct	10	8
	Nov	10	8
	Dec	10	8
year 2 -	Jan	$ 10	$ 8
	Feb	10	8
	Mar	10	8
	Apr	10	8
	May	10	8
	Jun	10	8
	Jul		8
	Aug		8
	Sep		8
	Oct		8
	Nov		8
	Dec		8
year 3 -	Jan		$ 8
	Feb		8
	Mar		8
	Apr		8
	May		8
	Jun		8
	Jul		8
	Aug		8
	Sep		8
	Oct		8
	Nov		8
	Dec		8
year 4 -	Jan		$ 8
	Feb		8
	Mar		8
	Apr		8
	May		8
	Jun		8

Figure 12-8

to the production process. Variable expenses are once-removed from the production process; direct costs are a part of it.

Fixed overhead includes all accounts which are not affected by changes in sales or direct costs. Unlike variables, they can be controlled directly by management.

The purpose of an expense budget is not to predict actual expenses. Rather, budgeting is a guide to management, a tool in controlling expenses. It is a standard against which you compare your results.

A budget for variable expenses is expressed as a relationship of sales volume to variable costs. This comparison tells you whether the variable expenses you have are effective.

A budget for fixed expenses should show monthly costs for each significant account category. Compare your budget to your experience to find areas where cost savings are possible. Only by acting on this information can you reverse unfavorable trends.

The distinction between variable (or selling) expenses and fixed overhead should be reflected in your company's general ledger. The subtotal for Selling Profit is an important figure which is available only if you maintain a well-defined classification system and if you carefully control posting. Any reports, statements, historical comparisons or summaries which are prepared from the expense accounts are only as good as the categories you establish and the posting you do.

The principal ratio analysis you need is that of variable expenses to sales. The trend of this ratio indicates your effectiveness as a manager. A comparison of fixed expenses to your budget for fixed expenses will help control those expenses.

There are three methods by which expenses are recorded in periods other than that in which they are paid. The first is by accrual: recording payable but unpaid expenses for goods or services in the current month. The second is by setting up a payment as a prepaid asset and taking the expense in an equal portion over a period of months. The period should be the length of time the payment covers.

The third method is by setting up a deferred asset. This asset is amortized over a predetermined period of time, like the method used for prepaid assets. The distinction is in the nature of the expense being handled. Prepaid assets are usually normal expenses; deferred assets tend to be one-time charges.

Chapter 13 Equipment Records

The largest single investment that most builders make is for equipment. Your equipment is a fixed or long term asset because it is usually held for more than a year. Your equipment is listed on your balance sheet at its cost and is depreciated over its estimated life. The act of setting up an asset in the general ledger is called *capitalizing*. Spendings for most expenses are taken in the month payment is due or the month paid. Capitalizing an expense sets up an asset which is taken or booked as an expense over a period of several years.

Tax regulations do not allow a business to write off large purchases of equipment in a single year. This is because the useful life of that equipment is longer than the year in which it is purchased. To take expense for a purchase over the life of the asset is to *depreciate* it. Generally you minimize your taxes in the current year if you depreciate equipment as quickly as the tax regulations permit. This is not always the best policy, as this chapter will show.

Fixed assets are not actually fixed, so the term *long term asset* is often used. This is more descriptive because it indicates an asset with a long useful life and distinguishes these assets from current assets such as cash and accounts receivable. This distinction is very important in the balance sheet because fixed assets are not as readily converted to cash as are current assets.

Fixed assets that you buy with borrowed money are listed at their total purchase price, even if the major portion of the asset value is owed to a bank. The liability is also shown, as a long-term debt. Any portion of the debt that is payable within twelve months is separated out and called a current liability.

Current assets are expected to flow into cash within a year; current liabilities are anything that is payable within one year. The entry to record a $15,000 purchase of a fixed asset with $12,000 financing (assuming payments of $100 per month) would be:

	Debit	Credit
Equipment:	$15,000	
Cash:		$3,000
Current note:		$1,200
Long-term note:		$10,800

This entry shows the full amount of the purchase price as an asset, the true value to the builder. The note is divided into current (the amount due within one year) and long-term portions. Any payments on the note reduce the long-term portion. The current portion stays the same each month. As each month passes, the current period is extended so that the current amount payable is always equal to twelve months total payments.

The well-organized builder should have a definite plan for fixed assets:

- A budget for acquisition of equipment and a complete knowledge of the planned use

- An established standard for return of investment in fixed assets. How many manhours will a new piece of equipment save and what will be the costs of owning and operating it?

- A study of alternatives. Would it be better to lease or buy new equipment?

- Dependable and complete asset records (serial numbers, costs, acquisition dates)

- Records of maintenance costs and idle time

- A policy establishing depreciation rates

- An identification system for all fixed assets, either by labels, tags or etched numbers

- A goal for maximum use of fixed assets including scheduling, minimizing idle time, and a maintenance policy.

- An estimate of the expected future market for the builder's services and the fixed assets required to meet that volume level

Have a clear picture of your own equipment needs and plan for them. Planning makes the purchase of a major piece of equipment a normal business process. Without a plan, the unexpected need for a replacement could delay your work. Investments tied up in unplanned major commitments can affect your ability to meet payroll and other obligations.

With a plan, you can compare your expectations about a fixed asset purchase to the actual results of having that asset. This should help you see whether the investment is a wise one. Equipment is a necessary part of your operation. But it should save manhours, make your operation more competitive and speed up completion of your jobs. Most important, the investment should not impair your ability to meet your obligations.

CLASSIFYING FIXED ASSETS

Your balance sheet should show enough detail to reveal the true position of your business. Lumping several categories of assets into one total obscures your investment in any one part. The following balance sheet categories are about what most builders should use to classify their fixed assets.

- *Office furniture and equipment* - Includes all desks, chairs, files, tables, typewriters, adding machines, and fixtures.

- *Trucks and autos* - Includes delivery vehicles, automobiles and any vehicle used in the general operation of the business.

- *Machinery and equipment* - Covers a wide range of assets including construction equipment, tools, specialty machinery, and equipment kits.

- *Small tools* - Many tools are included under machinery and equipment. Use this category for smaller, non-mechanical tools such as hand sets and small appliance tools.

- *Building* - Includes any structure owned by the company, excluding the value of land.

- *Improvements* - Includes any renovation or addition to structures owned or leased.

- *Land* - Includes the value of any land owned by the company. This asset must be kept separate from all others. It is not subject to depreciation.

All long-term assets (except land) are depreciated. The improvements classification is subject to amortization. This distinction is subtle. Depreciation is the recognition of the cost of an asset over a period of its useful life. Amortization is the process of spreading the cost. So, where depreciation uses up the valuer of an asset, amortization spreads the expense over a period of time. For example, you would

JOHNSON CONSTRUCTION COMPANY
FIXED ASSET SUMMARY
SEPTEMBER 30, 19___

Description	Gross Asset	Accumulated Depreciation	Net Asset
Office Furniture and Equipment	$ 1,280.00	$ 450.00	$ 830.00
Trucks and Autos	86,419.60	31,004.82	55,414.78
Machinery and Equipment	14,200.00	3,480.00	10,720.00
Small Tools	900.00	200.00	700.00
Building	62,500.00	21,800.00	40,700.00
Improvements	1,360.55	311.12 *	1,049.43
Land	35,000.00	-	35,000.00
Total	$201,660.15	$ 57,245.94	$144,414.21

* Amortization

Figure 13-1

amortize the cost of improving leased office space over the term of the lease.

There are several methods available for listing assets on a balance sheet. The least desirable is to lump all depreciation together after finding a subtotal. No distinction is made as to which assets are depreciated and to what extent:

Office furniture and equipment	XXX
Trucks and autos	XXX
Machinery and equipment	XXX
Small tools	XXX
Building	XXX
Improvements	XXX
Land	XXX
Total Gross Assets	XXX
Less: Accumulated Depreciation	XXX
Net Fixed Assets	XXX

Undesirable as this method is, it is commonly used. This is because the books of many companies are not set up to keep track of the accumulated depreciation by each asset type. Since accumulated depreciation is a negative asset (actually, a reduction of fixed assets), the yearly expense to depreciation is accumulated in this account. It is an easy matter to record depreciation by category. Rather than making a two-line journal entry, debiting the expense and crediting reserve, the entry should look like this:

	Debit	Credit
Depreciation expenses	XXX	
Accumulated depreciation: Office furniture		XXX
Trucks and autos		XXX
Machinery and equipment		XXX
Small tools		XXX
Building		XXX
Amortization expense	XXX	
Accumulated amortization		XXX

Figure 13-1 shows the most compact and informative way to report assets on a balance sheet. Totals are given for the total accounts, fixed assets, accumulated depreciation, and the gross and net totals in each category.

Equipment Records

JOHNSON CONSTRUCTION COMPANY
EQUIPMENT RECORD

Description _____

Company Number _____ Classification _____

Size _____ Model _____ Style _____ Engine No _____

New ☐   Used ☐   Serial Number _____

Purchased From _____

Date of Purchase _____ Terms _____

Location of Property _____

Cost information:

    Price

    Tax

    Delivery Charges

    Installation Costs

    Other _____

    Total

Date Sold _____ Sold To _____

Sales Price:

    Gross

    Tax

    Total

Machinery

Small Tools

Building

Figure 13-2

CONTROLS ON PURCHASING EQUIPMENT

You should establish a procedure for processing in new equipment. The procedure you follow should only take a few minutes. But it will provide all the information you need later. It's easy to forget to record these details when you buy new equipment and difficult to compile some of this information later when you really need it. The record you maintain for each fixed asset should include the following information:

- Identification (manufacturer and description)
- Assigned company number (if any)
- Classification of the asset (small tools, office equipment, etc.)
- Model, serial, and engine numbers
- Whether the asset is new or used
- The seller
- Date of purchase
- Location of the asset
- Cost information in detail

This equipment record should be maintained in a file or equipment record book and becomes the basis for analyzing fixed assets. It supports all general ledger entries for equipment.

Figure 13-2 illustrates Johnson Construction Company's equipment record. These forms should be kept in a file indexed by classification of asset. This provides both an easy reference and valuable historical data.

In addition to your record of equipment acquired, you should develop detailed records for:

- Equipment utilization
- Control of maintenance cost
- Depreciation

Your equipment purchase plan must include a number of key considerations. Insurance coverage must be adequate. The equipment must be adequately productive. It must be budgeted and affordable. On key pieces of equipment you should have a utilization schedule, assigning time to specific jobs. You should also have a secondary plan, since equipment can break down at any time.

Figure 13-3 is a schedule of equipment time. This should be prepared in advance as far as possible, with one schedule for each piece of high value equipment. Obviously, this kind of worksheet is necessary only if your operation owns heavy equipment. Separate schedules could be prepared for each job by piece of equipment.

UNIT COST OF EQUIPMENT

Every builder should establish an hourly cost for his high value equipment. This requires certain information about each piece of equipment:

- Estimated average hours of use and idle time
- Repair and maintenance time
- Repair and maintenance cost
- Cost of equipment less the estimated salvage value
- Estimated life of equipment
- Cost of storage, insurance and taxes
- Operating cost including gas, oil and accessories

Having this information will help you put together more accurate estimates and analyze project costs.

A large part of unit equipment costs is the cost of operation. Figure 13-4 is an equipment cost summary form. Make it a habit to keep a record like this on each key piece of operational equipment. The form breaks down the hourly operational cost and distributes that cost by job. Use both the schedule of equipment time (Figure 13-3) and the equipment cost summary, and you have about all the cost and bookkeeping information you need on the company's equipment assets.

Equipment Records

JOHNSON CONSTRUCTION COMPANY
SCHEDULE OF EQUIPMENT TIME

Unit_____ Company Number_____ Month_____

Day	Job No.	Scheduled Time	Hours Operated	Idle Time	Repair Time
1					
2					
3					
4					
5					
6					
7					
8					
9					
10					
11					
12					
13					
14					
15					
16					
17					
18					
19					
20					
21					
22					
23					
24					
25					
26					
27					
28					
29					
30					
31					

Figure 13-3

JOHNSON CONSTRUCTION COMPANY
EQUIPMENT COST SUMMARY

Unit_____ Company Number_____ Month_____

Date	Cost of Repairs		Tires	Cost of Fuel		Lube Costs		Total	Distribution by jobs				
	Labor	Parts		Gas	Diesel	Oil	Grease		#	#	#	#	#

Figure 13-4

In Figure 13-4, "Distribution by Job," record the amount of idle and repair time and the useful active time. Non-productive hours are as much a part of the cost of owning equipment as the productive hours. Naturally, the more non-productive time you have, the higher the unit cost will be. Well-maintained and scheduled equipment gives maximum use. Use the total hours assigned on the cost assignment worksheet as a basis of distributing costs by job. This is shown on Figure 13-5. Total monthly hours are summarized so that the total for repair and idle time can be seen, both as numbers of hours and percentages. These two categories are then assigned to jobs on the basis of operational percentage.

All costs of operation should be maintained as a percentage of total operational cost. In this way, all operational phases of the business are given a fair share of the total cost of operations for equipment.

Figure 13-5 shows an analysis for a four-week period, always taken as 160 hours. The period for analysis should always be the same length so that no allowance needs to be made for a greater number of hours in one month. While actual expenses occur on a calendar basis, they can be applied to jobs on a four-week basis. One common method of allocating costs over uniform periods uses 8 four week months and four five week months. For example, the first quarter would have 13 weeks. The whole 365 day year would look like this:

January	January 1 to January 28
February	January 29 to February 25
March	February 26 to April 2 (five weeks)
April	April 3 to April 30
May	May 1 to May 28
June	May 29 to July 2 (five weeks)
July	July 3 to July 30
August	July 31 to August 27
September	August 28 to October 1 (five weeks)
October	October 2 to October 29
November	October 30 to November 26
December	November 27 to December 31 (five weeks)

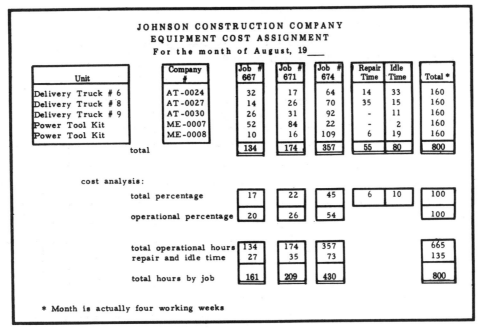

Figure 13-5

The equipment cost assignment worksheet provides an hourly summary by job and shows the relative efficiency of equipment. It also serves as a standard for all your equipment use. The idle and repair time figures indicate the degree of efficiency at which equipment is working.

DEPRECIATION

Depreciation is designed to allow businesses to "recover" (charge against income) the cost of buildings and equipment over the useful life of that property.

Writing off a portion each year spreads the cost over a period of time. Up to 1981, that period was referred to as "useful life." With passage of the Economic Recovery Tax Act of 1981, the idea of depreciation was replaced with "recovery" and the entire system of depreciation was replaced with the Accelerated Cost Recovery System (ACRS).

There are several methods for claiming depreciation. Under ACRS rules, you may use a prescribed method, or elect to use one of several straight-line alternatives. But assets you began depreciating under the old rules can continue under the old method until depreciation (or recovery) has been completed.

Questions you must ask before beginning to depreciate an asset are:

1) Is the asset depreciable? You cannot claim depreciation on land, inventory, or intangible property.

2) Are you entitled to a depreciation deduction? The tax law states that you are entitled to the deduction only if you suffer the economic loss as a result of decrease in value due to use. In most cases, the one who suffers the loss is the owner of the equipment. But under some circumstances you'll be allowed depreciation on equipment you lease.

3) What is the basis for depreciation? In most cases, the basis is the same as your cost.

4) When was the property placed into service? You begin depreciation during the year you begin to actually use property, not necessarily in the year it is purchased or paid for.

The ACRS system has been revised several times since its initiation in 1981. It now dictates the depreciation periods you are allowed to use for each specific class of asset. The specific periods, called class lives, specify the time over which you can claim depreciation for that class of asset.

Before ACRS, the concept of useful life was a determining factor in the number of years depreciation would apply. The true useful life of an asset was a primary factor in deciding how long a life the asset had for purposes of depreciation.

In many instances, property will actually appreciate in value. So the concept of useful life and depreciation doesn't truly reflect book value. If you own your own building, for example, it's probably worth more today than when you bought it — even though you have depreciated the structure down to zero.

You're allowed to use the prescribed ACRS method, which allows for a greater amount of depreciation in the earlier years, and a decreasing amount later on. Or you may elect in most cases to use an alternative straight-line method of recovery. You claim the same amount of depreciation every year for a specified number of years. At the end of that period, the asset will be fully depreciated.

In deciding which method to use, there are several factors you should consider:

a) What overall effect will your decision have? You must elect to use the prescribed or optional method for *all* assets placed into service for each class life within any one year.

b) How important is future depreciation to you? If you expect higher net profits in future years, you may come out better with straight-line depreciation. But if you need the expense this year, the prescribed method may be best.

c) Get the advice of your accountant. Use a trusted professional's advice in setting your policy for depreciation.

d) Obsolescence plays an important role in deciding how to depreciate assets. On the one hand, you want to consider the true useful life. On the other, the tax code restricts how aggressively you can write off your capital investment.

It's usually best to design a depreciation policy (within I.R.S. guidelines) that most closely reflects your use and replacement of equipment. If you use a lot of equipment that declines in value very quickly, accelerated depreciation is probably more accurate. If you replace equipment frequently because of increasing repair and maintenance costs, using the prescribed method of depreciation may be best for you. This relationship is illustrated in Figure 13-6.

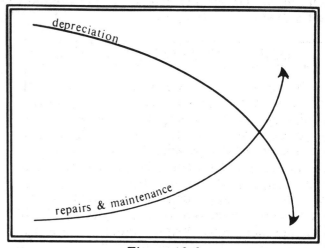

Figure 13-6

Once an asset has been fully depreciated (that is, the depreciation claimed to date equals the basis), it's either removed from the books altogether, or left in place. There are several ways for builders to handle adjustments for fully depreciated assets.

Removing a fully depreciated asset from your books is not good practice. Even though it has been fully expensed, it's still an asset of the company. Don't take it off the books until it's sold, traded or abandoned.

Once you do dispose of an asset, credit the value in the asset account and reverse the entry for your depreciation reserve:

	Debit	Credit
Accumulated Depreciation:	XXX	
Fixed Asset:		XXX

There are two ways to record depreciation. Both of these are acceptable methods of recording the expense. In the first, you make an entry for the part of the year you first own an asset. For example, an asset purchased on July 1 would receive exactly one-half year's depreciation during the first year. The second method would be used by builders who buy and sell a number of assets during a year. A full year's depreciation is taken on assets purchased between January 1 and June 30. Assets purchased between July 1 and December 31 are not subject to depreciation until the following year. The weakness of this method is that the recorded cost of doing business is distorted if purchases and sales of assets were not spread fairly evenly over the year.

OTHER DEPRECIATION RECORDS

Builders who have any substantial investment in fixed assets such as buildings, equipment or machinery are well advised to maintain their asset records as well as they maintain their equipment. Complete records for repairs and maintenance, hourly usage, insurance, cost, and depreciation are essential. Surprisingly, even established builders who have been in business for years have a difficult time breaking down their accumulated depreciation by units. As a result, fully-depreciated assets no longer in use remain on the balance sheet. Equipment asset accounts you carry should be backed up by subsidiary records to substantiate the components of each balance. Few small to moderate size operations can support their general ledger accounts. Yet this is a relatively easy set of records to maintain. It makes the difference between a clean set of books, and records that can not be explained at all. You can maintain good depreciation records in only a few minutes each month. Your records should include complete cost data from the purchase date of the equipment to the present. The depreciation record should balance with the total of the accumulated depreciation account in the general ledger and with the depreciation expense account for the current year to date.

Figure 13-7 is a depreciation record form used by Johnson Construction Company. One of these sheets is kept for each fixed asset the company owns. These forms are filed in a binder by classification of asset, similar to the filing system introduced earlier for equipment records. Each asset should have a separate page for recording depreciation. But small tools bought in bulk would have one page to cover the entire purchase. A second page could be attached, itemizing the various pieces that make up the one asset being depreciated. For practical purposes, each single purchase should be considered one depreciable asset.

A form such as Figure 13-7 allows you to compute depreciation far in advance. Simply check the "booked" column each month to find that month's figure for the depreciated value. Then, to control the journal entry, compute the total of all depreciation calculated for the current month by category of asset. This way you know that the current month's entry is correct.

If depreciation has been computed reliably and fairly, this record also serves as a signal when the assumed useful life of an asset is coming to an end. Analyze repair and maintenance bills to discover if the cost of operation is getting high enough to justify buying replacements.

DEPRECIATION METHODS

There are several ways to depreciate assets. Not all of these are acceptable for the depreciation of new equipment, although you may continue to use older methods on equipment that's held over from prior years. These methods include:

1) Straight line depreciation

2) Declining balance depreciation

3) Sum of the years' digits depreciation

Straight line depreciation is a method of spreading the cost of an asset evenly over its useful life. The same amount of depreciation is taken each year until the depreciable base is reduced to zero. An asset worth $8,000 with an estimated useful life of ten years and no salvage value would depreciate $800 per year. If the same asset had a salvage value of $1,000, the depreciable total of $7,000 would be depreciated at the rate of $700 per year.

You figure the amount of straight line depreciation by dividing the depreciable base by the useful life. In the example above, the total of $8,000 is divided by the useful life of ten years.

$$\frac{\$8,000}{10} = \$800$$

If the useful life was estimated to be eight years, the formula would be:

$$\frac{\$8,000}{8} = \$1,000$$

This can also be expressed as a percentage. A ten-year life can be expressed as: "Ten percent of the depreciable base." An eight-year life could be expressed as "Twelve and one-half percent of the depreciable base."

The effect of straight line depreciation is to recognize as expense an equal amount in each

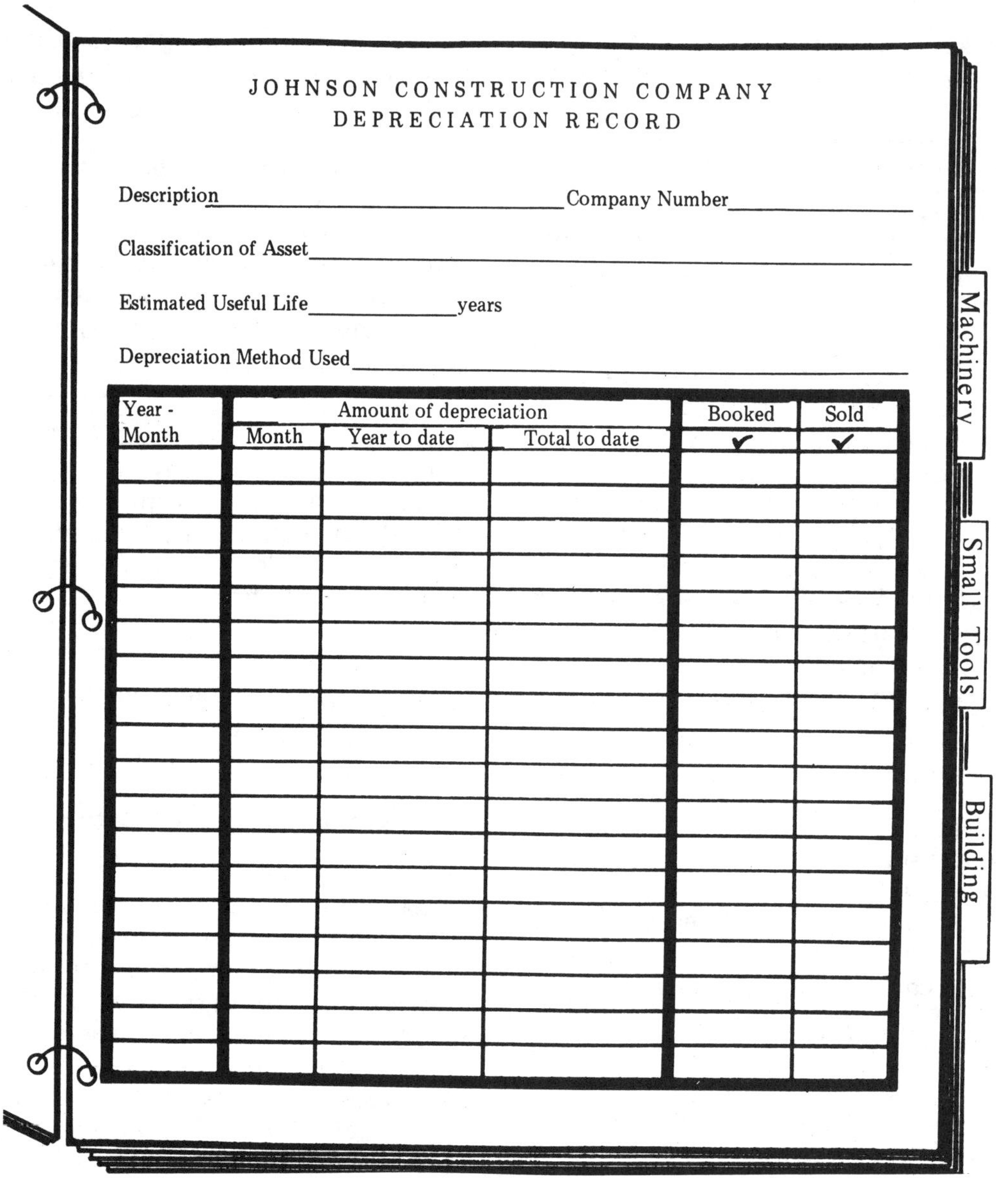

Builder's Guide To Accounting

```
          JOHNSON CONSTRUCTION COMPANY
                DEPRECIATION RECORD

Description_____ Company Number_____

Classification of Asset_____

Estimated Useful Life_____years

Depreciation Method Used_____
```

Year - Month	Amount of depreciation			Booked	Sold
	Month	Year to date	Total to date	✓	✓

(Tabs: Machinery, Small Tools, Building)

Figure 13-7

period of an asset's useful life. Accelerated methods recognize the greatest loss of equipment value earlier in the useful life on the assumption that the decline in value will be less in the asset's later years. The straight line method should be used for assets which are assumed to be as productive near the end of their useful life as they are in the beginning. Repair and maintenance cost and the cost of obsolescence must be considered. Equipment that will require progressively more maintenance as it ages should be put on an accelerated depreciation schedule.

An example of straight-line depreciation is illustrated in Figure 13-8. Assuming a ten-year life, each year's depreciation is based on the original depreciable amount. At the end of that period, the depreciable amount less depreciation is zero. Any value remaining is the asset's salvage value.

Declining balance depreciation is especially appropriate in the construction industry because most heavy equipment loses most of its market value fairly quickly. Automotive and mechanical assets depreciate right after purchase, even though the true value of that asset changes very little. You reflect much more closely the true market value of equipment and machinery when you use declining balance depreciation. Under this method, a large amount of the original value is lost in the early years of an asset's estimated useful life.

To calculate declining balance depreciation, the first step is to calculate the straight line depreciation, ignoring any salvage value:

$$\frac{\text{Full cost}}{\text{Useful life (years)}} = \text{Straight line depreciation}$$

STRAIGHT LINE DEPRECIATION

Useful life, 10 years

Each year's depreciation is 10% of the depreciable base

Depreciation:

first year	10% of base = depreciation	1st year
second year	10% of base = depreciation	2nd year
third year	10% of base = depreciation	3rd year
fourth year	10% of base = depreciation	4th year
fifth year	10% of base = depreciation	5th year
sixth year	10% of base = depreciation	6th year
seventh year	10% of base = depreciation	7th year
eighth year	10% of base = depreciation	8th year
ninth year	10% of base = depreciation	9th year
tenth year	10% of base = depreciation	10th year

Figure 13-8

The second step is to increase the straight line depreciation by the declining balance percentage selected. There are several percentages used with declining balance depreciation—125%, 150%, or 200%:

$$\text{Straight-line depreciation} \times \text{Declining balance percentage} = \text{Declining balance depreciation}$$

Under the straight line method, the depreciable base never changed. But under the declining balance method, the depreciable base is reduced each year by the depreciation taken up to that time. The adjusted value is used to compute the next year's depreciation:

$$\text{Full cost } less \text{ This year's depreciation} = \text{Adjusted depreciable base}$$

This method is illustrated on Figure 13-9. The 200% rate is most common with declining balance depreciation. Declining balance depreciation allows expensing amounts much higher than the straight line rates during the early years. The depreciation taken during later years is considerably less. You load most expense into the early years of an asset's life.

If assets are expected to be subject to higher maintenance costs in the later years of useful life, straight line depreciation would result in steadily increasing cost. Depreciation would remain level throughout the term, but higher maintenance costs in the later years would raise costs significantly. As a result, the unit cost of the equipment would increase as the equipment became less valuable.

Another accelerated method of depreciation is the sum of the years' digits method. It has the same effect as the declining balance method. The cost of an asset is recognized to a greater degree in the early part of its life.

The difference with the sum of the years' digits method is that the depreciable base of the asset remains at the original amount. To compute depreciation under this method, first figure the sum of the years' digits. A ten-year life would add up to 55:

$$1+2+3+4+5+6+7+8+9+10 = 55$$

The first year's depreciation would be:

Depreciable base x 10/55 = depreciation

The second year would use a multiplier of 9/55. The third year, 8/55, and so on until the tenth year, when a multiplier of 1/55 would be used.

The advantage of sum of the years' digits depreciation over the declining balance method is that salvage value can be computed at the exact amount wanted. Declining balance depreciation ignores the estimated salvage value.

The declining balance method always leaves some value and this is not necessarily the salvage value you select. Sum of the years' digits depreciation lets you compute your depreciation based on the estimate. See Figure 13-10.

There are advantages and disadvantages to each method of depreciation. At best, depreciation is only as good as the estimate of useful life and the proper distribution of the cost. And that depends entirely on your guess. There is no way to know the exact value of fixed assets. Averages must be used. But there are enough distinctions between the three principal methods that you should be careful to pick the right one for your particular application. The most appropriate method depends on the type of asset being depreciated. A comparative summary of the principal types of depreciation is shown in Figure 13-11. Assuming a ten-year life and a depreciable value of $60,000, the yearly depreciation is listed for straight line, sum of the years' digits, 200%, 150%, and 125% declining balance methods. Note that these figures assume no salvage value. If there were a salvage value, that amount would be subtracted from the $60,000 in the straight line and sum of the years' digits calculations.

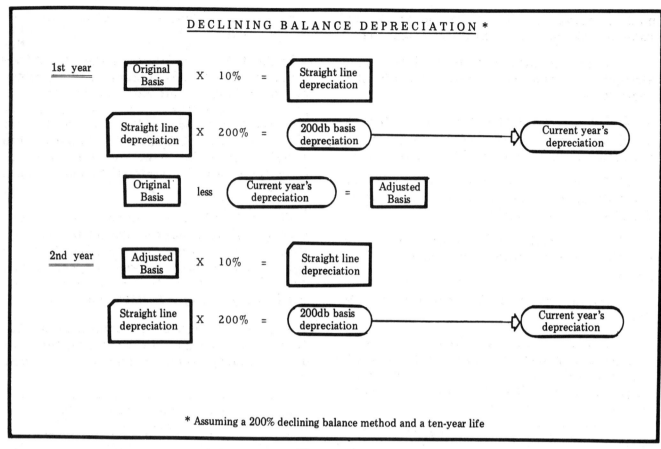

Figure 13-9

Assume that you sell the equipment depreciated in Figure 13-11. If the sales price is higher than the depreciated value you are currently carrying on your books, you have a *gain* that recaptures part of the depreciation. Your

SUM OF THE YEARS' DIGITS DEPRECIATION

Useful life, 10 years:

Sum of the years' digits is:

$1 + 2 + 3 + 4 + 5 + 6 + 7 + 8 + 9 + 10 = 55$

Depreciation formula:

Depreciable base for:

first year	x	10/55 = depreciation	1st year
second year	x	9/55 = depreciation	2nd year
third year	x	8/55 = depreciation	3rd year
fourth year	x	7/55 = depreciation	4th year
fifth year	x	6/55 = depreciation	5th year
sixth year	x	5/55 = depreciation	6th year
seventh year	x	4/55 = depreciation	7th year
eighth year	x	3/55 = depreciation	8th year
ninth year	x	2/55 = depreciation	9th year
tenth year	x	1/55 = depreciation	10th year

Figure 13-10

COMPARATIVE DEPRECIATION SCHEDULES

Year	Straight Line	Sum of the Years' Digits	200% Declining Balance	150% Declining Balance	125% Declining Balance
1	6,000	10,909	12,000	9,000	7,500
2	6,000	9,818	9,600	7,650	6,563
3	6,000	8,727	7,680	6,503	5,742
4	6,000	7,636	6,144	5,527	5,024
5	6,000	6,545	4,915	4,698	4,396
6	6,000	5,455	3,932	3,993	3,847
7	6,000	4,364	3,146	3,394	3,366
8	6,000	3,273	2,517	2,885	2,945
9	6,000	2,182	2,013	2,453	2,577
10	6,000	1,091	1,611	2,085	2,255

Figure 13-11

accountant should be able to advise you on the best method for your situation. But be aware that depreciation in excess of the actual decline in value often creates tax problems when the asset is sold.

For complete financial reporting, it is always a good idea to list in detail both the total value of assets in each category and totals for depreciation in each category. See Figure 13-1. Any significant difference in the various depreciation methods being used should be footnoted. For example, footnotes for **Figure 13-1** might be:

1) Office furniture is depreciated on a straight line basis, with a remaining life of 7 years (average).

2) Machinery and equipment and small tools are depreciated on a sum of the years' digits basis, with a remaining life of 4 years (average).

3) Trucks and autos are depreciated on a 200% declining balance basis, with a remaining life of 7 years (average).

4) Office building is depreciated on a straight line basis, with a remaining life of 32 years.

5) Improvements are being amortized over five years.

This kind of complete disclosure is rarely needed, but enough should be said in footnotes to indicate that a major portion of remaining depreciation for various assets will be realized in a specified period of time.

TODAY'S MACRS RULES

Today we use a *modified* system of depreciation (Modified Accelerated Cost Recovery System) from the Tax Reform Act of 1986 and later modifications. The following *class lives* are specified under MACRS rules:

- 3-year class— for assets that previously had a useful life of four years or less.

- 5-year class— including computers, typewriters, photocopiers, automobiles, light trucks, and telephone switching equipment.

- 7-year class— office furniture and equipment, for example.

- 10-year class— assets with previously categorized lives between 16 and 19 years.

- 15-year class— assets with lives between 20 and 24 years.

- 20-year class— assets with lives of 25 years or more.

- 27.5-year class— for residential real property including rental manufactured housing.

- 31.5-year class— nonresidential real property.

In the 3, 5, 7 and 10-year classes, the ACRS prescribed method involves using double-declining balance depreciation for the first part, then reverting to straight-line for the balance of the period. The 15 and 20-year classes employ 150% declining balance, then revert to straight line. And the 27.5 and 31.5-year classes are allowed to use straight line depreciation only.

Figure 13-12 is a summary of the percentages of depreciation allowed per year for the first six class lives. Note that the "half-year convention" is used in the first year in each case. This assumes that the asset qualifies for one-half a year's depreciation that year.

These percentages are prescribed by the Tax Reform Act of 1986. The I.R.S. could publish actual annual percentages that vary from these percentages.

Year	3-year class	5-year class	7-year class	10-year class	15-year class	20-year class
1	33.33	20.00	14.29	10.00	5.00	3.750
2	44.45	32.00	24.49	18.00	9.50	7.219
3	14.81	19.20	17.49	14.40	8.55	6.677
4	7.41	11.52	12.49	11.52	7.70	6.177
5	--	11.52	8.93	9.22	6.93	5.713
6	--	5.76	8.92	7.37	6.23	5.285
7	--	--	8.93	6.55	5.90	4.888
8	--	--	4.46	6.55	5.90	4.522
9	--	--	--	6.56	5.91	4.462
10	--	--	--	6.55	5.90	4.461
11	--	--	--	3.28	5.91	4.462
12	--	--	--	--	5.90	4.461
13	--	--	--	--	5.91	4.462
14	--	--	--	--	5.90	4.461
15	--	--	--	--	5.91	4.462
16	--	--	--	--	2.95	4.461
17	--	--	--	--	--	4.462
18	--	--	--	--	--	4.461
19	--	--	--	--	--	4.462
20	--	--	--	--	--	4.461
21	--	--	--	--	--	2.231

Figure 13-12

Two other major changes in the 1986 law include the elimination of investment tax credit and a change in the expensing provisions.

The investment tax credit was a provision that allowed you to claim a tax *credit* (a reduction of your final tax liability) up to 10% of the cost of qualified capital assets. Any portion of an asset that was used for ITC could not be depreciated. The ITC provision was repealed under the 1986 law, retroactively effective to January 1, 1986.

Expensing is a provision that allows you to write off a limited dollar total of qualified assets all at once during the year purchased. Under previous law, you were allowed to expense up to $5,000 in assets per year. The new provision, effective January 1, 1987, allows expensing up to $10,000 per year.

To qualify, the asset must be property used in a trade or business. If your total investment in capital assets is greater than $200,000 in any one year, your allowance for expensing is reduced dollar for dollar above that level. So once you buy $210,000 or more in capital assets in any one year, you have no expensing provision.

AMORTIZATION

Note that in footnotes improvements to the building are being *amortized* over five years rather than depreciated. Amortization spreads the cost of major expenses over a period of years. The cost of improvements is listed at its gross value in the same way as are depreciable assets. A Reserve For Amortization account is maintained just like the Reserve For Depreciation (Accumulated Depreciation) account. This is one type of amortization. Another type reduces the value of an asset. Instead of showing gross values, monthly expenses for amortization reduce the balance. This is used for accounts like organizational expenses.

A builder would use different amortization treatments for different types of expenses. In the case of improvements, there is a lasting, tangible value - the improvements themselves. As long as those improvements exist, their total value should be shown. But there is no lasting asset to be seen for organizational expenses. This would include any expense not taken when incurred and which would normally have been expensed - except that the reason for the expense was related to the formation or reorganization of the operation.

For example, suppose you had to spend the following to organize your operation:

Legal fees (to incorporate or to write partnership contracts)	$1,800
Accounting (to set up and advise on the formation of the business)	700
Rent and utilities paid for two months prior to moving in and beginning operations	1,380
Operating supplies (cleaning, etc.)	460
Total	$4,340

These are legitimate business deductions, but they do not apply exclusively to current operations. Instead of writing them off to expense as they occur, the correct treatment is to classify them as an asset and amortize them over a period of time. At the end of that period, there will be no identifiable, tangible asset, so the balance of the account should reduce to zero.

For improvements, the gross asset is offset by a Reserve for Amortization account. But for organizational expenses, the asset is reduced each month. If this $4,340 asset were to be written off over a period of three years, the account would look like this:

ASSET - ORGANIZATIONAL EXPENSES

	Debit	Credit	Balance
Balance forward			$4,340.00
January		$120.56	4,219.44
February		120.56	4,098.88
March		120.56	3,978.32
April		120.56	3,857.76
May		120.56	3,737.20

Because organizational expenses are not fixed assets, they are included in the Other Assets category.

LEASE OR BUY?

The decision to buy equipment can be a major step for any builder. It is sometimes to your advantage to lease equipment. You get the following benefits if you lease:

- Equipment needed for single jobs can be leased for that job only.

- An immediate tax deduction is available for the cost of monthly leases.

- No large investment or financing commitment is required.

- You have the chance to evaluate equipment performance before buying.

Leased equipment may be carried as an asset on financial statements if the offsetting liabilities are also included. This is the case only when:

- The lease includes a provision for purchase at a future date.

- The agreement provides that all or part of the payments can be applied to the purchase.

- The amount of the lease payments is higher than normal lease payments would be.

These points would normally be included in the terms of agreement on lease-option purchase plan. This type of plan allows a builder to lease equipment without a large cash investment. This may be a good plan if you intend to buy in the future. If you change your mind and decide to get out of the agreement, you lose only the higher payments you have made and the equity you have been building up in the equipment. This may be preferable to being stuck with unneeded or under-performing equipment. Under a lease-purchase agreement, the full value of the equipment is included as an asset and the full liability — less equity — is listed as payable, both currently and in the long term. Depreciation on the equipment can begin immediately as well.

The decision to lease or buy must depend on the equipment, its cost, the investment required, and the expected use. Will the use be regular or limited? What will the unit (hourly) cost be, and can you justify owning the equipment on that basis? Can you afford the investment itself?

Following is a comparison of the costs of leasing and buying equipment. Keep in mind that buying is cheaper for equipment that will receive regular use. All of the costs of buying equipment are built into a monthly lease total. Except on a short term, leasing is *not* cheaper than owning:

	Lease	Buy
Monthly payment	XXX	
Depreciation		XXX
Maintenance		XXX
Insurance		XXX
Property Tax		XXX
Interest on Financing		XXX

SALES AND TRADE-INS

When equipment is sold, the books have to be cleared of all entries relating to that asset. This includes taking out the gross value of the asset and the accumulated depreciation on it.

The other part of this entry is to book the gain or loss on the asset. These are not operating gains or losses, and they have to be kept separate from regular sales and gross profits. Otherwise, your comparison of operating results will be distorted. Besides, gains or losses from sales of fixed assets receive different tax treatment than ordinary income does.

If a piece of $25,000 equipment, depreciated to $7,000, is sold for $4,000, the loss on the sale is $3,000:

Original cost	$25,000
Less: accumulated depreciation	18,000
Book value	7,000
Selling price	4,000
Loss on sale	$3,000

The general ledger entry for this transaction would be:

	Debit	Credit
Cash	$4,000	
Accumulated depreciation	18,000	
Loss on sale	3,000	
Fixed asset		$25,000

A trade-in is recorded differently than a sale. For book and tax purposes, trade-ins are not subjected to tax treatment until the replacement piece is sold or abandoned. The tax gain or loss is deferred by adjusting the basis of the new asset. In the previous example, assume that the $25,000 fixed asset was traded in for a $14,000 replacement:

Original cost, old asset	$25,000
Less: accumulated depreciation	-18,000
Book value	7,000
Paid for new asset	14,000
Adjusted basis, new asset	$21,000

The general ledger entry in this case would be:

	Debit	Credit
(New) fixed asset	$21,000	
Accumulated depreciation	18,000	
(Old) fixed asset		$25,000
Cash		14,000

This defers tax on the gain until the new equipment is sold.

Chapter 14
Cash Budgeting

Builders must often make large investments in equipment and inventories, and they need short-term funding for construction in progress. Tremendous outlays of cash for these investments can result in low profits, and the remedies aren't easy to identify without a good cash budget. A builder needs a plan for both immediate and long-term cash funding that takes into account both profits and investments in the business.

A cash budget does not simply budget a certain amount of cash for a specific future use, or simply allow for reserves in the handling of income. Setting up a payroll account, for instance, and accumulating funds to pay payroll taxes is one type of cash plan. But this is a procedure, rather than a plan; a matter of habit after a while. A complete cash plan does much more. It allows for funds to be developed and used for business growth. It prevents errors in cash planning and direction by channeling cash to get the most for the sales dollar and more sales dollars from the cash flow. See Figure 14-1.

But a cash budget aims to do much more than this. Here is a list of what a good cash budget can accomplish.

- Makes cash available for day-to-day and month-to-month operations

- Plans for the use of excess funds when they are available

- Times operations for seasonal business decreases

- Plans adequate levels of inventory in advance

- Serves as a model for the control of receivables and collections

- Budgets loan payments

- Prepares for tax liabilities

- Takes advantage of discount terms

- Lets you improve and maintain a good credit rating and helps you qualify for needed loans

- Controls the purchase of materials and helps prevent over- or under-buying.

No builder can afford to ignore the problems

THE CASH PLAN

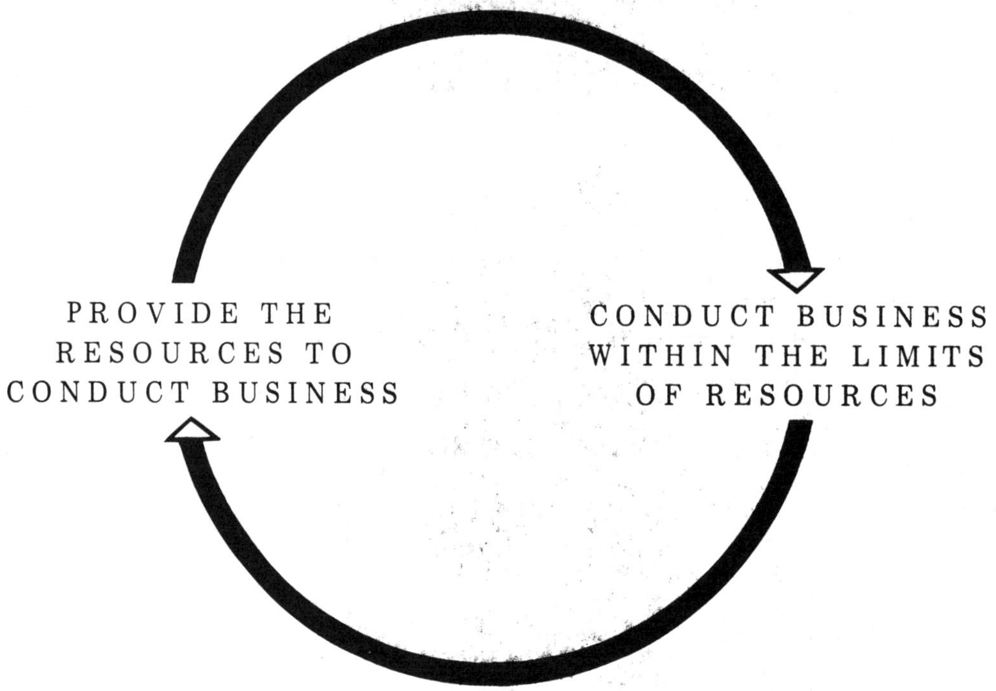

```
          ┌──────────────→┐
PROVIDE THE          CONDUCT BUSINESS
RESOURCES TO         WITHIN THE LIMITS
CONDUCT BUSINESS     OF RESOURCES
          └←──────────────┘
```

Figure 14-1

that result from poor management of cash assets. When lenders examine your financial statements, they look closely at your recent cash flow. If you are constantly struggling to maintain your cash position, lenders doubt that you can plan for loan repayments. They see the loan as a high risk, and are likely to turn it down. Cash flow problems put growth markets out of reach. Almost any expansion requires capitalization. A builder with serious cash flow problems at his current volume level lacks the reserves to generate further cash flow with equipment, labor, or inventory investments. His solution, of course, is to improve his current cash flow through budgeting. When he has built up his funds and improved his financial statements and credit rating, then he can think about expansion.

TESTING THE CASH POSITION

There are several ratios you can use to judge your cash position. They are called liquidity tests, and indicate the relative asset strength of a business. These tests are meaningless by themselves. You must look at them in the following ways to see the larger picture of your operation's cash position.

- In relation to each other. Each test measures cash position using different standards.

- As trends. The history of a business is as important as its current status, and is often more revealing.

The first test is called the *current ratio*. It compares current assets (cash, receivables, inventories) to current liabilities (current notes, accounts payable, and accrued taxes). Current means within one year. A business that has over-invested in fixed (long-term) assets may have liquidity problems. That is, it may not be able to meet current obligations from current resources. A business can be profitable and can still be hindered from making greater profits if its assets are tied up and unavailable for current use.

The current ratio is said to be *one to one* if assets and liabilities are about equal. If current assets are twice as large as current liabilities, the ratio is said to be *two to one*. Ideally, current assets should be higher than current liabilities. That is, the assets you have available to you currently should be able to pay off the liabilities that you currently owe and still leave you some funds. The current ratio is said to be *one to two* if assets are half of liabilities. This means you can pay half of what you currently owe with all of your current assets. Any time the current ratio is negative (current assets less than current liabilities) the potential for growth is negative as well.

When you prepare your financial statement for a lender before applying for a loan, you may be tempted to accumulate current assets for a couple of months to improve your cash position. But accumulating current assets always results in piling up current liabilities. Try instead to reduce your current liabilities rather than increase your current assets. Of course, when you reduce current liabilities you reduce current assets at the same time. But working on your liabilities first improves the current ratio. Assume that you accumulate your current assets for a couple of months; this allows your accounts payable to increase. Your total cash position looks like this at the end of two months:

Current assets	$74,000
Current liabilities	$52,000
Current ratio	1.4 to 1

Now you pay off $30,000 of (probably overdue) accounts payable. This reduces your cash by $30,000:

Current assets	$44,000
Current liabilities	$22,000
Current ratio	2 to 1

Paying off the current liabilities reduces both current assets and liabilities, but improves the current ratio. Now you have a solid current ratio to show the lender.

A second useful test shows your cash position over a period of time. Take your gross profits as a percentage of sales volume for a given period. Then compute this percentage for a number of periods to get a trend. This indicates the amount of real control you have over your costs. When gross profit decreases in relation to sales, either you are not controlling your costs or your prices do not allow enough markup over your costs. Other factors that affect the level of gross profit can be unexpected inventory losses, increases in idle time, and material theft.

The ratio of *expenses to sales* is the third way to judge your control of your business. Sudden large rises in expenses affect your cash balances. Good cash budgeting is impossible if you don't keep expenses in line with sales.

A fourth ratio, *income to net worth*, shows the overall yield on your total investments. Compare present and past ratios to find the trend of your yields. If you actively seek out profitable work and favorable yields for your risk, the trend will be favorable. But the trend will show the opposite if you begin to take lower-yield work and lose control over costs and expenses.

The fifth cash position shows the amount of *debt to capitalization*. Total debt includes all liabilities; capitalization is the worth or value of the business. The value is the total of assets less liabilities. This ratio is a good test of capital strength, as it shows the level of debt commitment against the worth of the business. Compute this ratio for several periods and compare the ratios to get a trend. A decreasing trend indicates that you are running your business on more and more borrowed money. This is an unhealthy direction for any business to take. Interest on borrowed money consumes profits and decreases the overall yield on investment.

PREPARING A CASH BUDGET

A cash budget must reflect business realities, since you use it to chart the future course of your operation. If you have never used a budget before, start by estimating your immediate future cash income. Figure out as closely as possible the amount of incoming cash for the next month. Once you know this you can time your known expenses to the availability of these funds. Most businesses have had to budget this way out of necessity at one time or another. But effective cash budgeting does more than allow you to pay your bills. It puts you on a controlled course of financial action designed to achieve long-range goals.

There are two principal methods of preparing a cash budget or forecast: the *cash movement* method and the *source and application of funds* method. Although there are other forecasting procedures, these two are commonly used because they are easy to prepare and at

JOHNSON CONSTRUCTION COMPANY
BUDGET OF CASH MOVEMENT
September, 19____

	September 1 - 7	September 8 - 15	September 16 - 22	September 23 - 30
Cash at Beginning of Period	$ 1,680	$ 1,280	$ 2,775	$ 1,475
Receipts:				
Trade Collections	600	6,400	8,600	1,400
Retainages	-	-	2,000	-
Cash Sales	2,200	2,400	2,400	2,200
Other Receipts	-	-	-	-
Total Receipts	2,800	8,800	13,000	3,600
Total Cash Available	4,480	10,080	15,775	5,075
Payments:				
Materials	900	3,000	200	-
Labor	1,200	1,500	1,200	1,500
Subcontractors	-	-	1,000	-
Selling Expenses	400	400	800	600
Fixed Overhead	700	1,600	200	100
Notes and Interest	-	-	-	450
Payroll Taxes	-	805	-	-
Income Taxes	-	-	-	-
Capital Assets	-	-	10,900	-
Total Payments	3,200	7,305	14,300	2,650
Cash at End of Period	$ 1,280	$ 2,775	$ 1,475	$ 2,425

Figure 14-2

least as informative as any other method. The cash movement method involves budgeting only the flow of actual cash in and out of your business. Do not list additions to accounts receivable, as no cash changes hands in charge sales. And do not include liabilities. But make an estimate of payments on account. Estimate payments on account over several months, taking the amount of your charge sales from your total income budget. Budget taxes and loan payments as they become payable.

This method is especially valuable for builders who have wide variations in their business volume from month to month. Such variations usually follow a predictable course. Building volume is affected by the seasons, zoning and political actions in your community, unusual weather patterns, the rate of community growth, and varying demands for the different types of work you do. Budgeting by the cash method allows you to handle directly cash flow changes that occur because of monthly volume variations. Figure 14-2 shows the form this budget should take. List each type of expected cash receipt and payment. This assures you that funds will be available to pay normal monthly expenses. Carry forward the amount at the end of the period on the last line to the beginning of the next period on the top line of the next column. Include in the cash budget the cost of any fixed assets you acquire, such as the $10,900 under Capital Assets.

You can use this form for yearly cash budgets by making a column for each month. Be sure to compare your actual cash flow to the budgeted amounts both during and after the

Cash Budgeting

JOHNSON CONSTRUCTION COMPANY
BUDGET, SOURCE AND APPLICATIONS OF FUNDS
For the Fourth Quarter, 19___

	October	November	December
Net Income	4,100.00	3,200.00	2,800.00
Plus: Expenses for affecting Working Capital:			
Depreciation	160.00	160.00	195.00
Amortization	37.00	37.00	37.00
Provision for Bad Debts	105.00	105.00	105.00
Working Capital Provided from Operations	4,402.00	3,502.00	3,137.00
Increase in Long-Term Debts	-	4,500.00	-
Sale of Fixed Assets	-	2,000.00	-
Other Increases in Cash	-	-	-
Total Cash Provided	4,402.00	10,002.00	3,137.00
Purchase of Fixed Assets	-	7,500.00	-
Increase in Long-Term Assets	-	-	1,000.00
Decrease in Long-Term Debts	3,500.00	3,500.00	3,500.00
Other Decreases in Cash	550.00	550.00	550.00
Total Cash Used	4,050.00	11,550.00	5,050.00
Net Cash Increase (Decrease)	352.00	(1,548.00)	(1,913.00)

Figure 14-3

budget period. You can improve on your budgeting technique by comparing your forecast to your records of where the cash actually went.

The advantage of this method is its simplicity. Only cash flow is forecast; you need no estimate of depreciation or accruals for income and expenses. You include no tax liabilities or any other non-cash item. But the disadvantage of the cash movement budget is that it makes forecasting assumptions about cash items that vary greatly with non-cash factors. A good example is receipts on trade collections. Trade collection receipts depend on your maintaining a certain number of charge sales. Thus you need to analyze your charge sales carefully when you prepare your cash movement budget. Too often, cash budgets are drawn up without studying true sales expectations. The result is an inaccurate and nearly useless budget based on only partial estimations of cash flow.

The source and application of funds method is more precise than the first method. It assumes that you will attain a certain level of net income. Then it adjusts for non-cash changes to arrive at the net increase or decrease in working capital. This form of budget does not let you compute the actual ending balance of cash, but instead budgets for the use of available funds. You maintain your cash balance within each period by timing your cash flow.

The form this budget takes on paper is similar to that of a standard financial statement that might be included in a package of company reports. Along with a balance sheet and an income statement, the source and application of funds statement makes a meaningful third main report.

Figure 14-3 is Johnson Construction's source and application of funds budget. The first line is the budgeted net income. Johnson derives this figure from two main sources: 1) the previous season's net income, and 2) the timing of its current jobs by stages of projected completion. It subtracts its costs and expenses from the gross income figure. To find out how much income it derives from its operation, Johnson Construction adjusts all non-cash entries from net income. The principal entries in this case are depreciation, amortization, and the reserve for bad debts. These do not reflect cash changing hands, so they are added back. The result is the amount of working capital provided from its operations.

Cash can come from loan proceeds as well as from jobs. November's increase of $4,500.00 represents the long-term portion of a new note, that part which is payable beyond the current twelve months.

Fixed assets sold for cash are not reflected in net income from operations. Yet any sale of fixed assets increases the available cash of a business. Note the $2,000 increase in November.

Any other increase of long-term assets will increase working capital in that period. The total of all such increases, plus the adjusted cash-basis income from operations, is the total cash available.

Cash used to pay costs and expenses has already been subtracted in the first line, Net Income. So the only costs and expenses Johnson lists in the Total Cash Used section are those other than from operations. These include the purchase of fixed assets. Note the $7,500.00 entry for November. Cash can also be used to purchase or pay for other non-current assets.

Decreases in long-term debts (such as loan payments) are generally cash payments. This is a use of cash just like any other payment, and it indicates how heavily committed Johnson's available cash is to debt repayment. In the budget in the figure, a more careful plan of action would have spread out the net decrease in cash (on the bottom line) over all three months. We assume that Johnson has adequate cash reserves to cover the decreases in November and December. The principal causes of the large drain on Johnson's available cash are the purchase of a large fixed asset in November and the payments of long-term debts in all three months. The fixed asset does not affect the cash flow by itself—the result of good cash budgeting. But the long-term debt payments seem to take up all available cash after other payments have been made. This indicates that the business is relying heavily on borrowed money.

The source and application of funds budget details the changes in current assets and current liabilities. Changes in net income, long-term assets, and long-term liabilities will equal changes in net current assets and current liabilities. This is shown in Figure 14-4. Current assets and liabilities change over a period of time. List the amount of change (increase or decrease) in current asset and liability accounts to see how your total cash flow will affect the availability and commitment of funds. These changes to current asset and liability accounts are referred to as the *components* of cash flow. Income and the changes in *all other* asset and liability accounts reflect the *detail*, or the real source and application of funds. The changes in all other asset and liability accounts show the same total change as that in current asset and liability accounts.

Plan thoroughly when you prepare a source and application of funds budget. Calculate billings and payments, material and labor needs, expenses, planned loans and interest. Include a blueprint for acquiring fixed assets. These backup schedules should support the cash flow conclusions in the budget.

FIGURING A BREAKEVEN POINT

You reach a breakeven point when your profits fall to zero. This can happen at any sales volume and with any cash flow. Know your breakeven point at any time, and calculate it as part of your overall business plan. Coordinate your sales planning with your budgets and estimates for cash flow, overhead, and direct costs, relying on high and low expectations of your sales volume and your overall costs to come up with profit figures. When you calculate the sales volume you need to break even, you know that your cash budget must allow you to use your capital to produce sales of no less than that volume.

The breakeven point has a very specific use. It does not distinguish between the various yields of the different jobs and job types you do. It cannot help you decide what kinds of work you need to reach a volume goal. A breakeven point does not encourage growth. It merely tells you the least amount of sales you need to break even

Cash Budgeting

BALANCE SHEET

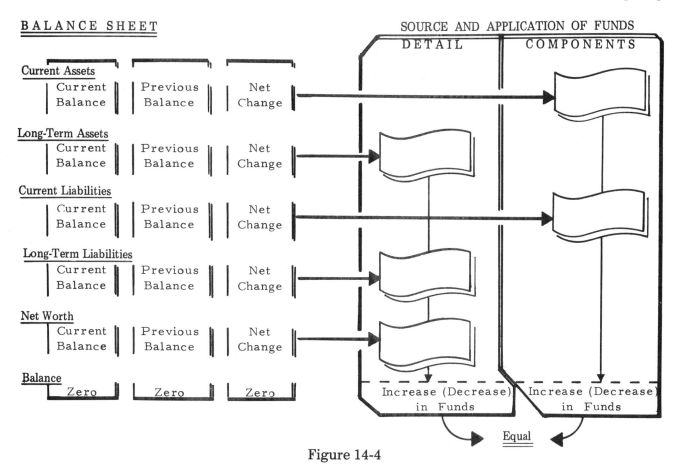

Figure 14-4

and serves as a reminder to maintain a strict cash budget to better that goal.

Compute your breakeven point by one of two methods. You can build a budget from the bottom up. Calculate your fixed overhead, then your sales expenses, and finally your direct costs. This amount will equal the breakeven volume of sales you need. Second, you can compute the breakeven point without starting from the bottom. Assume different volumes of sales, then subtract the direct costs and selling expenses you need to maintain those volumes. Next take out fixed overhead, then from that figure establish a breakeven point at each approximate volume by sliding the other figures up or down relative to it. Each approximate volume must equal direct costs, selling expenses, and overhead for you to reach a breakeven point. For example, on a volume of $150,000 with a direct cost of 60%, selling expenses of 22.5%, and fixed overhead of $28,000, you could project a loss of $1,750.00. Thus the breakeven point is slightly higher than $150,000. On a volume of $200,000, with projected direct costs of 60%, selling expenses of 25% and fixed overhead of $28,000, your profit would be $2,000.00. The breakeven point would be somewhat less than the volume of $200,000.

The value of all this is that a builder preparing his cash budget knows the minimum volume he must find to avoid a loss. This assumes that the levels of costs and expenses remain in line with previous experience. Volume is limited within reasonable bounds by your ability to invest in inventories, receivables and fixed assets. The result of this cash budget planning is a set of realistic goals. And a plan that starts from realistic goals for growth of volume, profits, and markets has a better chance to succeed. In a good plan you buy new major assets with the knowledge that the purchase won't strain your assets. You won't over-borrow if you know the limits of your volume and available cash.

CASH CONTROLS

You need to control your cash as it flows through your office on its way to the bank. Cash is the most often lost asset you have. This is

153

because it is easy to lose control of any group of balances when volume goes up during seasonal fluctuations. Cash, unlike a company truck or piece of heavy equipment, is not identified as yours. Anybody can use it.

The rest of this chapter shows how your cash can be lost and how you can avoid this problem with proper office controls. Daily, weekly, and monthly summaries of cash income and expenses keep you informed of your cash status. Balance your bank accounts every month. Above all, be concerned about your cash. A builder who closes his eyes to how his books and records are handled invites theft and loss.

Cash can be stolen in a number of ways. The following list should give you ideas about how to prevent cash theft in your business.

While most employees are honest and trustworthy, there are those who can not resist temptation. Too many builders tempt their employees by themselves demonstrating how easily cash can be taken.

- *Lapping* Cash is taken from the business by moving outstanding balances from one customer account to another. The thief can write off the amount of cash taken as a bad debt or he can adjust the amount as a "reconciling" item to an out-of-balance system. The way to prevent lapping is to insist on monthly aging lists of all outstanding accounts. Compare this total to a reconciled accounts receivable total. Review all writeoffs and adjustments personally.

- *Unofficial borrowing* An employee can take cash for periodic "emergencies" and intend to pay back what he borrows some day. But the employee is often tempted to take more and more until it would be impossible to repay. Prevent this by establishing office rules on petty cash and by establishing stringent controls on all funds. Carefully monitor your employees who have access to checking and savings accounts and cash awaiting deposit.

- *Failing to record cash sales* Sales made for cash are only reported if the sale is recorded. An employee who receives the cash and who also records the sale has the opportunity to pocket the whole amount. Discourage this by setting up an effective method of accepting cash. Use invoices for billings and orders, and fill out sales receipts for cash sales. File a copy of the invoice or receipt with the cash when you receive it. Keep and be able to justify all voided receipts and invoices.

- *Adjusting the check book* An employee with access to the check book can change the cash balance to absorb checks he writes and does not record. The voided stub is ignored as an unused check. Prevent this by keeping the voided checks themselves. You encourage this form of theft if you allow the person who balances the bank account to also be responsible for the check book.

- *Falsifying discount amounts* Protect against this form of theft by knowing the discount terms of your suppliers. A monthly discount summary allows you to verify the terms used by your employees against those quoted you by your suppliers.

- *Writing off accounts as bad debts and taking payments* Approve all writeoffs yourself, never delegate this job to others. Also prepare the bad debt analysis yourself for greater security.

- *Falsifying sales records or deposit slips* Keep your cash, deposits, and control records secure by limiting access to them. High-volume businesses use deposit controls involving several employees. Owners of small operations handle all the cash and deposits themselves. A secure method for mid-size builders is to count the cash yourself before turning it over to an employee to record and deposit. You can compare your total to a duplicate bank slip. You retain control over your cash without spending a lot of time on the procedure.

- *Double use of petty cash receipts* You may think you can control your petty cash by checking each set of petty cash receipts yourself. Employees can take funds by reusing receipts from previous batches. Prevent this by initialing each receipt in ink in the same place on the receipt every time. You can tell at a glance if the space for your initial has been tampered with in order to run through a used receipt a second time.

- *Paying personal bills* Some builders pay

accounts on their books that are never checked or balanced. An employee can make a journal entry to increase the balance of the cash account by making an offsetting credit to the uncontrolled account. He can then write unrecorded checks on this overbalance or remove cash from deposits up to the amount of the overbalance. Prevent this by balancing and checking all your accounts periodically.

- *Overbalancing* Many builders have several bills with a single check. The employee staples these bills together with one or two of his own in the middle, and readies the check for your signature. Look over such payments carefully to make sure that the bills you pay are all your own.

- *Submitting false invoices* This is a rather elaborate scheme for petty theft. The employee establishes a false vendor to send invoices to the builder. Prevent this by knowing what all checks pay for. Inspect all invoices carefully before signing checks, and be certain that you have approved all payments.

- *Double payments* This is another idea that requires elaborate planning and, often, a conspiracy between two people. Know who you are paying, what you are paying for, and who you have paid already.

- *Changing check amounts after signature* This kind of theft is relatively easy to prevent. Be sure checks are imprinted before you sign them. Compare the written amount to the imprinted amount and the amount on the invoice.

- *Cashing unclaimed checks* An employee may be tempted to take a check he knows has been lying around unclaimed for several months. Know how many checks are long-outstanding, and for what amounts. After a predetermined period of time (such as 90 or 180 days) stop the checks and reissue them. If checks are returned to your office, know what happens to the money.

- *Forging checks* Keep close tabs on all checks to prevent the loss of blank or unnumbered checks. Check numbers are the most important check control you have. Keep a numerical log of all checks you sign. Include the numbers of all the voided checks you keep.

- *Forcing bank accounts* Bank account reconciliations can be falsified to cover up the theft of funds recorded as deposits or not recorded as checks. You can minimize the chance of bank account thefts by having the bank account balancing done by two different people.

- *Outright thefts* You invite this kind of theft when you are careless around your employees with cash, books, and records. Take at least minimum security precautions when you handle your cash. Keep books and records close at hand. Show concern for all transactions in the office.

The best, most foolproof methods of preventing theft can't take the place of having trustworthy employees. A weakness in the cash control system often brings on theft; but a strong system can also challenge someone to beat it. All you can do is develop the best cash controls possible within the limits of your time, resources, and employees. But you can screen before you hire those to whom you entrust the handling of your cash. Check their references. Make sure their backgrounds and reputations are clean. You can also bond employees who handle large amounts of cash.

A series of related cash controls often discourages theft by simply not encouraging it. Well designed cross-controls are usually more effective at this than sophisticated, tempting rituals for cash security.

Keep in touch with your cash flow and update yourself on the status of actual cash balances with a cash flow summary. See Figure 14-5. This summary should list all incoming and outgoing amounts. Prepare it as needed daily, weekly, or monthly. But a cash flow summary is most valuable when issued on a daily basis. You can see high and low balance days within the month, and this helps you time payments against collections. A monthly cash forecast and your records of actual monthly cash flow cannot do this.

A series of daily cash summaries compared over several months can tell you on what days you can expect a strain on cash and on what days you can count on an excess of available funds.

```
                JOHNSON CONSTRUCTION COMPANY
                    DAILY CASH SUMMARY
                       Date _____



    BALANCE FORWARD                              _____

    ADD: Receipts                                _____


                                                 _____

    LESS: Payments                               _____

    ENDING BALANCE                               _____


    Ending Balance by Location:

            General Account A                    _____
            General Account B                    _____
            Payroll Account                      _____
            Tax Account                          _____
            Petty Cash                           _____
            Cash on Hand or in Transit           _____

            Total                                ==========


    Prepared By:_____ Date _____
```

Figure 14-5

SUMMARY

Budgeting for cash requires an understanding of future needs, a clear plan of acquiring fixed assets, building inventories, and collecting receivables, and a method of cash controls.

A good budget plans the use of resources required for conducting business. Within that broad area, the benefits of a good budget are many and varied. An organized and logical budget insures that the balance between investment and expected yield is maintained at a realistic level. Cash flow problems are common to builders because the construction business requires large investments. It's a fast-changing business, and a cash budget that works is essential.

There are several liquidity tests available. The trend of some account relationships indicate whether the builder has successfully controlled cash in the past, and in what direction his liquidity is moving. An essential part of analyzing these trends is to view them together.

The control of cash is closely related to the control of receivables, inventories, direct costs and expenses. Without those controls, a budget for cash flow will never work because the parts that make it up cannot be safely projected in advance.

Besides the long-range plan, cash budgeting requires good day-to-day systems for recording cash—including the control against theft. Cash is the most available of a builder's assets, and too often the man in charge doesn't want to become involved. Cash controls are easy to maintain, even if there are several in effect. A cash forecast depends on the availability of receipts. Thefts of cash (whether prior to its going to the bank or by falsifying checks) can have a very negative impact on even the best cash plan.

A quick daily cash summary is one of the most practical and useful cash controls available. The value goes well beyond mere cash control. This report may be the only source of information on your daily cash balances, important information for timing payments and collections, receipt of discounts, and payment of payroll, notes, and taxes.

Chapter 15

Cost and Expense Records

An efficient and practical cost system is a high priority requirement for every builder. Too often, cost recording systems are nonexistent or inadequate in small and moderate size building operations. Many builders see a cost system as too complicated for day-to-day maintenance and effective application. In fact, a practical, streamlined cost system provides a wealth of information that is available from no other source.

Because you are usually involved in several different jobs, you need a flexible, efficient and detailed cost procedure. Keeping records by job can be efficient and precise or it can be time-consuming, uninformative, and a complete waste of effort. The key to good cost recording is streamlined reporting that collects maximum information with minimum effort on a daily basis. This is possible without getting bogged down in an overly complicated file of forms, lists, and calculations.

Cost accounting for contruction work doesn't require specialized training. You know what information you need and should be able to produce your own reports in a short time. And knowing how the information was produced makes using that information to control costs and expenses much easier.

Good cost records make it easier to control costs on jobs in progress, find out where costs have exceeded estimates, and estimate future jobs based on your own known costs.

Each job in a builder's operation can be thought of as a cost center. The total of all cost centers are the areas of the business that produce profits. The activities in your office are not cost centers. Neither are your overhead expenses. These costs are allocated to cost centers to record total costs. Allocating all non-profit-producing costs and expenses to jobs requires that you include your office expense and overhead in each job. When this is done reasonably and consistently, each job you have carries a fair amount of non-profit-producing expenses. All costs and expenses of doing business are either direct costs or assigned to a cost center. When you have done this, your operation is working within a cost center format that makes estimating and cost analysis both easier and more realistic.

OBJECTIVES OF THE COST SYSTEM

Finding the best way to keep track of job costs and expenses is not easy. Every builder's operation is unique and the right way is different for everyone. But knowing the objec-

JOHNSON CONSTRUCTION COMPANY
WORKSHEET, COST AND EXPENSE DEFERRALS
For the month of _____, 19___

Description	Total	Job	Job	Job	Job	Job
Percentage of Completion						
Materials						
Direct Labor						
Subcontractors						
General Expense (Allocated)						
Total						
Completed Contract						
Materials						
Direct Labor						
Subcontractors						
General Expenses (Allocated)						
Total						

Figure 15-1

tives of a cost system should help you design procedures that will work well for you. A cost system should do the following:

- *Provide a control of job costs* Keeping within a job schedule requires that you monitor the completion of work and control the use of material and labor. Otherwise, spending is not under control. An easily prepared report can tell you what your actual costs and expenses are by job. This report is then compared to the schedule commitments.

- *Serve as a guide to future estimates* Many builders do not use past cost records when preparing estimates. Your cost records are as valuable as your own knowledge of current labor and material costs. An organized cost system is your best reference for avoiding past estimating mistakes. Good cost records make it possible to estimate the scheduling and amount of your costs. They make it less likely that you will underestimate or overlook major parts of the cost of completing a project.

- *Allow for varying labor rates* Labor costs are the key to profitability on most jobs. A small percentage error can turn a healthy profit into a loss in a short time. Your cost accounting system should help you stay within the time and cost estimates established in the original budget.

- *Control material purchases and the level of surplus inventory* Your cost system should help you control the timing of material use. You have a large investment in materials on most jobs. Material use should be checked by job on a regular basis. Many material cost increases are the result of extra work not originally contracted for—work you should be paid extra for if you note the extra cost in time and discuss it with the owner promptly.

- *Judge the quality of job force productivity* Labor productivity can be evaluated from the time required to complete each phase of a job. It is often difficult to motivate people to achieve company goals because the outlook of employees is never the same as that of the owners and managers. The only way to control productivity is through a regular check of labor efficiency.

- *Analyze control problems unique to each job* Even though every project is special and no two are identical, a cost control system can work for every builder. Many contractors believe that their costs in a modest operation can not be controlled because every project is special. An effective system will identify and control the differences in each job.

- *Identify overruns promptly so corrective action is possible.* A procedure that tells you that you had cost overruns last month, or even last week, is of little value. But one that flags problems as they arise can be a real moneysaver, justifying the investment of time and effort required. You should be willing to make cost control one of your primary functions. Managing the day-to-day costs of your projects gives you (1) direct involvement in your projects, (2) a means for controlling costs, and (3) a growing familiarity with present and future cost requirements of your jobs.

An efficient cost system is based on summary reports that are simple to maintain yet highly informative. These reports should serve as guidelines for estimating future jobs and should be designed to conform to the bookkeeping system you use. This will help you avoid duplication of effort. The information has to be timely. It must be analyzed promptly, before expenses get away from the schedule established for the job.

Figure 15-1 shows the form used each month by Johnson Construction Company to record costs from five jobs during the month. These costs are recorded as deferred debits. Detail job costs and expenses come from the job cost card, which is explained in detail in Chapter 16. The worksheet shown in Figure 15-1 is used to summarize the current month's total costs and expenses by job, and to provide support for the balance sheet account which accumulates deferrals.

General expenses which do not apply directly to jobs are allocated among the work in progress. And a fast method for applying deferral totals to a large number of general expense accounts is explained in Chapter 16.

SCHEDULING COSTS AND EXPENSES

Your estimate is your budget for costs and expenses on a job. Costs and expenses should be scheduled over the life of the job. The construction budget and schedule of costs are your guide to controlling job costs.

Figure 15-2 shows headings for five forms that help you record and document costs and expenses on a job. Break the job into phases, defining each part carefully for your reference. If you have a clear division for each phase of the job you should have no trouble dividing your costs into distinct phases.

The form Total Costs and Expenses in Figure 15-2 includes a column for a revised estimate. This estimate should be revised on a regular basis to keep your deferrals current. Otherwise, the amount of deferral will be off by the cost or expense variance. If deferrals are off, your financial statement is inaccurate. Revising the estimate will help you forecast eventual profits and losses.

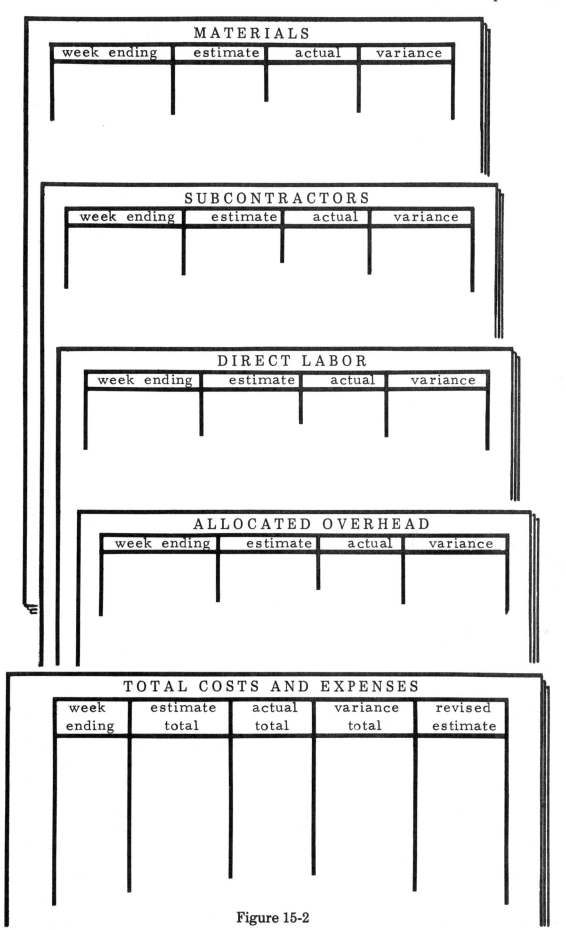

Figure 15-2

JOHNSON CONSTRUCTION COMPANY
LABOR PERFORMANCE STANDARDS

Job Description **Downs Housing** Job Number **214**

Phase of Project **Interior completion (Specs. section 9)**

Week ending:	Type of work	Labor hours Actual	Labor hours Standard	Ratio
4-15	Masonry	14	12	1.2 / 1
4-15	Carpentry	63	60	1.1 / 1
4-15	Painting	6	4	1.5 / 1
4-15	Sheetrock	3	3	1.0 / 1
4-15	Clean up	7	9	0.8 / 1
4-15	Finishing	3	6	0.5 / 1
4-22	Masonry	11	12	0.9 / 1
4-22	Carpentry	71	60	1.2 / 1
4-22	Painting	5	4	1.3 / 1
4-22	Sheetrock	4	3	1.3 / 1
4-22	Clean up	8	9	0.9 / 1
4-22	Finishing	6	6	1.0 / 1
4-29	Masonry	10	12	0.8 / 1
4-29	Carpentry	68	60	1.1 / 1
4-29	Painting	4	4	1.0 / 1
4-29	Sheetrock	4	3	1.3 / 1
4-29	Clean up	9	9	1.0 / 1
4-29	Finishing	7	6	1.2 / 1

Figure 15-3

PERFORMANCE STANDARDS

One of your most difficult but also most rewarding tasks is to interpret the information this cost system provides. For example, if it looks like labor will cost more than estimated, can you determine where the higher costs are originating and what should be done to bring costs into line? The degree of control you have is measured by your ability to act on useful information. Control means changing the course of events from what would normally occur to what you feel should occur. Your system is a tool only if you use it to exercise control.

Setting performance standards is one way to control costs. Assume you believe that the cost of labor on one job is likely to be higher than expected. To reverse that trend, you could establish a performance standard for the completion of each part of the project. A performance standard for direct labor is measured in hours. Figure 15-3 is a ratio analysis of direct labor. It compares actual time to a standard or estimated time. The standard for each phase of each project is the amount of time in which a particular task should normally be completed. A comparison such as that done in Figure 15-3, will help you locate problem areas. Then you can take steps to correct the problem. It may be that you have a less experienced crew doing fairly difficult work. Perhaps your crew size is too large or your crew leaders are not organizing the work well. Whatever the cause, the cost-conscious builder will try to bring the costs back into the range of the standard.

CONTROL OF DEFERRED COSTS

Income deferral accounts were discussed in Chapter 2. Under percentage of completion and completed contract accounting systems, income is deferred until the time comes to recognize it. Your records of costs and expenses are largely lists of deferred debits.

Percentage of completion billings are partially deferred to the extent that billings exceed earned income. For example, a 50% billing on a job only 40% complete would require a deferral of 10%. Under the completed contract method, all income is deferred until completion of the job.

The same rules apply for cost recording on your projects. Only the costs and expenses to be recognized are shown on a statement of income. All others are deferred.

It is easy to find the amount to defer. The contract price and the partial billings to date are exact amounts. But with costs and expenses, the amount to be realized and the amount to be deferred depend on the accuracy of the original estimate and your estimate of the work completed. For that reason, your cost recording system must be able to handle the deferral of costs and expenses.

In a well-controlled procedure, you can relate costs and expenses to income to see that the work is progressing as assumed in the estimate. If you are able to stay within your estimate, deferrals of costs and expenses can be based on your estimated totals. However, anticipated variations from the estimated totals require that you adjust the total to find the deferred amount.

You find the amount to be deferred by comparing actual costs and expenses to estimated totals each month. When variances show up, revise the original estimate to forecast a new actual cost and expense total. You may find that totals for some jobs are constantly revised as the job progresses. Following costs like this puts you in firm contact with progress on your jobs and the profit you expect. It also lays the foundation for better estimating in the future.

Deferred income is neither a current nor a long term liability. It is a deferred credit, a third, distinct classification on the operation's balance sheet. Like income, costs and expenses are neither current nor long-term. They form a separate balance sheet classification found in few areas other than the construction industry. This classification is called *deferred debits*.

Many builders include these deferrals as current assets. In practice they are not available in the current period, as are cash, accounts receivable, and inventories. Deferrals are not liquid assets and should not be considered or classified as such. To do so only distorts the current ratio (current assets to current liabilities) and presents an unfair picture of the operation's status.

Considerable detail is needed to back up the totals in the general ledger accounts for deferred debits. This requires a set of detail records maintained off the general ledger. The general ledger is not the place to include analyses of accounts, only summaries.

Figure 15-4 is a form for calculating and collecting entries to deferred debit accounts. There are two principal general ledger deferred sections, each with four sub-accounts:

Deferred Costs and Expenses:
   Percentage-of-completion method
      Materials
      Direct Labor
      Subcontractors
      General Expenses (allocated)

   Completed contract method
      Materials
      Direct Labor
      Subcontractors
      General Expenses (allocated)

The general ledger entry, which would be made in a special journal and supported by these schedules, would be:

Builder's Guide To Accounting

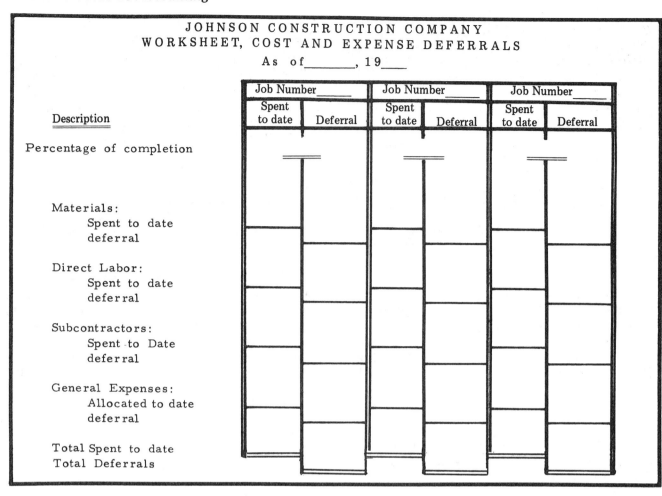

Figure 15-4

	Debit	Credit
Deferred Costs and Expenses (in detail as above)	XXX	
Materials		XXX
Direct Labor		XXX
Subcontractors		XXX
General Expenses		XXX

The entry to General Expenses can be made to a summary control account in the expense section of the ledger, or to each detail account, as explained in Chapter 16.

Since entries going into deferral accounts will eventually be reversed, the journal to take out deferrals would be the opposite of the example above. In most cases, the monthly summary total of reversals will not be the same as the previous month's recorded amount. For that reason, good detailed records should be maintained so you know the status of each job's deferred costs and expenses at any time.

A second way to support deferred debits totals is to reverse the previous month's entries every month, and calculate the current deferral each time the entry is prepared. The advantage of this method is that any errors which have been recorded previously are eliminated each time the deferral is recalculated. None are carried forward. Under this method, all deferrals from the previous month are reversed. Then a worksheet is prepared to calculate deferred debits as of the end of the month. This is a much simpler way to support general ledger totals. See Figure 15-4. The balance in the accounts at any time is new—it has been calculated this month.

Notice that Figure 15-4 covers only percentage of completion contracts. No special worksheet is necessary for projects kept under the completed contract method. The balance in the deferral account would always be equal to the total on the job cost cards. This is because neither expenses nor costs are recognized until completion.

It is obvious that bookkeeping under the completed contract method is much simpler because less calculating is required. It takes less time and you maintain more control.

A third method of handling deferred costs and expenses for percentage of completion jobs is to post all costs into the deferral account and remove an estimated percentage each month. The percentage to be removed is equal to the estimated percentage of completion excluding costs which exceed the budget for any category. Because costs and expenses can vary from the original estimate, this third method is not always as accurate as it might seem.

Chapter 16
Accounting For Costs and Expenses

Your cost accounting system does not have to be complicated. An efficient system can use a simple, easily maintained procedure. The sooner information is available, the more useful it is to you. The information is available sooner if your procedure is simple and as automatic as possible. This chapter shows you how to use your books and records to allocate costs and expenses to jobs. A system that does not permit prompt analysis won't be worth the trouble it takes to maintain. But one that reveals vital information about a job's profits and losses, cash flow, and use of material and labor will save you money.

Analyzing labor productivity is essential to controlling costs and making accurate estimates. It helps you spot lost time and make the correction required while the job is still in progress.

Controlling labor cost is probably your most difficult control problem. There are two areas of labor control to consider, your office sales and supervisory force, and your construction crews.

Idle time increases the cost of labor and reduces profits by the same amount. If you have to include idle time in your estimates, you become less competitive. Controlling labor cost should be among the top priorities of every cost conscious builder. Your bookkeeping records can tell you as much about labor productivity on your jobs as a professional labor cost analyst... and at a much lower cost.

The only way to analyze labor hours is by the job. A total picture is important in its own way, of course. But only detailed analysis for each part of each job will help you understand your true labor requirements.

Your first step is a good method for collecting costs on your jobs and assigning them to specific projects. Once you know what your true costs are, you can compute and compare gross profits, selling profits (the amount of profit before fixed overhead), and net profits.

While direct labor costs can be identified by job, office, selling, labor and all overhead expenses cannot. You need to find some reasonable method for assigning these expenses to each of your jobs. Your assignment method won't be perfect, no matter how precise. There is no absolutely fair and dependable method of making sure each job carries its share of overhead. You can only be as consistent as possible, realizing that the best estimate of true costs and expenses is better than no cost system at all. Specific methods for applying overhead expenses to your jobs in a reasonable and efficient way will be discussed later in this chapter.

ACCOUNTING FOR DIRECT LABOR

The goal of labor cost analysis is to determine the actual cost of completing each portion of each project. Most high labor cost jobs have inefficient work schedules that result in excessive idle hours. Jobs run with a large crew and only one foreman cost more to complete. Invariably several workers will be idle for a substantial percentage of the time. Most jobs that use only a few workers and one foreman will have a higher productivity rate, especially if the crew includes a highly skilled craftsman, one lower wage apprentice and one laborer. Small crews with a good mix of skill and pay levels are usually most productive.

The payroll register is the best place to perform a weekly analysis of labor cost. Posting and analysis should be done on completion of each week's payroll. This way the cost information is usable right away. Checks are written and recorded on the left side of a pegboard check register. This includes the date, name, hourly rate, total hours, gross pay, deductions, and net pay. The right side is available for cost analysis. The first step is to break down the hours by task or job, depending on the detail you need. This is not difficult if you have a properly-designed time card. The company time card can include a task breakdown by job for each worker. Each man, under supervision of the foreman or crew leader, notes the time spent on each major task. The time can be approximated to the nearest half hour, but every hour of the work day should be accounted for and noted on the card at the end of each day. Once your tradesmen are accustomed to this procedure, it should take only a minute or two to record each day's work.

The hourly breakdown on the time card is converted to dollars of labor cost in your office. Each man's hours should be converted to costs individually because varying rates result in varying costs.

Figure 16-1 shows the right-hand side of the payroll register. Gross wages are broken down by task or job number. The employee on the first line, for example, has divided his hours for the week between three columns. Number 437-P shows 24 hours of work and a payroll total of $252.00. This includes only your payroll cost, not your tax, insurance and overhead cost. Your other costs will be added later. The column for task number 437-P might be used to collect all costs for framing a roof on a particular job. The column numbered 432-R might cover other rough carpentry on the same job or a different job. With a little ingenuity you can create job and task categories that correspond to the type of work you do. Naturally, the accounting categories you plan to use on any job should be available to your work crews so they know what categories of time they should record. It will probably be enough to post a list of the account numbers you use for each job near where the work crews record their hours at the end of each day. The tradesmen can then record their hours by code number rather than description. This saves time both in recording and posting the hours to your analysis worksheet.

Recording and posting direct labor hours is easier if you assign numbers for certain types of tasks consistently. In Figure 16-1 the column 437-P might be task 437 on a job that has been assigned the letter P. Task 437 would always be roof framing regardless of the job number. All labor and material costs on job "P" would be posted to an account with the P. If you have more than 26 jobs under way during a year or if your crews do more than 999 categories of work, you would need a longer combination of numbers and letters.

You will need some numbered categories to record idle time, driving time, cleanup and maintenance time and other time spent not working on any particular job. These hours are allocated to one or more of your jobs with other overhead expenses. Be sure to leave enough flexibility in your numbering system so that your tradesmen have no trouble knowing what number to use in recording their hours.

The material costs you compile should be kept under the same numbering system used for labor hours. In Figure 16-1 the cost of lumber for framing the roof would be recorded on a separate material cost worksheet under the number 437-P. Naturally, some of the lumber ordered for roof framing may have been used somewhere else in the job, but your estimate of the materials required will probably be fairly close to the actual amount used. If the job

Builder's Guide To Accounting

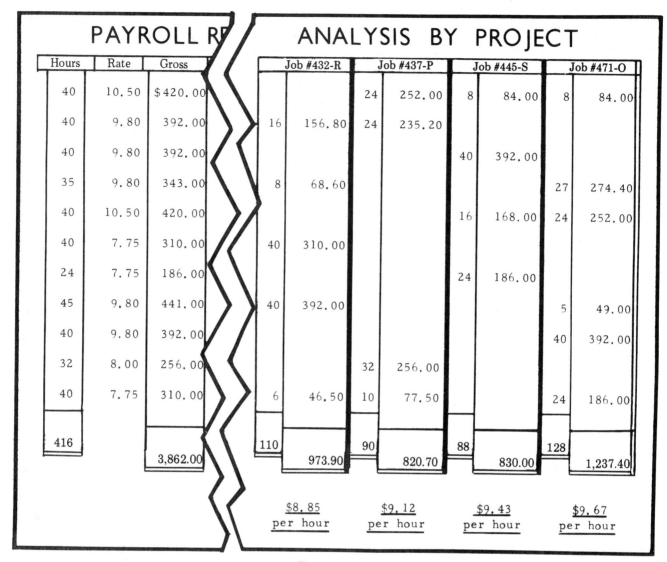

Figure 16-1

requires more lumber than you estimated, you should find out where the loss occurred or where your estimate was wrong. Then assign the additional cost to the appropriate category.

If you have more than a few jobs under way at a time and more than 10 or 15 work categories, your payroll register won't have enough columns to record all your crew hours. Any ruled pad with vertical columns will work well for this purpose. Keep the completed sheets in a loose leaf binder. They become important cost records that you will use many times when estimating new jobs. At the end of a year you will have recorded several hundred hours of work under a number of categories. A quick tally on the number of square feet of roof framing this represents will tell you on the average the manhours required for each 1,000 square feet of roof your crews framed. Even better and easier, divide the roof framing hours by the board feet of lumber you used for roof framing and you know on the average how many hours are required per 1,000 board feet of lumber. Of course, some jobs take more time per thousand board feet and some take less. There are many factors that affect productivity. But you should be able to tell in advance which jobs will be harder or easier than your average and how much harder or easier they will be.

Note in Figure 16-1 that eighty of the 110 job hours for job 432-R were worked by employees earning $7.75 per hour. The balance, 24 hours, were at a rate of $9.80 per hour. Other projects had a mix of hourly rates ranging up to $10.50 per hour. Keep this average hourly cost in mind when comparing different jobs. The average hourly cost in Figure 16-1 ranges between $8.85 per hour and $9.67 per hour. Obviously, the demands of different projects are not the same. This is an important consideration when compiling estimates. Some projects require the constant presence of a foreman while others may not. Some projects require a higher ratio of more highly skilled craftsmen. Most of your specialized work is probably contracted out, but your cost per labor hour will vary widely on your jobs.

While the special needs of each job will vary, you should determine whether the average hourly labor cost is reasonable on each of your jobs. If a rate varies considerably from week to week or remains at a high level, there is a need to investigate. It could indicate that too much supervision or too many highly paid tradesmen are on the job.

The direct labor analysis shown in Figure 16-1 is practical because it provides valuable information with a minimum amount of work. The information is available right away in a format that is easily understood. The components of the hourly rate can be seen in detail, but each employee's time for the week is listed on one line. This kind of analysis establishes clear trends for each job. Most important, it provides invaluable information for use in future estimates.

ACCOUNTING FOR MATERIALS AND SUBCONTRACTS

Since payments for materials and subcontracts are readily identified by project, the cost recording can be done when payments are made. A complete bookkeeping method would include coding for both the general ledger account and the job. This method is commonly used by companies that have their general ledgers on a computer system that can handle a large number of accounts. You get the maximum value from automated accounting procedures if you can generate a number of useful reports from a single data input. The entries for general ledger purposes are coded so that they can be resorted for project analysis. A typical computerized system might have a code number like this:

04 - 6320 - 4800 - 61004

The first two digits indicate that the code is for a direct cost. This would flag this entry as being subject to cost analysis. The second group, 6320, is a project number. The third group, 4800, is the general ledger account for materials purchased. The program might also include all accounts beginning with the digits 48 in an inventory control listing. The last group, 61004, could be the number assigned to the vendor to whom payment was made.

From a single set of account numbers a computerized recordkeeping system could provide several types of reports. The cost of a computer to do this work has become quite reasonable in the last few years. But the program or *software* required by the computer must be very specialized for your type of business. The same information can be developed at modest cost in a smaller organization with manual procedures. All material purchases are coded with a double set of numbers—one for the general ledger and one for the project and task. A typical manual account might be 632 - 480. The number 632 could be the job number and 480 the account number. Even in a manual system, the uses of coded information are unlimited.

On a pegboard ledger, cost recording for materials and subcontract work is done at the same time as the payment of bills. Double post all material purchases and subcontract payments to a general ledger account and to a job cost ledger card. This is explained later in this chapter.

ACCOUNTING FOR OVERHEAD

Unlike payroll and materials, overhead cannot be assigned to specific jobs. For this reason, the allocation must be done on some other reasonable basis.

Finding a good, uniform way to assign overhead is difficult. It can't be based only on income because income from jobs varies with various phases of completion. In the meantime, overhead is a constant and recurring part of that job's cost. Material purchases vary as well, not only by the phase of completion but by the type of job. The only reasonable way to assign overhead is in proportion to total direct labor hours. This can be justified by a few considerations:

- The percentage of hours spent on any one job can be said to represent the degree of commitment to the project or the emphasis placed on it.

- Management's involvement with a job usually varies directly with the labor

JOHNSON CONSTRUCTION COMPANY
DISTRIBUTION OF OVERHEAD EXPENSES
For the month of September, 19___

Description	Total	Job #432	Job #437	Job #445	Job #471	All others
total hours	416	110	90	88	128	36 *
percentage of total	100	26.4	21.6	21.2	30.8	
selling expenses (list attached)	$3,260.18	$860.69	$704.20	$691.16	$1,004.13	
fixed overhead (list attached)	$2,209.60	$583.33	$477.27	$468.44	$680.56	

* not included in total - idle time, sick leave, vacation, and shop time

Figure 16-2

hours. As labor hours increase, there usually is a corresponding increase in the need for controls, supervision, and planning.

- Several large overhead expenses vary with labor hours—union welfare, payroll taxes, use of trucks and autos, and insurance.

- No other method of allocation comes as close to truly representing a fair distribution of overhead expenses as does total labor hours.

There are several ways to assign overhead expenses to jobs on the basis of direct labor hours. Remember, you probably have 20 or more overhead expense accounts. In fact, you probably have more overhead expense accounts in your general ledger than all other types put together. It would be too time-consuming to assign a portion of each expense account to each job. It would be meaningless as well. A detailed breakdown would take hours to complete. There is an easier way.

Assignment of overhead is a guess. A detailed distribution would have little or no value. The practical and efficient solution is to total your overhead expenses each month. This is illustrated in Figure 16-2. Your books are posted at the end of each month. Overhead expenses are available in total at the same time.

Figure 16-2 shows total hours listed on the first line, including 36 hours of time not related directly to jobs. This 36 hours is not allocated. All you are concerned with on this worksheet is the direct labor by job. All costs other than direct labor on specific jobs will be allocated among the jobs by a formula. The second line shows the percentage of hours for the week on each one of the job classifications. These percentages are applied to total selling and overhead expense to find the additional cost each project should carry. When this has been done, the increased general ledger balances for the current month can be assigned to jobs and recorded on the job cost ledger cards.

JOB COST LEDGER CARDS

Accounting for the costs and expenses of several jobs at one time can be a burden,

Accounting For Costs and Expenses

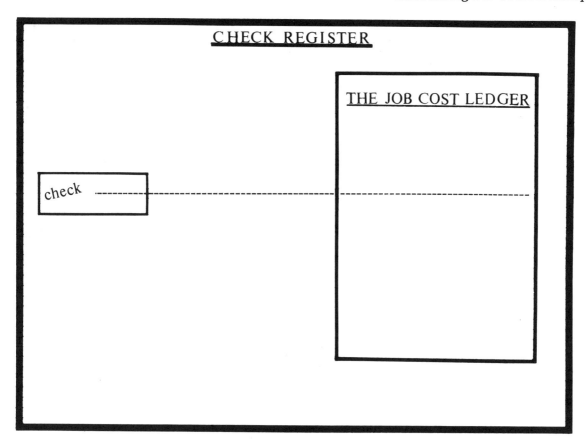

Figure 16-3

especially if it involves going over all entries twice—once for the general ledger and once for cost analysis.

Figure 16-3 diagrams the usual pegboard system for checks. The distribution to accounts for all payments occurs between the check and the middle of the board. The job cost ledger card is placed on the far right. In this position, the amount of the check is entered again, providing complete cost analysis. This method is intended to ensure that all costs are entered on a job cost ledger. For that reason, at least one card must be used for all jobs. A card for all other expenses must be used as well. This would include miscellaneous expenses, one-time work and any job categories which are not accounted for separately.

In the pegboard writing method, the total of all columns posted to the job cost ledger card should equal all columns for distribution to the general ledger. The total of all checks will then balance to the distribution columns, and also to the cost control columns.

Materials and subcontract payments are posted directly to the job cost ledger, but payroll and expenses are not. Figure 16-4 shows the method for posting the payroll and expense totals. The labor totals can be posted on a weekly basis and expense totals on a monthly basis.

If you keep cost records as described here, all of your costs and expenses are expressed on a cost center basis, giving you a tidy and comprehensive summary of business. The complete system is easily maintained, requires a minimum of calculation, and handles assignments of indirect expenses in a logical manner.

Total monthly and project-to-date costs and expenses can be determined from the job cost ledger cards. The total system is practical and provides useful information with little added effort.

ACCOUNTING FOR COMPLETED CONTRACT EXPENSES

Accounting for general expenses on jobs kept under the completed contract method presents other problems. All costs and expenses

171

Builder's Guide To Accounting

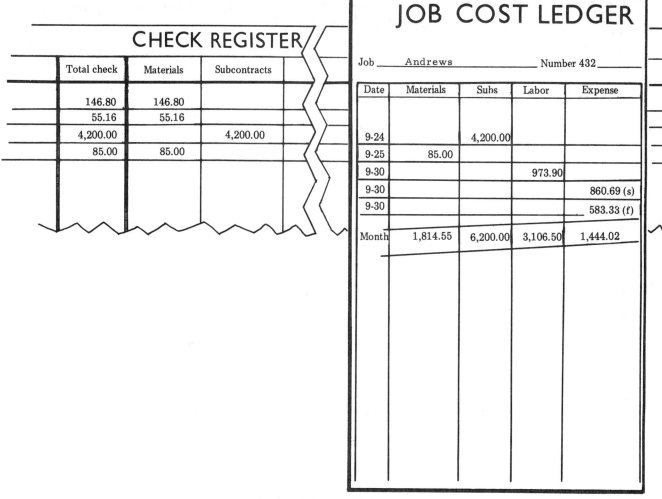

Figure 16-4

are deferred, and general expenses assigned to completed contract projects should be treated the same way. But how do you defer specific expenses when the allocations are done in total? Deferring a percentage of each expense account would be time-consuming and meaningless. Entries to defer parts of each expense would obscure the analysis of overhead, making it impossible to control selling expenses and fixed overhead. Too many accounting entries would have to be recorded to each account.

But selling expenses and fixed overhead can be estimated in total. The answer is to estimate deferrals in the same manner. Each broad section of expense in the general ledger (such as "selling expenses" and "fixed overhead") can include a special account for deferral of expenses on projects kept under the completed contract method. You should control these accounts with great care and always know what makes up the balance.

Each month's total expenses are divided between projects. The completed contract totals would be deferred by the following entry.

	Debit	Credit
Deferred Debits	XXX	
Selling Expense deferrals: Job # 432-R		XXX
Job # 445-S		XXX

These expense account reductions would continue through the life of each project, reducing selling expenses in total each month. Fixed overhead entries would be as follows:

	Debit	Credit
Deferred Debits	XXX	

	Debit	Credit
Fixed Overhead deferrals:		
Job # 432-R		XXX
Job # 445-S		XXX

These, like the selling expenses, would continue through the life of the job.

When these projects are completed, all income, costs, and expenses are recognized at the same time. The job cost ledger has been accumulating the totals for each account. The entry to recognize the completed contract account is:

	Debit	Credit
Deferred Credits (Income)	XXX	
Income from Contracts		XXX
Materials	XXX	
Subcontractors	XXX	
Direct Labor	XXX	
Selling Expenses	XXX	
Fixed Overhead	XXX	
Deferred Debits (costs, expenses)		XXX

By this entry all deferrals are reversed and the income, costs, expenses, and profit or loss are recorded on completion of the job.

If you have no uncompleted work accounted for by the completed contract method, the balance of the following accounts would be zero:

Deferred Debits   (Asset)

Deferred Credits   (Liability)

Selling Expense deferrals (Expense)

Fixed Overhead deferrals   (Expense)

A complete income statement can be put together at any time by using the job cost ledger system. Even for completed contract accounts, all profit and loss information for each job is readily available in one place. If you use these controls, you can proceed with confidence, knowing that all costs and expenses have been assigned and there will be no surprises at the end of the job.

The advantage of the job cost system is that completed contract accounts, which are actually not included in profit and loss, can be analyzed off the general ledger.

SUMMARY

Since material and subcontractor costs are easily identified by job, the method for recording them by code number is simple. Each account entry should include a job number as well as a general ledger code.

Overhead expenses should be assigned to jobs in proportion to the actual labor hours on each job. This method is only a guess, but it is the fairest and most reliable method available for assigning indirect expenses.

Job cost ledger cards should be maintained for each job. All costs and expenses are assigned to the various profit-producing jobs.

Costing for jobs accounted for by the completed contract method is more difficult because none of the income, costs, or expenses are recognized until completion of the job. Some builders tend to ignore completed contract projects in their cost accounting procedures. But this forces you to operate without control or a picture of your anticipated profit or loss. The chief benefit of any good cost accounting system is that it keeps you informed of your costs. This information gives you the opportunity to control the direction of your operation. The information must be available quickly if it is to be useful. The time required to prepare the figures must be reasonable. The information you develop must be put to use or your effort is wasted.

Simplifying cost accounting should help you guide your business toward a reasonable profit on every job you take. A well-planned cost system compliments estimating records and can be valuable to any builder in planning for profits and growth.

Chapter 17

Petty Cash Funds

Like most businessmen, builders need ready cash around the office from time to time to pay for miscellaneous expenses. But the haphazard way this cash is usually kept can result in losses of funds and legal tax deductions. You don't want to let small amounts of cash bleed out of your cash flow and build up to big losses over a period of time. What you need is a well documented and controlled office fund large enough to make incidental cash payments. The fund must be easy to use. Yet it should give you all the accounting data you need to deduct from taxes the business expenses you pay out of it. Keeping undocumented cash in a drawer or box invites pilfering. And you can't deduct from taxes what you can't document.

Petty cash funds are set up in one of two ways: 1) cash is withheld from a bank deposit, or 2) a check payable to "Cash" is made out and the amount is recorded in the check register. When you withhold cash income from bank deposits to set up and supply the fund, you make it impossible to account for all sales by using deposit records. Recording all sales on deposit records is one of the best ways to clean up your recordkeeping. When you withhold sales dollars from the bank you blur your profit picture and all your sales-related analyses. And you can't keep records of business expense deductions as expenses are paid from the fund.

Books and records should be set up so that all entries can be traced to their origin. But you can't make accurate entries to the general ledger if you make payments from a fluctuating cash fund that is the product of checks held back from deposits.

Some builders write checks payable to "Cash" whenever they need funds. Here again, no accurate accounting entry can be made to verify and balance this kind of office cash fund. You can't code the check accurately because the entire amount is not immediately deductible. And since you don't know how all of the cash will be spent when you record the check in the check ledger, you can't accurately distribute the amount of the check among your various expense accounts.

THE IMPREST SYSTEM

The best method to use for petty cash is called the *imprest system*. It accurately controls the flow of cash payments and documents them fully on your check register. It also lets you record your deductible petty cash expenses in the general ledger. An account is set up in the ledger as a separate cash fund. The balance of this ledger account does not change at all unless the balance of the fund is increased or decreased.

The actual balance of the imprest petty cash fund will vary as money is spent. But confine the entries in the general ledger to the total amounts paid out of the fund and paid back to it from an expense account in the period of the entry. If you keep a $300.00 petty cash fund and you pay out $50.00 in expenses one week, the actual fund balance is $250.00. The only time you make a general ledger entry is when you write a check from expenses to bring the fund back up to $300.00. The check is a reduction of cash for expenses, and this is recorded in the ledger as an increase to one or more expense accounts. Thus the general ledger does not adjust each and every payment you make from petty cash, only the amount expensed to reimburse the fund. Cashing the check brings the fund back up to $300.00.

The imprest method has these advantages:

- It is easily maintained.
- Controls are built in—the fund can be balanced each time it is reimbursed.
- The fund is isolated from your checking accounts.
- The documents supporting the petty cash checks also verify the increases and decreases to petty cash.
- Entries to the general ledger do not require special effort.
- Cash shortages are cut down.

THE SIZE AND USE OF THE FUND

Good business practice dictates that you handle most business expenses by check. But realistically, every business must make some payments in cash.

The amount of the fund depends on the requirements of your business. Petty cash payments should not be large compared to those in the rest of the operation. The average payment is for a miscellaneous expense, not a major one. But the fund should be adequate to pay the following kinds of expenses:

- Expenses that cannot be paid by check
- Expenses that are more realistically paid in cash
- Expenses that are too small to justify a check
- Expenses that have been paid on your behalf, and that you are paying back (usually to an employee)

Estimate the amount of cash you need to cover payments in these categories. But keep two points in mind when you decide on the size of your petty cash fund: Too high a petty cash balance leaves some of your cash idle indefinitely. This invites theft. Too low a balance means that you must issue checks to pay back the fund too frequently. You are likely to run out of funds often. Most operations need to keep at least fifty dollars in petty cash. You may require much more, depending on the size of your payments, especially if your creditors have placed you on a C.O.D. ordering status because of collection problems, or if you pay many of your bills in cash. But try to eliminate any such cash-payments and reduce the size of your petty cash fund to less inflated amounts. Too much cash payment in an operation makes overall controls more difficult.

To *increase* the balance of your cash fund, issue a check to the fund above the amount needed to pay it back. Then make the following entries in the general ledger: 1) raise the fund balance by adding the amount of the increase to the old balance, and 2) reflect the increase in the account on which the check was drawn. To *decrease* the fund balance, let the fund dwindle to your intended balance by not writing checks to replenish it. Then 1) lower the fund balance in the ledger by subtracting the amount of the decrease from the old balance, and 2) reflect the decrease in the account by the amount by which you lower the fund. Or you can deposit a portion of the fund into the general bank account, decreasing the fund by that portion. Make only one entry in the ledger in this case.

CONTROLLING PETTY CASH

Controlling the flow of information for cash payments is not difficult with an imprest fund. All payments from the fund are supported by a voucher or receipt. Figure 17-1 shows a petty cash receipt that serves several purposes.

- It is a signed document verifying that someone received cash from the fund.

Builder's Guide To Accounting

```
Amount $_____      Number_____
PETTY CASH RECEIPT
                           _____19____
Description_____
Charge Account Number:_____
Approved_____Received_____
```

Figure 17-1

- It is a support document for accounting entries, coded for proper classification.
- It is a document to support unreceipted expenses (such as repayment of mileage on company business).
- It is a uniform way to describe and identify payments from the fund.

All funds taken from the cash box should be replaced by a completed receipt. Make one person responsible for the fund. That person

```
         JOHNSON CONSTRUCTION COMPANY
                PETTY CASH SUMMARY
                   March 31, 19____


    Balance, 3-1-__                          $300.00

    Paid from petty cash:

         office supplies    21.60
         postage            30.00
         bridge tolls        4.50
         C.O.D. expense     11.45
         coffee              3.60
         donations          15.00

         total paid out     86.15

         cash shortage       3.11

         total reduction                       89.26

    Balance in fund                           210.74

    Payment to petty cash:

         3-31-__, check 412                    89.26

    Ending Balance                           $300.00
```

Figure 17-2

should either complete the receipts himself or see that the form is filled out before funds are removed. Make sure the person responsible knows that, at any time, the total of all cash plus the total of all petty cash receipts should equal the fund balance.

A petty cash summary as shown in Figure 17-2 should be prepared at least once a month. Making up this summary is like balancing a bank account because you are accounting for the flow of cash in and out of the fund. The report also summarizes your accounting entries for each payment into the fund. Expect small unexplained amounts over or short of the amounts on the summary if the volume of your payments is high. Keep these reports in a small binder, or attach them to the collected receipts and file them as support for pay-back checks.

The imprest petty cash fund is a form of budgeting. You estimate the amount of the fund based on your expected cash payments for a specific period of time—one month, for example. If your actual cash expenses begin to exceed this estimate, look for a way of reducing the frequency or the amount of some of your cash payments.

It might seem like an unimportant detail to develop such a rigid system for small expenses. But considering the level of federal (as well as state and local) taxes, and their effect on your profits, it's time well invested.

For example: Let's say your monthly cash expenses add up to $85 and your effective overall tax rate is 32 percent (the combined effect of federal, state, and local taxes). That means the value of having a system to document these expenses is:

$85 per month x 12 months = $1,020.00

Tax benefit (32%)   = $326.40

The value of tracking expenses is $326.40. That's a significant enough number in itself. But there's more. Having the fund in place helps you to capture and document expenses you might miss without the fund. For example, you might spend a few dollars each week on legitimate business expenses. But without a fund for reimbursement, you tend to lose receipts or forget what they were for. The imprest petty cash fund helps you to claim all of the deductions to which you are entitled.

Chapter 18

Balancing The Checking Account

The only way to know your true cash balance at any time is to balance your check account once a month when you make your monthly general ledger entries. Trouble starts when you miss a month or when you do the job inaccurately or incompletely. Save any adjustment memos sent to you by the bank that month. At the end of the month sit down with the bank's statement and this chapter as a guide. Many builders view their checking account as a mystery. There is no mystery to account balancing. The bank statement shows a different balance than your check book only because the bank records entries at a different time than you do. You increase your balance for deposits and decrease it for checks as they occur. But the bank records deposits and checks only when they find out about them. Here are ten steps to balancing a bank account. Follow these steps completely and accurately and your account should balance every time.

1. Start with a "good" balance for the previous month. A balanced account for the month just past is the first step in balancing your account this month. Otherwise you risk carrying mistakes forward and never finding them. This means that you must balance your account every month. If you have never bothered to balance your account you'll have to construct a balance for the first time. Constructing a balance is discussed later in the chapter.

2. Account for all *cancelled* checks you have received from the bank in this month's statement. Place a mark in your check register next to each check listed on the statement. This isolates any unmarked items for later examination. After marking off all the received checks on the statement, put the checks in numerical order.

3. Find out whether the previous month's outstanding checks have cleared the bank this month. Most previously outstanding checks should have cleared by now. To find out, look at the list of all outstanding checks. Making this list is part of each month's procedure. The check numbers and the amounts of uncleared checks should be listed each month. Delete from your list all previously outstanding checks that have cleared. List any checks that were out last month and that still have not cleared.

4. Find this month's starting point—the point at which you did last month's balancing.

This gives you the balance after the last check from last month was either received or listed on the previous outstanding checks list. This is also the point at which the last booked deposit was shown on the previous bank statement or was a deposit in transit (a deposit recorded in your checkbook but not on the statement). Make each month's starting point logical and consistent by balancing your account on the last day of each month. This makes your accounting simpler because the month-end general ledger balance is supported by the end-of-the-month bank account balance.

5. Prepare the current month's list of outstanding checks. Include all outstanding checks from last month that have not cleared the bank, plus all outstanding checks in the current month. Make this list on the finishing point or the last day of the month—next month's starting point—to catch all outstanding checks. This step is a likely place for making errors, so double-check all amounts and math.

6. Determine that last month's deposits in transit appear on this month's statement.

7. Compare deposits listed in the checkbook to deposits reported on the bank statement. If there are any amount differences, make an adjustment to your checkbook balance. These differences can occur because of errors in addition on deposit slips. You hope you catch this kind of mistake before you take deposits to the bank. But errors do slip through sometimes. Place a mark next to each deposit listed on the bank statement as you find it in the checkbook. This procedure is just like comparing the cleared checks to the recorded checks; here you are accounting for all deposit entries in the checkbook and the general ledger with the bank statement and adjusting every deposit amount that is off.

8. Were there deposits made this month that are not on the bank statement? Deposits not listed on the statement should be deposits made at the end of the month. The bank closes its daily accounting before recording your final month's deposit. When this happens, record the amounts as deposits in transit.

9. You have been marking each item on the bank statement as you go along. Soon nearly every item on the statement is checked off on the check register or your deposit record. What

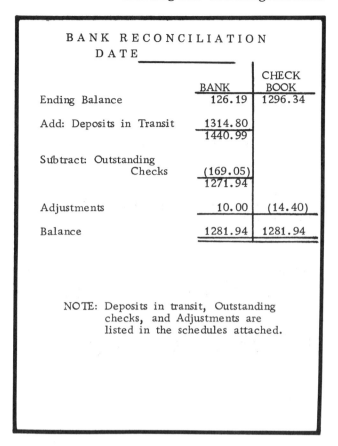

Figure 18-1

is left over are the bounced checks, bank charges, automatic loan payments and the like. Such items reduce your checking account balance. All such adjustments should be listed on the bank reconciliation page or on an attached schedule.

10. Once you have accounted for all checks, deposits, and adjustments on the bank statement, and you have used the statement to check your entries in the check book, find out if you are in balance. See below.

Figure 18-1 shows a typical bank reconciliation form. Both the bank's balance and the checkbook's balance are adjusted to an agreed balance. Most of the adjustments on the bank's side are for timing differences. Deposits in transit are added to the balance, and outstanding checks are subtracted from it. A ten dollar adjustment in the checkbook's favor is listed to show a bank error. Adjustments on the checkbook side may be for checkbook errors or for items on the statement you see for the first time. All adjustments made on either side of a reconciliation form to arrive at a reconciled

OUTSTANDING CHECKS
DATE_____

Check Number	Amount
127	$ 12.00
128	126.00
131	18.94
133	12.11
Total	$169.05

DEPOSITS IN TRANSIT
DATE_____

Date of Deposit	Amount
9-29	$ 600.00
9-30	714.80
Total	$1314.80

ADJUSTMENTS
DATE_____

<u>Adjustments</u>: Bank errors (call the bank and straighten it out.)

Example:
Check recorded $10.00 over amount of check .... $ 10.00

<u>Adjustments</u>: Check book (Adjust the check book balance.)

Examples:
Returned check .... $(15.80)
Service Charge .... (2.60)
Math Error 4.00
Total $(14.40)

Figure 18-2

Clearing Last Month's Balancing Entries

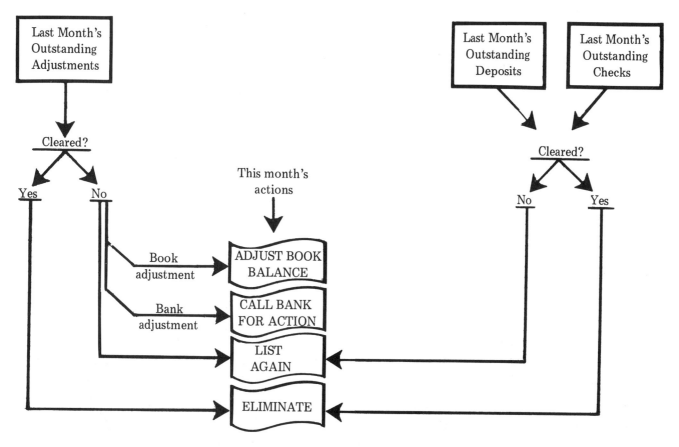

Figure 18-3

balance are shown in Figure 18-2. Outstanding checks and deposits in transit are always adjustments to the bank's balance. Miscellaneous adjustments must be made to either the bank or the checkbook, depending upon which side is incorrect.

A false way to summarize the balancing of a bank account is to show how the bank's balance becomes the checkbook's balance:

Bank's ending balance	$126.19
Deposits in transit	1,314.80
Outstanding checks	(169.05)
Net other adjustments	24.40
Checkbook's ending balance	$1,296.34

The problem with this method is that it doesn't reveal the true balance of $1,281.94. Do not use this short-cut method. You must have a true balance of the statement and the checkbook to support the monthly ending balance in the general ledger asset account for Cash.

Figure 18-3 summarizes the flow of balancing entries from one month to the next. Think of outstanding checks as blocks of cash that the bank has not yet taken from your account — a timing difference. Think of outstanding deposits — deposits in transit — as your money, deposited but not recorded by the bank at the time monthly statements were prepaid. Recognize adjustments as business you must act on promptly. Correct bank errors by calling the bank and pointing them out. Checkbook errors require that you adjust the checkbook balance. You will never have a true balance if you don't make adjustment entries and try only to balance

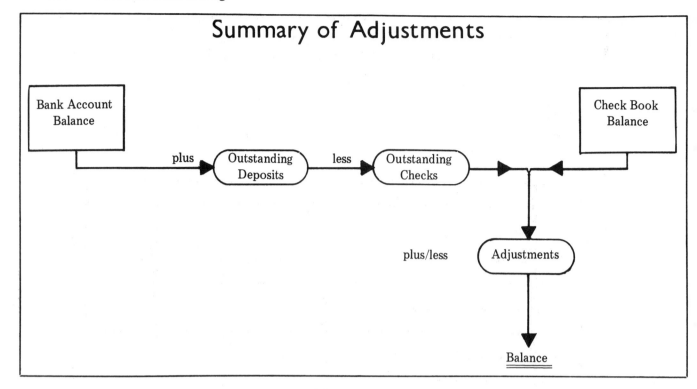

Figure 18-4

deposits and checks alone. Figure 18-4 summarizes the correct method for balancing your checking account. The bank statement and the checkbook will never agree unless your account has no activity for most of the month.

Adjustments can be made to either side, the bank or the checkbook, depending on the nature of the adjustment. Only adjustments for timing entry differences or bank errors adjust the bank's side. Errors in math, special charges, and automatic loan payments are the principal adjustments to the checkbook. Other checkbook adjustments relate to deposits—deposit slip errors, returned checks, or totals incorrectly listed in the checkbook.

A CHECKLIST FOR BALANCING YOUR ACCOUNT

Once you have accounted for every item in the checkbook and on the bank statement and made all the adjustments to both balances, see if the checkbook balance and the statement balance agree.

If the balances don't agree, check your entries again to make sure you haven't made any incorrect adjustments. If you are still not in balance, go through your statement, checks, and checkbook again. Is the math in your checkbook correct? Is the math correct on the outstanding checklist? Are cleared checks recorded at the same amount on the bank statement as they were in your checkbook? Compare your checkbook amounts to the imprinted number in the lower right-hand corner of each check. These are entered on each check by the bank, and hard-to-find errors can slip through. If you still can't find the error, check everything again. Look over the bank statement very carefully. Are all the entries accounted for? Look for math errors on the reconciliation page itself. Do not give the bank the benefit of the doubt about anything on the bank statement. People in banks can make mistakes just as easily as you can. Their accounting systems catch most errors, but no controls are foolproof. Mistakes do come through.

Below is a checklist of errors that can cause your account to be out of balance:

- An incorrect or nonexistent starting point

- Unadjusted corrections from previous months

- Bank errors in recording check totals

- Math errors on outstanding check lists

Balancing The Checking Account

JOHNSON CONSTRUCTION COMPANY
Record of Bank Balance

Midfield Bank
General Account   Month of _____

Date	Deposits	Date	Check Numbers	Amount	Balance
Balance forward............			....	......	
1					
2					
3					
4					
5					
6					
7					
8					
9					
10					
29					
30					
31					
total					

Figure 18-5

- Checks listed for the wrong amounts on outstanding check lists

- Previous months' outstanding checks still outstanding and not listed again

- Previous months' outstanding checks that have cleared but that have been listed again by mistake

- Deposit errors not recorded

- Deposits in transit not recorded

- Returned checks not recorded or recorded twice

- Adjustments made to the wrong side (doubling the amount out of balance for those adjustments)

- Unrecorded adjustments for service charges, bank-provided services (such as printing your checks) or automatic loan payments

- Errors in recording the amount of deposits in the checkbook

- Trying to balance to the wrong finishing point
- Math errors on the bank reconciliation summary form

Many other kinds of errors and omissions can occur. One-time or very unusual circumstances often arise which are especially difficult to find because they are unusual.

KEEPING TRACK OF YOUR BALANCE

In a manual check-writing system you carry the balance forward with each deposit and check. Some pegboard systems also allow space to keep the balance. But for many check-writing systems, including some pegboards, there is not enough space to allow for this function. You must keep track of the balance and if your pegboard does not allow it you must do so separately. The merit of keeping a separate daily balance is that it provides you with a math control for your checkbook. Keep a separate balance with a record of bank balance form as in Figure 18-5. The math control helps you balance your account at the end of the month. By adding the columns for deposits and for checks you can quickly determine the totals for the month. The balance forward, plus deposits and less checks, should equal the last amount in the Balance column. If it does not, there is a math error somewhere on the form.

Add and list each day's checks in the check amount column. If you write several checks on one day, add them together. Make sure there are no duplications or omissions by indicating the range of checks you write that day in the check numbers column: "#'s 110-117" for instance.

List each deposit in the deposits column on the line for the day the deposit is made. Add today's deposit to the balance for the previous day in the balance column and subtract today's checks in the amount column to compute the new balance forward.

Include on the record of bank balance all adjustments to the check book such as previous months' errors, bank charges, and returned checks. Isolate these adjustments. If possible, put them on a separate line so that they are easily traced. Handle the current month's adjustments in the same way; record any adjustments made during the month by the bank whenever they send you a adjustment memo. A notice to increase is listed on a bank *credit memo* and a notice to decrease your balance is listed on a *debit* or *charge memo*. These forms are shown in Figures 9-2A & B.

HOW TO CONSTRUCT A BALANCE

Chances are that if a bank account has been out of balance for months, a good starting point does not exist in that year's account activity. Find the most recent good balance and work your way forward, balancing each month until you get to the present balance. Your other alternative is to give the bank credit for its accuracy and rely on its balances. Banks have good controls for giving accurate service to their customers. They compete with one another for your business, and accurate, professional statements and support documents are one way to do this. The law requires banks to maintain certain controls and records, so their reports to you are likely to be accurate most of the time.

You can construct a balance where there was none before by determining your outstanding checks as of the end of the month. Go back through several months' cleared checks and find out what checks have not cleared. Next, figure out whether there are deposits in transit at the end of that month. Then adjust the bank's balance by adding any deposits in transit and subtracting all your outstanding checks. The result is a constructed balance. Enter it in your checkbook at the beginning point, the end of the month. Balance your account at the end of each month from then on.

OLD OUTSTANDING CHECKS

Most balancing adjustments are for timing differences. You write checks on one date in the month and the bank clears them on another date, sometimes in that month and sometimes in the next. But from time to time a deposit in transit or an outstanding check does not clear the following month.

Call the bank when a deposit is lost and compare your duplicate deposit slip to their records. The deposit may have been applied to someone else's account, or it may have been posted late. Make sure the bank corrects its error in time for the current month's statement.

When outstanding checks become too old to keep on your list, contact the payee. The check may have been lost in the mail. Call the bank if that is the case and issue a *stop payment* order.

Most banks charge for this service, but it is the only way to protect yourself against a possible double payment. Once the bank determines that the check in question has not been cashed, reissue the payment. This keeps check adjustments up to date.

After the bank issues a stop-payment on the check, record the amount of the check as an increase to your account. At the same time, remove the check from the outstanding list. Reissuing the payment once again decreases the balance in your checkbook by the amount of the stopped check.

KEEPING YOUR PERSPECTIVE

You should spend as much time balancing your checking account as needed to reassure yourself it's been done correctly. But you also need to decide when enough is enough.

This raises a question. Let's say you're out of balance by three cents, and you've checked everything three times. You've invested three hours. Is the three cents worth the effort? That's not as easy to answer as you might think. The problem is that there could be a number of things adding up to the out-of-balance condition. For example, you could be wrong by $1,000 in one direction, and $1,000.03 in the other direction.

Some guidelines for dealing with very small amounts:

1) Check everything twice. Once you're certain that the most likely problems have been checked, you can probably assume the small amount is not worth the time required to investigate.

2) Check and compare only the cents (everything to the right of the decimal point) when your out-of-balance problem is less than a dollar, or when the amount includes odd numbers of cents. This will help you find the problems more quickly.

3) Write off very small amounts, once you've checked everything, to the cash over or short account. Don't waste valuable time looking for insignificant problems.

Chapter 19
Accounting For Estimates

Accurate bidding is a result of having the facts you need and putting those facts together so they make sense. The successful estimator gets information from many current sources: estimated wage rates, current material price listings, and subcontractor quotes. But these sources, important as they are, leave out the most important—the cost records of jobs completed. With good job cost records, a builder can see very quickly his costs on various types of work. He can analyze a list of final and complete costs and see where estimates varied from actual costs.

No accounting system in the world is a substitute for an intimate knowledge of your own business. But your accounting records can be a valuable source of information and guidance in preparing current estimates, and an aid to organizing bids so that none of the parts are left out. Builders who win bids because they forget to include costs for some part of the job can not expect to prosper.

JOB COMPLETION SCHEDULES

The first part of every estimate should be a schedule of proposed job completion. While you need detailed information and support for your material, labor and overhead costs, the schedule is a framework in which your estimate is referenced.

The schedule is in itself an estimate, because no one knows for sure exactly how long a job will take. Too many uncontrollable variables exist to pin down an exact completion date. But your cost estimate must be based on a rough estimate of when the work will be done.

Once you determine a fair completion time, arrange the phases of the job into a flexible schedule that allows for favorable and unfavorable time variances. This should give you both earliest and latest completion dates. Early completion will often result in higher profits because overhead is allocated at least partly on the length of time required for completion.

Figure 19-1 is a sample job completion schedule. With a plan like this, you can forecast the cash flow, commitments to subcontractors, assignment of labor (not just for this job, but for all jobs), and the purchase of materials. Since

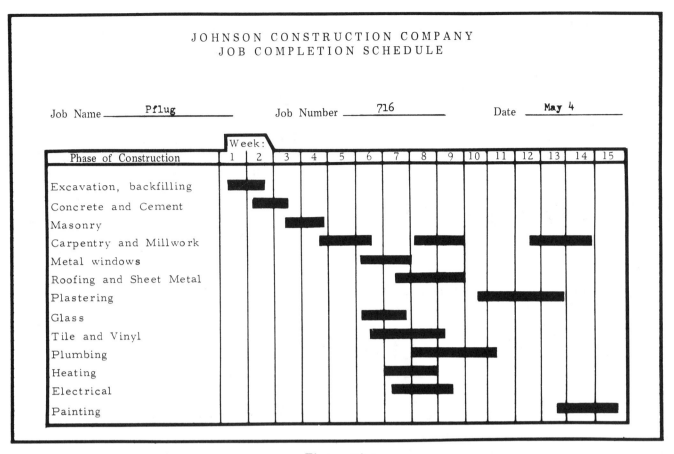

Figure 19-1

there are many factors that are beyond control, the schedule must allow for reasonable variations in each phase.

Many phases of a job can overlap one another. Extending the beginning and ending period for each phase of a job may give you a clearer and more realistic schedule. The plan illustrated in Figure 19-1 is designed to be flexible. Notice that the later phases are longer than the earlier phases. The builder in this example does not expect that plastering will take three weeks to complete. But the three-week bar indicates the period during which plastering must be completed. It is easier to judge the start and finish time for the initial phases of a job. But as time goes on more variables enter the picture. So the builder in the example has extended the time for later phases to allow for complications in the job schedule.

If all goes well, this job could be completed in twelve weeks. If the job is right on schedule at the end of the ninth week, most of the work will be completed. If there are delays, the completion schedule must be stretched out. The more generous completion schedules for plastering and painting allow for those delays.

A schedule like this will help you deliver the profits you expect and cuts costs. Your schedule will be more realistic if you review completion times for similar jobs to see where delays have most often occurred. Careful planning will allow for similar delays that are beyond your control.

THE ESTIMATE

Once you have developed a job schedule based on the specs, plans and your own familiarity with all the building components, make an estimate of costs. A consistent, well-organized approach is the most likely to succeed. As you make your estimate, remember that the document you are preparing will be used beyond the bidding phase. Many builders view the estimate as a document prepared to get a contract. But a well-prepared estimate also serves as a control over the job once it is in progress. Comparing estimated to actual costs shows the trend of the project. If costs begin to exceed estimates, you have to work to bring them back in line to preserve your estimated profit.

It is a good idea to have an analysis of jobs both by field and office personnel. Foremen or superintendents should give you status reports regularly. Direct costs can be watched from the office. But actual progress occurs at the jobsite.

The estimate for each part of the job is made up of various components. These may include:

- Materials
- Labor
- Subcontracts
- Equipment use / rental
- Other selling expenses (in detail)
- Fixed overhead
- Profit

MATERIALS

The builder who has kept up a job cost card on all previous work and has made a good material take-off should have no trouble estimating the true material cost. Prepare a complete list of materials for each phase of construction. This list documents your material estimate.

Don't assume that there will be material losses that require a reserve contingency. Any real losses will be covered by insurance. Building contingencies into several places in your estimate creates a safety cushion but also makes it more difficult to estimate the true cost. Too often, builders guess at material costs, assuming that they may have left something out. You should include a contingency allowance in your estimate. But contingency should cover unknown conditions, not estimating errors. Add contingency at the end of your estimate just before you add in the profit. Don't clutter your estimate with little guesses that allow for errors. This will help you keep your cost figures clean and precise and make them most useful when used as cost reference data months later.

LABOR

Some men work faster than others, perform better quality work, and make fewer mistakes. Some have a better attitude about their work than others. Know what tradesmen or crews you will use when you make your estimate. Many builders who have worked on their own jobs make the mistake of estimating labor based on their own experiences. "I could do that work in fourteen hours," you might say. But would it be reasonable to estimate fourteen hours for the same work to be done by others? Instead, rely on educated guesses and calculated averages to estimate your labor costs. Some men could do the work in fourteen hours, others would take eighteen. A mixed crew might be able to finish the job in a different amount of time. You need to estimate a fair time requirement based on an appraisal of your own work force. This makes labor the single most difficult part of your costs and expenses to estimate.

The hourly wage you pay for labor is not your hourly cost per man, of course. Payroll taxes, insurance, and employer-paid union benefits must also be included:

Labor, $20.00 per hour	$20.00
Liability insurance (2.5%)	.50
Social security (FICA) (7%)	1.40
Unemployment insurance (4%)	.80
Unemployment insurance (federal) (.7%)	.14
Union benefits (pension, vacation, etc.) (10%)	2.00
Workers' compensation insurance (5 to 15%)	2.00

This works out to $26.84 per hour, not simply the base rate of $20.00 per hour. The percentages above are approximate but taxes, insurance and benefits will add about ⅓ to your cost.

Review your recent payroll analysis records such as shown in Figure 16-1 to estimate the hours required to do each phase of the work. Estimate the hours required in the greatest detail possible. The more detailed your analysis, the more accurate your direct labor estimate is likely to be.

SUBCONTRACTS

Builders often run into cash flow problems because they don't anticipate payments to subcontractors. Especially important to cash flow is matching payments to be received against payments due your subs. You can't anticipate payments to subs unless you follow progress of the work. In a small organization this is no problem. But larger projects require weekly progress reports. These weekly reports serve many purposes, and one of these is aiding forecasts of when subcontract payments are going to be done.

EQUIPMENT COSTS

The builder who maintains good equipment cost records and understands the capability of each major piece of equipment will have no trouble estimating equipment costs. Set up a procedure to compute the hourly cost of equipment and keep that cost current. You need to know what your equipment costs, whether the equipment is owned or leased.

Control equipment cost with good scheduling to minimize idle time. A realistic estimate of productivity should be based on your experience in previous jobs. The hourly cost estimated for equipment should allow for idle time, repairs and maintenance.

OTHER EXPENSES

Expenses which are usually a percentage of direct costs but are not themselves direct costs can be estimated fairly precisely. Payroll taxes and union benefits, for example, will be a certain percentage of the direct labor cost. If your estimate of the direct labor cost is accurate, your estimate of taxes and insurance should be just as good.

ALLOCATING FIXED OVERHEAD

Overhead costs must be allocated to each estimate you prepare. As discussed previously, the allocation of overhead is never precisely accurate. Each estimate may become a portion of your total business. But it is seldom certain that any estimate will become a job.

Assigning a portion of overhead expense on the basis of estimate dollar volume is a poor method because jobs can produce widely differing yields. Many builders add some fixed percentage to the estimate to cover overhead. But this ignores the fact that some jobs require a heavy commitment of your assets and labor while others are largely subcontracted and involve a much smaller commitment. A heavy mark-up on a largely subcontracted job is likely to give the work to your competitors. Subcontract and material costs are generally passed on to your client with a much lower overhead burden. These costs by themselves have little to do with fixed overhead.

Once again, the reasonable way to allocate overhead is on the basis of total labor hours. This is true even though there is no perfect way to allocate fixed overhead. No matter how it is analyzed, overhead expense bears no exact relationship to an individual contract except that it is a very real part of your cost of doing business.

Most fixed overhead expenses vary with the cost of labor, either as a result of labor-related factors, or for the same reasons that labor varies. Most important, direct labor can be identified easily by job. This gives you an easily-defined base from which to assign fixed overhead.

To allocate fixed overhead by direct labor hours, first use your estimate of the duration of the project. Next, figure what your annual overhead expense will be. You should be fairly certain about how many direct manhours the job requires and your annual fixed overhead. The next step requires an approximation. Estimate how many direct manhours will be worked on all of your projects during the next 12 months. Divide this figure by 12 to find the average monthly direct manhours. If the job in question will require 20% of your total direct manhours while it is under construction, it should carry 20% of your fixed overhead for that period.

Fixed overhead expenses are more easily estimated than any other cost you have. In most cases, the past year's costs can be projected ahead one year fairly easily. While overhead is "fixed," it can increase due to raised rents, inflated prices of supplies, higher telephone bills, and so forth. Consider increases you expect in these costs.

A contract estimated to run for twelve months should be allocated a portion of the next twelve months' projected fixed overhead expenses. But a contract that will run about four months should be allocated its share of the average fixed overhead for a period beginning four months before the start and ending four months after completion. This should prevent overloading fixed overhead on one or two jobs during a low volume period.

Estimating total labor hours a year in advance can be difficult because you do not know what your building volume will be. You can't base your estimate on past performance because the total commitments in the next year probably will not be the same as in the past year. Still, your volume for the past year is usually the starting point for an estimate of the coming year. Estimate labor for the period of the contract on the assumption that you will win the job you are bidding. This total can be revised and restated on an annual basis. To compute labor for a four-month period, annualize the total by multiplying it by three. For an eighteen-month contract, annualize the figure by multiplying total labor hours by 67 percent. In other words, express total labor hours as though for a one-year period. If you are estimating that a job will take fifteen months to complete (as in Figure 19-1), annualize total labor hours as follows:

$$\frac{\text{Total labor}}{15} = \text{average per month}$$

Average per month x 12 = annualized labor

Once annual labor hours have been estimated for one contract, you should have no difficulty deciding what your labor hours will be over a shorter term. Consider the contracts in progress, normal changes in activity to be expected during certain seasons of the year, and assume that this new contract will be won. Your estimate should be fairly accurate unless something unusual happens during the coming year.

Once you have annualized labor on the contract you are bidding, you are ready to compute a fair allocation of fixed overhead:

1) $\frac{\text{Annualized labor}}{\text{Company's total annual labor}}$ = this contract's percentage of total

2) Percentage of total x total annual overhead = this job's amount of annual overhead

3) $\frac{\text{This job's amount of total overhead}}{12}$ = monthly average

4) Monthly average x 15 = overhead for total contract

You now have documented your fixed overhead costs and have a number to include in each estimate. You also have a detailed estimate of average monthly fixed overhead. Compare actual amounts allocated to estimated allocations. This helps you revise estimated profits on the job if the contract is awarded.

YOUR ESTIMATE SUMMARY

All totals in your estimate summary should be supported by detailed calculations and lists. This lets you identify cost changes as the job progresses. Also, it gives you a checklist of problem areas to flag for revision when estimating similar jobs.

Figure 9-2 shows a sample final estimate, prepared in enough detail so that you can identify specific costs by job phase.

Using the methods described in this chapter should require no more time than preparing estimates with other methods. But the method described here should improve your batting average. This is especially true if you do the following:

- Use past performance to temper present estimates

- Schedule the completion of each phase and of the total job

- Document each part of the estimate

- Control costs of jobs on a regular basis

- Avoid adding contingencies without good cause

- Establish a methodical procedure for estimating

SUMMARY

The builder who controls job expenses and costs will have a higher success ratio than one who does not. Controls are not difficult, even in a small struggling operation, if the books and records are informative and allow the builder to make detailed comparisons.

The schedule of completion is the foundation of a well-prepared estimate. Flexibility within that schedule should help prevent overruns that kill profits.

A consistent approach should help you improve the reliability of your estimates. The consistent approach requires intelligent use of all available information—including past performance compared to past estimates. Evaluate

JOHNSON CONSTRUCTION COMPANY
FINAL ESTIMATE

Job Name _____ Company Number _____ Date _____

Description	Quantity	Unit	Labor unit	Total	Material unit	Total	Subs	TOTAL
General Conditions								
Foremen	7	wks.	$400.00	2,800				$ 2,800
Permits				700				700
Temp. Bldg.						1,200		1,200
Washing							900	900
Rentals	3	mos.			3,000	9,000		9,000
Total				3,500		10,200	900	$14,600
Carpentry								
Materials								
Finish floors								
Shelving								
Roof nailer								

Figure 19-2

shortcomings in previous estimates to improve current estimates.

Most estimating systems require consistency but do not need to be complicated or time-consuming. In fact, a good procedure simplifies estimating. Too many builders have no consistent procedure at all. In those cases, profitable contracts are won more by luck than by skill.

Detailed listings for materials, subcontracts, labor, and overhead expenses are required to document the estimate, control the job when it is in progress, and create cost reference material that will help you win profitable jobs in the future.

Controlling costs begins when the estimate is complete. Once the job is awarded, you must produce the expected profit by controlling the costs projected in the estimate. Your estimate may be the best-prepared of all submitted, and the best documented. But if you fail to control costs once the job starts, the estimate will be meaningless and the profits nonexistent.

A well-documented estimate makes control easier because each individual phase of the job has been assigned a budget. You control costs within each phase of construction. This tightens control. When an estimate isn't prepared by job phase, overruns are often hidden until all costs have been tabulated. This rarely works to the builder's advantage. Cost overruns should be stopped as they begin to occur. Looking back at a problem doesn't solve it. Successful builders control their costs during each phase of the work.

Section III

Financial Statements

- Recording Before The Event
- Financial Statements
- Using Financial Information • Financial Ratios
- Putting Together a Statement
- Comparative Period Statements
- Restatements By Accounting Methods
- Statements By Job
- Statements For Loan Applications

Chapter 20
Recording Before The Event

Accrual and deferral entries record real events—exchanges of money that have not yet taken place or that occurred previously and are being spread over a period of time for accounting purposes. Recording before the event is a way of equalizing timing differences between work you do and payment you know you will receive later. It also equalizes differences in timing between lump-sum prepayments you make in one month and the resulting services or benefits you will receive over many months to come for that single payment. Accruals and deferrals give you a truer picture of your business than if you base your accounting only on cash transactions as they occur. No matter what the size of your contracts, completed contract accounting procedures require that all business be deferred. Percentage of completion contracts are accrued or deferred in part whenever billings do not match receivables. And accounts receivable, or income booked and not yet received in cash, is often a significant portion of total assets—and certainly the largest current asset.

You record accruals other than for contracts to show accounts and taxes payable, prepaid assets, and any other non-cash exchange that you expect will result in cash changing hands in the future. Your accruals should reflect the true financial condition of your business, regardless of the timing of cash exchanges. Don't misuse accruals by creating on paper the payment results you hope will occur.

A distinction should be made between journal entries and accruals. While accruals are always made by journal entry, all journal entries are not accruals. Below are some examples of non-accrual journal entries:

- Entries to correct coding errors
- Entries to record automatic loan payments
- Entries to record interest or payroll taxes paid
- Depreciation and amortization entries

This chapter explains how accruals and deferrals are recorded and how they fit into a builder's operation. This and the following chapter, which discusses the individual accounts and their meaning, should help you read and construct a financial statement and make these special accounts work for you.

ACCOUNTS RECEIVABLE

All sales on account should be accrued because no cash changes hands. Make an entry to Receivables instead of to Cash when you make sales on account. When sales on account are paid, make a cash entry to reverse the accrual you made earlier to Receivables.

Your accrual entries should always record events in the period in which they occur. Your reversal should always be recorded in the period in which cash actually changes hands. Record the total accounts receivable for each month in a single entry as a portion of the sales journal totals. If this entry could be isolated from sales, it would look like this:

	Debit	Credit
Accounts Receivable (asset)	XXX	
Sales (income)		XXX

When customers pay their accounts, their portion of accounts receivable in the monthly sales totals would be reversed.

	Debit	Credit
Cash (asset)	XXX	
Accounts Receivable (asset)		XXX

The net effect of these two entries is to cancel the accounts receivable. Since the original charge is reversed in a later month, the remaining two entries in your general ledger will reflect this inflow of cash:

	Debit	Credit
Cash (asset)	XXX	
Sales (income)		XXX

This entry is identical to the one made for cash sales. The accrual system and the accumulation clearing account, Accounts Receivable, adjust income timing differences. To wait until you receive cash before you record the income would distort the true business picture of your operation, since you probably do most of your work on account.

PREPAID EXPENSES

You don't usually pay all your expenses in the current month; you accrue them for payment in later months. And some of the expenses you do pay apply to more than a single month's service. You may pay three year's insurance coverage in advance, for instance, in the first month of the policy period. It would not be accurate to record the entire thirty-six months' expense in the first month. The entire amount should be classified as a prepaid asset and assigned to an expense account. Then month by month the asset should be removed from your general ledger in equal installments over the period of the policy. The monthly portion of the insurance asset is *reversed,* since the whole asset was entered in the first month.

You can do this for any expense that applies for longer than one month. But it is not practical to set up and amortize expenses for only a few months. Reserve this procedure for expenses that 1) apply over twelve months or more, or 2) apply for less than twelve months but belong partly in the next tax year.

For example, if you pay six months' insurance coverage in the eleventh month of a tax year, it would be proper for you to set up four months' payments as a prepaid asset. This would recognize two months' coverage this year (for the eleventh and twelfth months), and four months' coverage next year.

Several types of expenses are commonly accounted for as prepaid assets. These are insurance, printing expenses, office supplies, interest, and contract services.

Consider office supplies and bulk printing purchases as part of inventory. They are used over a period of time and replaced as needed, just like any other items in your yard. The difficulty in accounting for these expenses is that you must estimate the length of time you will have these supplies on hand and thus the number of months' prepaid asset you need. Unlike insurance policies, which cover defined periods, office supplies and printed stationery can last for an indefinite number of months. You must make a fair estimate of the "life" of these expenses before you establish a prepaid asset for them. Base your estimate on the prepaid asset's expected use—the average monthly consumption, the budgeted months of supply or the like.

There are two ways to record interest paid on bank loans. Since the interest applies over the period of the loan itself, you can recognize that amount over the same period. Banks sometimes supply a scaled interest summary sheet showing the portion of each payment applying to interest and the portion applying to principal. Use this information when it is available to recognize the exact interest amount per month. The second method is to recognize equal amounts of interest per month. Under this method a thirty-six month loan repayment period would call for recognizing 1/36th of the total interest each month. Or a loan with thirty-six equal installments and a balloon payment would call for recognizing equal portions of the interest with some interest removed from each portion to be recognized as the balloon payment at the end of the period.

Say you spread the interest equally over the period of loan repayment and recognize these equal portions. There are two methods of booking this interest expense. In the first, you recognize the interest as a portion of each installment. In the second, you amortize a pre-established prepaid asset total. The first method would begin with an entry to record the loan:

	Debit	Credit
Cash	3,000	
Note payable (current)		1,000
Note payable (long-term)		2,000

The drawback of this method is that you make no entry to reflect interest. The monthly entry to record note payments would be included with the check register totals. If the interest entry could be isolated, it would look like this for one month:

	Debit	Credit
Note payable (long-term)	$83.33	
Interest expense	16.67	
Cash		100.00

In the first method, you recognize the interest expense only as each month's installment is paid. You establish a prepaid asset account in the second method. Total interest is recognized as an asset and amortized over the repayment period. The entry to record the loan would be:

Recording Before The Event

	Debit	Credit
Cash	3,000	
Prepaid interest	600	
Note (current)		1,200
Note (long-term)		2,400

The monthly check-register breakdown, if isolated, would show note payments as follows:

	Debit	Credit
Note (long-term)	100.00	
Cash		100.00

The monthly journal entry to *reverse* the monthly portion of the prepaid interest and *record* the interest as an expense would be:

	Debit	Credit
Interest Expense	16.67	
Prepaid Asset		16.67

The method you use to record interest for prepaid assets is up to you. Your method should reflect the total prepaid interest as an asset, allowing you full reporting on financial statements. Documentation in the original note entry should provide a trace of the loan's history.

Contract services such as legal fees, engineering, or special business consulting can apply to future periods as well as current ones. These future expenses should be spread over the whole period to which the services will apply. You may have to estimate this period. Then amortize the expenses over the life you have estimated.

Some types of expenses require special treatment involving both deferral and accrual. Figure 20-1 shows a twelve-month schedule for property tax expenses, payable in December and April and applying to the period of July 1 through June 30. Since Johnson Construction reports on a calendar year, it wants to report one half of the expense in each period. But to show December and April as the only periods of taxation would distort the financial picture through November of the first year, for the first quarter of the second year, and for May and

JOHNSON CONSTRUCTION COMPANY
ACCRUAL SCHEDULE
PROPERTY TAX EXPENSE

Month of Entry	Payments	Monthly Expense	Accumulated Expense	BALANCE OF: Prepaid Asset	BALANCE OF: Deferred Liability
July	$	$150	$150	$	$150
August		150	300		300
September		150	450		450
October		150	600		600
November		150	750		750
December	900	150	900		-0-
January		150	150		150
February		150	300		300
March		150	450		450
April	900	150	600	300	-0-
May		150	750	150	
June		150	900	-0-	
Total	$1,800	$1,800	*		

* Expense is split between the two years - $900 in each six-month period.

Figure 20-1

June. The company wants to show $150 expense in each of the twelve months. From July through November, Johnson makes this entry each month:

	Debit	Credit
Property Tax Expense	150	
Deferred Expense (liability)		150

Johnson's December's entry, when the tax is finally paid, is as follows:

	Debit	Credit
Property Tax Expense	150	
Deferred Expenses	750	
Cash		900

The entry for the first three months of the second year is the same as the previous year's entry in the deferral journal. In April, Johnson pays the entire second installment and makes this entry:

	Debit	Credit
Property Tax Expense	150	
Prepaid Asset	300	
Deferred Expenses		450

Johnson thus recognizes taxes paid in advance as being prepaid. The payment of the expense is not applicable until later. The entries for May and June are:

	Debit	Credit
Property Tax Expense	150	
Prepaid Asset		150

At the end of the period in June the prepaid asset account and the deferred liability account are both zero, and Johnson has reflected the property taxes as $150 each month. Johnson could use this procedure for contract services that apply to a period both before and after the actual payment date.

PREPAID EXPENSES AND DEPRECIATION

Amortizing an expense over a period of time is not at all the same thing as depreciation, although the two seem similar at first glance. *Depreciation* is the recognition of the cost of new and permanent fixed assets over the useful life of those assets. These are generally large-cost purchases which have tangible value to the builder and often last beyond the depreciation period. *Amortizing* expenses over a period of months or years does not indicate that there is a tangible value. You are not recognizng a cost over a useful life. Rather, by prepaying and amortizing an expense you spread the expense over a period of time. You do this so that you do not misrepresent the true financial condition of your business for the month in which you prepay.

DEFERRED ASSETS

You can treat expenses by paying them and deferring them in your books until a later period. Unlike prepaid asset accounts which recognize expenses by amortizing them over a period of time, deferred asset accounts are set up for payments you now make that will be recognized at a later date.

There is an important distinction between deferred liabilities and deferred assets. Figure 20-1 illustrated how to treat expenses by setting up a liability or a payable account for expenses booked and not yet paid. On the other hand, deferred assets are those items that are paid first and then coded to the asset account to be deferred later. Suppose you pay for having letterhead and envelopes printed with your new business address, but the actual move isn't scheduled for another three months. The proper treatment of this expense would be to establish a deferred asset account and record the entry as follows:

Recording Before The Event

	Debit	Credit
Deferred Assets	XXX	
Cash		XXX

Once the move is completed and you begin using your new stationery, remove the expense from the deferred asset account and do one of two things, according to the expected life of the stationery: 1) you can write the amount directly to expense, or 2) you can set up the amount as a prepaid asset and amortize it.

You can also defer assets such as deposits you make against future expenses. These deferred deposits may be for conventions or lectures coming up in six months. Or you can defer your last month's rent payment if you lease rather than own your property. Be sure to distinguish between security deposit payments and payments for last month's rent. A security deposit is refundable, and should be placed in an asset account. The last month's rent is a true future expense and is deferred until the lease expires.

Defer expenses that refer to future sources of income. A builder might pay for permits and licenses for a job this month knowing that the income from that contract will not be generated until next year. Such items are deferred assets. Be sure to account for expenses at the same time you recognize the corresponding income.

PROVISION ACCOUNTS

Occasionally you must establish some provision for expected future losses. This is especially true when a comparison of actual and estimated costs shows that a net loss will result. The loss may have a significant impact on operations when it occurs, and it should be spread over the life of the contract.

A bad debt reserve is a good example of a provision for future losses. Record a monthly expense in the general ledger by setting up a provision against the accounts receivable asset. Offsetting entries show a monthly bad debt expense, even if no bad debts are recognized that month. You record actual losses against the provision you have set up.

Establish a provision against your assets—either receivables or deferrals—any time you know of a future loss. In some cases, such as losses in contract agreements, establish a liability account for future losses.

Set up provision accounts with caution. Habitual use of a provision procedure can distort the true state of your business as shown by your financial reports. Allow provisions only for certain losses. Estimates must have a high degree of accuracy. Base your estimates on solid current data. Use your past records for comparison if you are sure of their accuracy and relevance to the current situation.

Identify specifically what losses your provision is for when you set it up. You might want a provision for increased bad debts at rising volumes of business or for early overruns on contracts. These are supporting reasons for setting up provisions. But provision accounts should never be established as contingencies against the mere possibility of future unexpected losses.

CONTROLS

Each time you accrue an expense or establish a provision, set up a corresponding subsidiary control account. These sub-accounts control the flow of accrual entries. This keeps the general ledger a summary report as it should be. At the same time these control accounts document the reasons for the special accrual or provision treatment.

Know at any time exactly what items are in the Prepaid Assets, Deferred Assets, Deferred Liabilities, and Provision for Future Losses accounts.

The accounts payable and receivable controls were discussed in previous chapters. The general ledger should contain an entry to show the monthly flow of taxes payable, especially payroll taxes. Consider these accounts as strictly clearing accounts.

Any special journals you use to record and reverse accruals should contain full explanations of the entries. Keep reversing and recording journals separate. When you reverse a previous entry, make your explanation refer to the original entry you are backing out. This lets you maintain control over your accounts yet reflect the true status of your business when you prepare your current financial statements.

THE CASH ACCOUNTING METHOD

Builders commonly keep their books on the accrual method because of the nature of the construction business. This requires that financial statements reflect events in the period of activity, not simply when cash changes hands.

The other method of accounting is called the "cash method." In this method, as we have seen in earlier chapters, only exchanges of cash are recorded and financial statements show only cash exchanges. You can use this method for income tax returns, but only if you do so every year. Cash only is much simpler to account for, as no records need to be kept for contract progress, accounts receivable, or accounts payable.

But accrual records make for more accurate financial reporting and also give you better control over your business. These controls minimize cost and expense, improve efficiency, and result in healthy cash flow and profits. Without controls, your chances of staying in business are cut considerably. The cash method of account never reflects the true status of business. While accrual methods present their own special problems, cash accounting is a poor alternative if you want to improve the quality of your financial reports and controls.

Chapter 21
Financial Statements

Financial statements summarize your operation's activity for a certain period, usually a month, a quarter, a year or more, and show its status at the end of that period. Data for financial statements is taken from the general ledger, but only after the ledger is proven to be in balance on a worksheet called a *trial balance*. The general ledger is an important record, but it is not useful in revealing financial data. It is really no more than a listing of account balances. Financial statements arrange the general ledger accounts in meaningful groups to reveal the financial truth about your operation. There are three distinct financial statements; each them serves a specific function: 1) the balance sheet, 2) the income statement, and 3) the cash flow statement.

The *balance sheet* is a summary of the existing conditions of the business. It is called a balance sheet because its line-by-line details show the balances of general ledger accounts. The balance sheet lists the assets, liabilities, and net worth of a business as of a given date. Assets are the value of the properties owned, liabilities are the company's debts, and net worth is the owner's equity or the net value of the assets less liabilities. The relationship between assets, liabilities, and net worth is shown in this formula:

Assets *less* Liabilities *equals* Net Worth

The balance sheet is broken down into two parts—assets on the debit side and liabilities and net worth on the credit side. Debit accounts generally represent assets, and credit accounts generally represent either debt or worth. Liabilities here are debt and equity is worth. The balance sheet will always be in balance because:

Total Assets = Total Liabilities + Net Worth

The *income statement* summarizes business operations within a period of time. The ending date of that period is always the same as the closing date given on the balance sheet. The period covered by an income statement is always specified on top of the report. For example, an income statement for the second quarter would read: "For the quarter ended June 30, 19__."

The income statement is broken down into more or less standard divisions. Except for variations tailored to the needs of specified individuals and business, here is the format for a standard income statement.

Gross Sales

less

Direct Costs

equals

Gross Profit - (Income before operating expenses)

Gross Profit

less

Operating Expenses

equals

Income from Operations - (Income before Income Taxes)

Income from Operations

less

Provision for Income Taxes

equals

Net Profit

The income statement allows for flexibility in how you want income broken down for the period. The details for operations can be shown in a summary form in one column. Income can be compared by month, quarter or year to income in another month, quarter or year. The statement can be broken down by job, redefined by another accounting method, expressed in dollars or percentages. You need this flexibility because the income statement is usually subjected to closer and more detailed analysis by lenders than are the other two statements.

The *statement of cash flows* is a summary of the sources of and uses of cash—where it came from and where it went for the same period of time covered by the income statement. The statement of cash flows has two sections:

1) The fixed or long-term provisions of funds, which includes net income

2) The change in current assets and liabilities

The bottom line of this report shows either an increase or a decrease in funds.

The *sources* of funds are generally cash-basis net income, sales of assets, increases in liabilities, and decreases in other assets. The *applications* or uses of funds include the purchase of long-term assets, the decrease of liabilities, and the payment of taxes. Any time you change the sources or applications of your funds you affect your current assets and liabilities.

For example, paying off a bank loan is an application of funds. You use money to decrease a liability. At the same time, you decrease the balance of cash. Every source and application of funds has an offsetting effect on the components of the cash flow change—your current assets and liabilities.

The change or the net increase or decrease in funds is expressed in two ways—as a change in the source and application of funds, and by the change in liquid position (current assets and liabilities). Thus a change in your cash flow is a change in:

1) Your current cash position, and

2) The source and application of your funds on long-term assets and liabilities

The statement of cash flows will be in balance if the general ledger and the balance sheet are in balance. You can check your accounting accuracy in the following way. The net increase or decrease in your funds should be equal to:

Current Assets		Current Assets
less		*less*
Current Liabilities	*less*	Current Liabilities
(End of Period)		(Beginning of Period)

In the above formula an increase from the beginning to the end of the period means an increase in funds for the period, and a decrease from the beginning to the end of the period means a decrease in funds.

RELATIONSHIP BETWEEN STATEMENTS

Each of the three statements has a distinct function, and these functions are interrelated. Any one statement by itself can appear promising without revealing the true status of your operations. Combined, they paint a fuller picture.

1) The balance sheet is a general indicator of the strength of your operation—how well the business is able to finance its commitments, how heavily financed it is, how weighted its assets are against its liabilities, the degree of debt commitment versus gross value.

Financial Statements

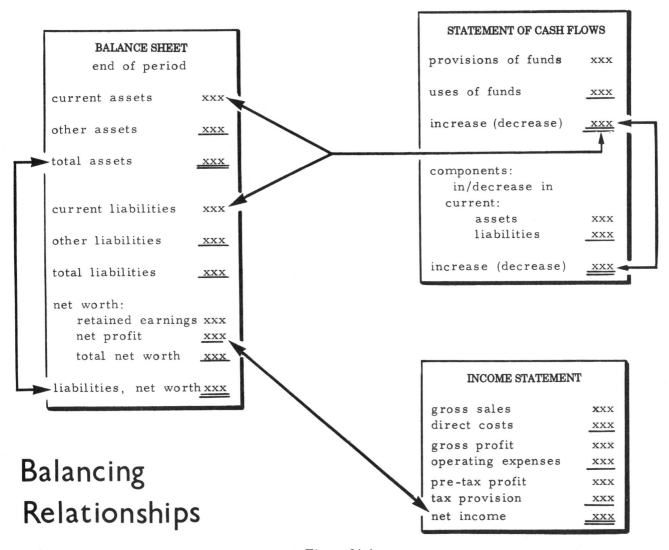

Balancing Relationships

Figure 21-1

2) The income statement shows how well you are controlling costs and expenses, the yield on your operations, and the volume of business you are doing. The income statement also divides the information into categories which indicate where you are having problems.

3) The statement of cash flows tells you whether your operation is handling cash properly. It also points to cash trends. Know how well you can meet your present and future cash requirements by studying this statement carefully with the others.

The three statements should always be reviewed together. While a balance sheet could present a healthy picture, an income statement could show costs far out of control and yields that are much too small. Or an income statement may show good volume and profits, but the cash flow may indicate that all the cash is tied up in uncollectible accounts.

Comparing an income statement and a statement of cash flows might also show that a seemingly profitable business has overbought fixed assets, thereby seriously draining its capital resources. On the other hand, a low-profit or loss status on the income statement may look negative alone, but in fact the business may have a better potential for growth and profits because of an excellent cash position shown on its statement of cash flows. Consider the value of this last status when, for example, your slow season is ending and business is about to pick up. You want to be ready to commit your funds in contracts that will produce income down the road. No one report can show you a complete profile of your operation. You or a lender must analyze all three statements together along with your plans for the future, to come to any realistic lending or planning decisions.

The three statements have a descriptive relationship to one another. That is, all three statements describe your business better than any one of them could. But they also have a balancing relationship. Here are the principal balancing relationships of the three statements. See also Figure 21-1.

- Change in current assets *less* change in current liabilities *equals* net increase or decrease in funds

- Increase or decrease in net worth *equals* income or loss from operations

- On the balance sheet, total assets *equals* total liabilities and net worth.

THE TRIAL BALANCE

After you post your ledger at the close of the period, the first step to take in drawing up financial statements is to put together a *trial balance*. See Figure 21-2. This worksheet proves that the ledger is in balance. In all cases, debits should equal credits. So the total of all debits on one side of the trial balance must equal the total of all credits on the other side. Otherwise, none of the financial statements will balance. A trial balance can also show at a glance the distribution of accounts and can thus isolate the profit. Because some entries are made on the general ledger side and others on the profit and loss side, find the profit by adding each side separately in the order the accounts are listed.

The general ledger is in balance when all entries into it have been made correctly. The way to prove this is to add up the totals of all accounts. Modern-day general ledgers are maintained by the double entry system. This means that every entry is made twice—one debit and one credit. Debits are positive numbers and credits are negative. Because every entry includes a positive (debit) and a negative (credit), a correctly posted general ledger will add up to a net of zero. For example, to record a payment you debit an expense account and credit cash. To record a charge sale, debit accounts receivable and credit income. Double entry bookkeeping was discussed in Chapter 1.

Look at the summary trial balance in Figure 21-2. Each section of the general ledger is shown in summary. An actual trial balance would list each and every account in detail. Total assets, liabilities and net worth balance with a net profit of $47,062.66. The same amount is needed to balance the profit and loss sides. This proves that the general ledger is in balance.

When you finish the trial balance, proceed to make the closing adjustments. Adjustments are usually entries made only on a worksheet to bring the statement up to date. Make these entries for the general ledger once a year when the books are closed out. But make the adjustments for interim financial statements against the trial balance.

SUPPLEMENTARY SCHEDULES

The trial balance adjusted for closing entries forms the basis for the financial statements themselves. Financial statements are summaries; sometimes they must be explained further. The three-page package of financial statements is designed for uniformity and compactness. But a particularly unusual item or high balance requires a supplementary schedule. Most comprehensive statements include explanatory notes and schedules. These brief notes are usually included on the statements themselves. For example, a notation might read:

Accounts Receivable (Note One)

"Note One" would provide all the detail necessary to explain the balance of receivables.

Notes on supplementary schedules might explain the accounting basis on which income is recorded in the statements. This is particularly true for builders, whose accounting basis for income is usually more complicated than that for most other businesses. Include notes to explain or elaborate on any item you think should be detailed, such as the various valuation methods used and the ways some accruals have been made.

BALANCE SHEET ACCOUNTS

The balance sheet contains the following account categories in the order they appear on the statement.

Current Assets Assets available to the builder within one year. These accounts include cash, accounts receivable, inventories, and other current assets. This category is especially

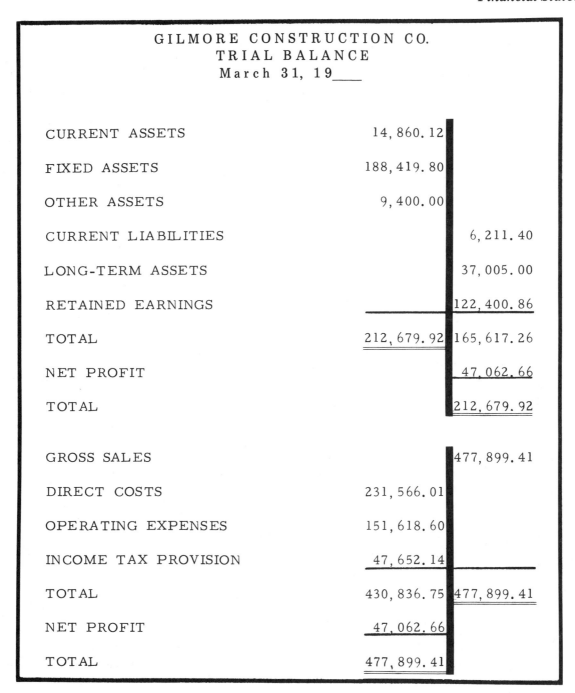

Figure 21-2

important when it is compared to current liabilities payable within one year. It is an indicator of the general health of a business.

Current assets are liquid assets. This means that they could be turned quickly into cash. Accounts receivable and inventories fall into this category because if they are currently receivable and usable they generate cash within a year.

Current assets are the guideline for a builder's immediate and future financial decisions. If delinquent and uncollectible accounts are on the books without an adequate bad debt reserve, or if you include dead inventory in current assets, you may not be able to judge your own ability to fund future operations.

Assets such as equipment are not included with current assets because they generally represent a long-term investment. Besides, no builder can be certain he would be able to sell his equipment at book value within a year—and he would not normally make decisions based on such a move anyway.

Fixed Assets This accounts category includes the gross value of equipment and machinery, small tools, improvements, furniture, fixtures, structures and land, *less* a reserve or an accumulation for depreciation that has been expensed throughout the history of the assets. These assets are fixed because they are not available to be turned into cash within one year.

Include in this category an account called Reserve for Depreciation or Accumulated Depreciation to show the to-date total of all depreciation of assets still on the books. Each year's depreciation is booked as a debit to expense and a credit to this account. Since fixed assets have a debit balance, the accumulation of yearly depreciation will eventually bring the total of fixed assets to zero. But since builders replace some of their equipment from time to time, a zero balance is never really shown.

When you sell or abandon equipment, remove from the books both the gross value and the accumulated depreciation on that asset. This means maintaining detailed records for depreciation on each piece of equipment.

Other Assets Include in this accounts category all assets that do not properly belong in other categories, such as suspense debits and prepaid and deferred assets and deposits. Also include here an account called Intangible Assets. *Intangibles* are those assets for which no material property exists. Cash, fixed assets, and prepaid assets are all identifiable specifically and have tangible value.

Goodwill is an example of an intangible asset. When a business is bought, the seller often adds a goodwill value to fixed assets, inventories, and other tangibles. You might decide, for example, that by selling your business you are adding to the company's reputation in the community and that this reputation will generate business in the future for the new owner. You and the buyer agree on a value for this reputation and call it goodwill. The buyer pays you for goodwill as well as for the established values of material or tangible assets.

Another type of intangible asset is a "covenant not to compete." The buyer of a business might want assurances that the seller will not open a similar business in the same area and pull away old customers. Part of the buying agreement would include a covenant by which the seller promises not to compete within a defined geographical area for a certain number of years. The buyer pays an agreed sum of money in return for the covenant.

Goodwill cannot be amortized or depreciated and it remains at its original value indefinitely. But a covenant can be amortized over the period of the agreement, as long as it was purchased materially from the seller.

Current Liabilities Like current assets, which are liquid and convertible within one year, current liabilities are payable within one year. Accounts payable, taxes payable, notes, contracts, and deposits to be refunded are all examples of current liabilities.

Say you obtain a four-year note from a bank. Thirty-six months' payments are long-term debt and twelve months' payments are current debt. This distinction is important, because you should be able to see current assets and liabilities in relation to one another. As you repay the note the last portion you remove from the books is the current portion, because at any time during the first thirty-six months the balance sheet should reflect twelve months' of current debt. During the last twelve months, the current debt decreases each month.

Long Term Liabilities Some liabilities exist on a builder's books for a long time. If these amounts are not payable within one year, keep them apart from current liabilities. The most common type of long-term liability is a long-term note or a contract payable. You may have other liabilities on a long-term basis, such as advances from officers or partners. These loans may be sizable and may never be repaid in full. The loans may have been made out of necessity originally, and may not become current for many years.

Other Liabilities Include deferred credits or unusual items here rather than with Long Term Liabilities. Long term liabilities will eventually become current liabilities, but deferred credits are likely to be clearing or "wash" items that enter this account and are soon cleared back out again.

Contingent Liability This is a debt that does

not occur now but may in the future. Because it is only a possibility, the amount is not added in with the other balances of liabilities. But be sure you disclose it on financial statements. A possible debt from a pending lawsuit or a potentially unrecoverable loss are examples of contingent liabilities.

Equity Account Partnerships have an account called Partners' Equity while corporations have an account called Capital Stock. This account represents the original investment of the owners of the business.

Capital stock accounts are only changed by the addition of new capital or the removal of old. Partners' equity accounts are adjusted each year. Profits are split and added to the equity accounts. Amounts withdrawn called *draws*, are deducted from them. This provides a yearly running balance of each partner's investment balance in the company. These accounts should remain at about the same amount for each partner in an equal partnership. Some equal partnerships have only one working partner. It would be unfair to deplete his equity account by showing his draw, with no draw taken for the other partners. The working partner should be paid a salary so that all the equity accounts increase or decrease in fair proportion to the percentage of ownership.

Retained Earnings Corporations have a retained earnings account in addition to a capital stock account. Once the corporation is operating each year's profits and losses end up in this account. The balance of the account is the to-date accumulation of profits, losses, and non-deductible expenses.

Partnerships have a similar account into which all profits and losses flow. But it is closed out each year as the partnership accounts are updated.

Corporations sometimes give out portions of the retain earnings as dividends to their investors or stockholders. Partnerships are more flexible and can simply increase the amounts drawn by each partner. Draws are not taxed as are corporate dividends. Each partner pays tax only on his share of profits.

Members of partnerships often express surprise when they discover that their taxable income is higher than the amount they withdrew that year. This can happen in a very profitable year, and the reverse can occur in a period of poor performance.

Net Profit Each year's net profit or loss is recorded and placed into an earnings account when the books are closed at the end of the year. This account may be called a *retained earnings* account (corporations) or an *equity account* (partnerships). Profits are taxed in any corporation, but the amounts drawn or distributed in partnerships are never taxed. This account contains net profit funds placed there after draws are made.

The net worth of an organization increases or decreases directly with profits and losses. An organization that loses money every year will eventually deplete its net worth into a negative number. Owners or partners are then pumping their own money back into the business to keep it alive, or depending on outside financing.

When the net worth is a negative number the liabilities are larger than the assets and the company may be insolvent. Bankruptcy often occurs in such cases because the owners overdraw against their profits.

INCOME STATEMENT ACCOUNTS

The income statement contains the following account categories in the order they appear on the statement.

Gross Sales This includes all recognized income in the period covered by the statement. Recognized income is usually not the same as cash received. First, receivables represent unadjusted income. Second, the accounting method you use determines the amount of income you book. Under the completed contract method, no income is booked until the contract is complete. Income on such accounts is recorded as a deferral. Percentage of completion accounting recognizes income as the contract progresses, regardless of the billings or timing of payments. Billings are often prepared and mailed to reflect the correct percentage.

You can use a hybrid method to recognize income. Short-term contracts are accounted for by the completed contract method and long-term contracts are accounted for by the percentage of completion method. This hybrid method is the most realistic and practical to use. But whatever method you use to account for income, include on the income statement a supplementary schedule and note, explaining what the amount of gross sales represents.

Returns and Allowances This category of accounts decreases gross sales to net sales. Sometimes income is returned or allowances are made for special provisions. The amount of gross sales is lowered by the amount of returns or allowances. This account is less commonly used by builders than it is by other businesses like retail shops, where large amounts of purchases can be returned for cash. But returns occur in your yard from sales you make to the public. List the amount of these returns separately.

Direct Costs This account shows the total of all direct expended items, adjusted by the change in the inventory.

It is important to distinguish direct costs from operating expenses. Find the gross profit figure, a significant indicator on income statements, by subtracting direct costs from net sales. The percentage and the amount of gross profit indicate your profit and yields potential. Direct costs, because they relate directly to sales, vary directly with increases and decreases in volume. Make a yearly comparison of gross profits.

Direct costs are for materials, direct labor, payments to subcontractors, and other costs. The amount of direct costs also reflects the change in inventory level as shown in the following:

Direct Costs:	
Inventory, 1-1-19__	XXX
Add: Materials Purchases	XXX
Add: Direct Labor	XXX
Add: Other Direct Costs	XXX
Total	XXX
Less: Inventory, 12-31-19__	(XXX)
Direct Costs	XXX

Materials purchased and used to increase inventory decrease the total of direct costs. Materials used from inventory increase direct costs.

Operating Expenses Include in this account all indirect expenses. Operating expenses are sometimes split into selling expenses and fixed overhead for a more detailed breakdown. You can also break down operating expenses into production and administrative expenses.

Report operating expenses in one or two lines and explain these expenses on a detailed summary schedule. This keeps the income statement itself to a minimum of detail and allows the reader to scan it without examining the itemized listings of relatively small expenses. You can also show the expense totals for several large-balance accounts and lump the smaller expenses together in one All Other category.

Tax Provision This account category should include federal, state, and local taxes, even if the amounts have not yet been paid or deposited. Establish each tax account in the period of the tax liability so you know how much of an increase in your net worth to expect.

This provision allows you to show a profit before income taxes, and then list the tax provision by itself. In this way, the reader of the financial statement can view the profitability of your business without considering the tax consequences of operations.

List no tax liability for partnerships, which are only shells created for the convenience of each partner. Partners file and pay their own income taxes, and the partnership is not affected by the liability. But corporations pay income taxes regardless of the profitability of the investors. For tax purposes, a corporation is an individual with its own tax liability like anyone else. It is a form created for the participation of several investors.

Chapter 22
Using Financial Information

For a modest-size builder success depends upon his ability to control the finances of his operation himself. Well-prepared financial statements provide a basis for financial control that can't be had in any other form. Of course, common sense and a talent for building have no substitutes. But being able to apply financial statements to current conditions and managing your operation with long- and short-term goals in mind is just as vital to a growing, healthy business.

Many builders have an oversimplified concept of financial statements. They tend to take bits of information out of context and apply them randomly to the business and prejudge it. For example, a company's income statement might show a $20,000 profit but the builder has no cash. He may ask, "I made $20,000; where is it?"

The answer isn't always simple, of course. The builder might assume that his information is faulty. But he may simply not understand how to apply his data. Profit is not an indicator of cash flow control—it merely shows the yield on sales volume. Several factors could contribute to absorb the builder's $20,000 profit.

- An increase in the amount of outstanding accounts receivable

- A decrease in the amount of accounts payable

- Heavy commitments to payment of notes (not taken into consideration in profits)

- Purchase of new equipment

- Partnerships with draws above the amount of profit

A combination of some or all of the above would certainly affect cash flow. So the builder should to refer to his cash flow statement in addition to his income statement. No one statement by itself can tell the whole story. And no piece of financial data alone makes sense out of context.

APPLYING FINANCIAL INFORMATION

Any business depend on the ability of its officers to control finances by applying their analyses and raw data to their practical knowledge. You can't control your cash flow if you can't understand the statements or know how they relate to operations. Any successful builder understands that numbers by themselves don't tell the story unless he knows what those numbers mean to every phase of his business.

Controlling the purchase and disposal of fixed assets intelligently depends on careful long-term planning. Too often, builders make large equipment investments only when cash is available. Others buy equipment only if they can obtain financing from a bank. These aproaches makes it impossible for a builder to enter into contract commitments that depend on his supplying the heavy equipment. Bidding on such jobs would mean taking too many chances. As a result, the builder stays small.

Start your expansion plans by regularly analyzing a well-prepared, complete set of financial statements. This allows you to develop an overall financial plan whereby you place your operation in a position to afford the new equipment you need. Buying the equipment when it is needed becomes a goal that develops from a cash flow and profit concept you formulated months before.

Financial statements can also help you plan for debt repayments, for new financing, and for the uses of those loans. Covering accounts receivable, building an inventory, buying new assets, and getting through a seasonal slump with cash to commit in the busy season are reasons for controlling cash flow by analyzing financial information. Make decisions to obtain loans, cover accounts receivable or seasonal slumps, and increase inventories with a full understanding of the financial impact the moves will have on operations. Only with knowledge based on financial reports can you know you're making the right decisions.

Financial statements are invaluable in helping you control costs and expenses. You can't possibly know whether your efforts are successful unless you see the results each month on paper and compare them to previous statements. Monthly statements along with budgets and cost and expense forecasts are only common sense to any good builder interested in raising profits and volume.

Without this essential information you don't really know whether you are making adequate profits on your investments in equipment and inventories. Statements drawn up only once a year are too far apart to be useful in a constantly changing business climate. And you can't wait till next year to find out you could have raised your profits this year if you'd only had statements for a critical month.

SETTING STANDARDS

Previous chapters discussed the need for establishing financial standards. Standards are rudimentary budgets. You need standards in order to increase building volume, the type and quality of that volume and the expected yield from it. A standard may be expressed as a percentage, a ratio, an amount, or a level in relation to some other factor (such as bad debts to accounts receivable or expenses to direct labor.

Builders may base their standards on other operations, on an idea of what they think they can achieve, or on financial statements that have been analyzed practically and with the future in mind. Standards based on other operations are dangerous and untrustworthy. Many builders run operations similar to yours, but every businessman and every operation is different. The personality of the builder affects the conduct and philosophy of his operation, and every successful builder has found his own way of achieving that success. The mix of his labor force may be different than yours. The geographical location or the year or quarter his operation was established can also make a difference in standards. The location of office and shops, the degree of initial success of a new operation, the capitalization available and the kinds of work performed all influence the nature of a business and make each operation unique. Because there are so many variables, standards cannot be taken from one operation and applied to another without extensive modification.

Setting and enforcing standards is a continuous job. You can't merely set a standard once a year based on the financial statements alone and sit back to let it enforce itself. Modify your business assumptions as you obtain new information.

And keep your eye on your overall goal. This goal—it might be growth, expanded markets, new geographical areas of operation and the like—should be the point toward which the standards face.

Be sure to establish the degree of control needed to reach your goals. A well-organized operation has controls and standards that are interdependent and rest one upon the other like a pyramid. At the top is the overall goal of the builder. Below that are a few well-defined general standards. Still below those are a variety of controls, needed to bring about the standard.

There are four major areas in which standards must be set. These do not include minimum accounting standard and taxes and record-keeping duties required by law.

1) *Planning* A portion of your time should be devoted to the planning for the future. Some successful businessmen claim that more than half of their hours are spent in developing goals and future concepts. This is not always practical for modest sized builders who need to budget their time carefully. Heavy direct commitments, don't allow much time at all for planning. Yet the only way to implement your standards to reach your financial objective successfully is through planning—and that takes some time out of every day.

2) *Analysis* Analyze your results in order to define your goals. No business condition stays the same for long, and the builder who ignores well-prepared analysis documentation suffers by losing touch with the direction his operation is taking.

3) *Control* To reach goals and carry through on the conclusions drawn from intelligent analysis, become directly involved in control. No matter how enthusiastic key employees are, they can never approach company problems with the intimate concern you have for your own shop. That means operational and financial control is *your* job, and you must set control standards for the rest of your personnel.

4) *General matters* You spend some time each day deciding on many incidental planning matters such as which jobs to bid, how much insurance coverage to buy, when to start preparing for the heavy season and the like. Without well-organized schedules on how much time to spend in each decision area you can easily get bogged down in details.

Find time for these four areas every day. They may at first seem secondary in importance to bringing in profits, but in fact they are essential. In the middle of the pressures and deadlines it might not seem practical to spend all that time with the books, feet on the desk,

Using Financial Information

planning the future. But take an hour or two each day and follow up on planning matters. An hour a day is vastly more effective than a Sunday catch-up session once a month.

While complete accounting records and well-prepared financial statements are only one of many tools you use to set standards and acheive goals, they let you make intelligent planning decisions based on objective data. This is not the same as practical experience; but if you are totally involved in operations you make it very difficult to direct the company toward broad, long-term goals.

ENFORCING STANDARDS

The four areas of standards and controls—planning, analysis, control and general functions—need constant definition and enforcement: Standards only work when you following through on them and continually question their validity. Ask yourself the following questions when you evaluate which plans to enforce and which plans to scrap as out of date.

- Is the plan realistic? If not, it won't work. If it is, proceed with confidence.

- Do other business factors support the plan's basic idea? For example, is it realistic to expect a large increase in profits in light of the need for increases in volume? Will the market demand support such growth? Does a subsidiary plan exist to increase the labor force to meet this objective? Will funding be available?

- Have changes in the economy been considered in putting together the plan? Is the plan conservative enough to weather these changes?

- Will the plan bring in profits?

- Does the plan fit into the overall strategy of the operation?

- Is the plan expressed in terms of a goal? That is, is there an outside limit or deadline by which the plan is to be realized?

Coordinate analysis standards with planning standards. Analysis and paperwork must serve some needed purpose or it is a waste of effort. This can be one standard for any new paperwork you take on: does it contribute to higher profits, cut time, or have some practical application?

Analysis should "pull its own weight": it should produce profits directly or indirectly. Analysis without a useful end result is frustrating and wastes time, energy, and management. Together, financial analysis and control can keep business on steady keel. When your operation develops a problem, isolate the problem area and all other areas that are affected by it. Then analyze the problem area for all the possible solutions that reverse the trend or eliminate the problem. Choose the solution that has the fewest adverse side-effects to the other areas of the operation.

Some types of controls call for only occasional checking, while others need daily attention. These regular controls becomes time-consuming if they are not managed properly. Planning based on direct labor hours requires monitoring the field constantly to make sure that idle time doesn't eat up profits. A large part of this function can be assigned to superintendents or foremen; but try to work directly with the initial scheduling phases and with the contingencies for direct labor.

Results of past controls or of uncontrolled jobs can serve as guideline for realistic enforcement in the present and future.

Budgeting the time factor in your own work day is a good place to control enforcement. Many builders' time is totally consumed by administrative functions: arguing about discount and freight terms, buying insurance, hiring new employees, terminating or reassigning existing employees, consulting with estimators about the small details of bids, and fending off salesmen. Allow only so much time per day for these duties. Make the problems fit the schedule. This gives you time to enforce your overall standards and a chance to think about long-term goals.

USING FINANCIAL INFORMATION

Financial statements are too often merely numbers. They are often prepared only to comply with a procedure, to fill out a file, or to apply for a loan. These are only functions of paper flow, not useful in themselves. Financial data must be acted on to be effective; from it you should be able to direct sales efforts, control costs and expenses and plan profits.

Rather than issuing financial statements in preestablished formats, highlight exceptionally noticeable items on your reports. This way, recurring problems are not buried in a mass of numerical columns. Control depends upon flagging the exceptions and reversing the trends that those exceptions indicate. Highlight these problem areas by preparing the reports in standard formats but adding footnotes or extra worksheets to explain or analyze special situations. Then summarize the information on another sheet, omitting unnecessary details.

CONTROLLING CASH FLOW

A statement of cash flows is issued to report on the company's recent cash activity. Unfortunately, by the time the report is issued everything on it has already happened. The builder has documentation, but no control over the events documented.

Get around this problem by analyzing your cash flow as often as necessary. Look at your cash flow against a budget. Make up a cash flow report once a week or even once a day, and compare it to budgeted amounts for that period. On the cash flow report analyze and summarize receipts and payments, carrying the balance forward to the next day. This way, you can continuously monitor your actual cash flow using estimated figures.

Time large spendings, such as for new fixed assets, according to the cash flow budget. Know your needs well in advance of a purchase date and know where the funds will be generated to meet those needs. By the time a cash flow statement is issued the weekly or daily budget controls will have paid off.

The statement of cash flows points out in a summary form the areas of cash problems in the immediate past. It lets you modify cash flow budgets in the future and maintain preestablished performance standards.

Nothing motivates a builder to control receivables like not having enough cash on hand. But controls resulting from actual necessity are often too late to be fully effective. By the time a regular procedure is established to collect old receivables, for example, many of those accounts have become uncollectible. The proper way to control receivables is to start with a credit check prior to granting credit. Other examples of improper cash control include suddenly discovering that your inventory is much too high or contains many "dead" items or damaged and unusable parts. A costly cash flow problem results when you invest in an asset that is not really needed. Purchase fixed assets only after evaluating the need for it and the yield expected from the investment. Design a cash flow arrangement that allows the purchase of the asset with the least strain on your budget. Don't let cash availability problems become chronic. Several cash flow mistakes can compound the effect on your operation.

One of the most common losses due to bad cash management is that of discounts on purchases. Discounts you take from suppliers on a large inventory volume can represent a significant portion of profits. Too often, builders don't have the cash available by the 10th of the month to take advantage of the discount terms. They may be fully solvent, but due to timing problems the cash just isn't available. This can result in losses of thousands of profit dollars per year. Planning your cash flow against a budget, using cash flow reports to check current cash status, and analyzing your cash flow statement and comparing it to past statements can all help alleviate cash availability problems.

SHORT-TERM GOALS

Goals of one year or less are what budgeting is all about. You can apply a budget to any aspect of business to reach some short-term goals. Sales volume, costs and expenses, receivables, inventories, fixed assets and cash flow can all be controlled if a realistic budget is established and monitored.

Because the future depends on the present, planning for long-term goals means concentrating on current conditions. The dependability of a long-term goal decreases the further in the future you plan. Such long-term standards are best expressed as general ideals for the direction of the business.

On the other hand, short-term goals can be realized in the immediate future. They can be prepared realistically and then implemented and controlled in the business climate you know best—the present. A budget is the best way to maximize short-term results, since you can compare past vs. present performance for blocks of time that are short enough to be practical and long enough to let you see how your controls are working. Following through on short-term budgetary goals is more effective in the long run than trying to realize long-term goals or concepts with a single year's profits and cash flow.

Financial information can be most helpful in setting your short-term goals. But remember that these goals are only general guidelines for your own use — they're not set in concrete.

Any estimate is flawed, because it depends on the past. While historical information is useful to a degree, there are many factors that can make it less than dependable as a predictor for the future. But as flawed as it is, historical financial information is probably your most dependable source for developing short-term goals. You can use the past to establish immediate standards for the future.

When your goals don't work out, remind yourself that setting them was only one method for monitoring your progress and developing a method of control. Without the goal itself, you have no means for monitoring progress, for grading your own performance, or for identifying areas needing more work.

Chapter 23
Financial Ratios

This chapter explains the purposes of ratios in financial statement analysis, shows how ratios are applied, and defines the commonly used account ratios.

It is much easier to interpret financial information if a large amount of the work can be expressed in some kind of summary format. Because of their size, numbers tend to become obscured when several summaries are examined together. Ratios are easier to analyze because they show the relationships behind the numbers concisely and simply.

Ratios can be expressed in several ways, depending on the type and context of the analysis. Here are the three most common expressions of ratio relationships:

1) The two factors or numbers expressed as *x to 1*. The second factor is given a value of one and the first factor is compared to it.

2) The two factors expressed as a percentage, one to the other — *x/1*

3) The two factors expressed as one factor to another as a component of *1 — x per sales dollar*, or *x per day*.

These are clear and understandable ways of communicating financial and operational relationships. For example, the *current ratio* (current assets to current liabilities) is normally expressed by the first method — x to 1. Avoid thinking that "current assets are $266,000 and current liabilities are $133,000" by saying that *the current ratio is 2 to 1*.

Ratios let you boil down information into a usable form. Rather than merely present financial data, ratios interpret and draw attention to significant conditions and trends. Ratios can be used two ways:

1) To indicate a single business condition. Ratios are used alone to express a relationship between two factors at one point or period in time.

2) To indicate a trend. Two ratios from different periods or points in time are compared to discover the amount of change between the first one and the second.

When ratios stand alone as in 1) above, they serve as general indicators of conditions, strengths and weaknesses in your business at a given time. Ratios used this way must be compared to some standard outside your own company to have any worth to you. Lists of average expense ratios can be obtained from the Internal Revenue Service, and Dun and Bradstreet publishes the vital ratios of selected companies in all fields. The current ratio, for example, is used universally to measure a company's ability to meet current obligations from current assets. A 2 to 1 current ratio is the minimum current assets to current liabilities considered healthy for a business. A 2.5 to 1 current ratio would be better, as explained later.

Two ratios from different periods or times can be compared to one another to show how your business has changed over that time. A trend is established when a ratio changes, indicating a change in the business condition the ratio describes. You find the trend by figuring the ratio in one period and the ratio in the second period, and by comparing the two to get the amount of change. Done several times with ratios in successive periods, you plot a more reliable trend than if you calculate from only two ratios in different periods.

Here is an example of how a trend shows you more about business than a single ratio. "Charge sales to total sales are .95 to 1" is not a particularly significant statement by itself. It is more meaningful to say that "the ratio of charge sales to total sales has *increased* from 65% last year to 95% this year." It indicates a large increase in that portion of sales made under credit terms, and it prompts you to ask the following questions:

- Is this a true increase, or are the numbers somehow distorted?

- Is this a good trend or a bad one?

- How does this increase compare to the increase or decrease in volume of business?

- What change has occurred to affect cash flow in the average number of days that receivables are outstanding?

- How has this change affected the availability of cash?

- What is the trend in the ratio of bad debts to charge sales?

- Should this trend be reversed or decreased?

- What controls should be actuated on accounts receivables and charge sales?

- Are there many customers now participating in new charge sales, or are there few? Are they good credit risks?

These questions are all inspired by one ratio calculated at least twice, once last year and once this year, at the start and at the finish of the period under study. The answers to them provide valuable information on the various aspects of the operation affected by the change in account relationships.

Ratios give specific kinds of information that you can use to analyze your operation. There are four main reasons for doing ratio analyses:

1) To check existing conditions and relationships as indicators of business health, so that you can control and change them in the future.

2) To spot trends for the purpose of establishing standards and producing controlled yields within these standards.

3) To check on the effectiveness of existing controls.

4) To measure the levels of various accounts (such as Receivables, Inventories, Funded Debt and the like) in relation to other similar accounts.

Ratios can be calculated based on two balance sheet accounts, on one balance sheet account and one income account, and on two income accounts. In addition, ratios can be created which compare non-financial to financial information. The rest of this chapter gives examples of how to form these ratios, what they mean, and how to use them to better analyze and manage your business.

BALANCE SHEET RATIOS

Ratios that involve accounts found only on the balance sheet are called *balance sheet ratios* or *financial ratios*. The current ratio is that of current assets to current liabilities. Generally, a 2 to 1 current ratio is considered the satisfactory minimum. This means current assets are double current liabilities. Compute the current ratio as follows:

$$\frac{\text{Current assets}}{\text{Current liabilities}} = \text{Current ratio}$$

The current ratio is always expressed as x to 1. This ratio is favored by loan officers and credit departments as a good indicator of financial health. Since current assets are those you could convert to cash within one year, they represent the liquid assets of the company. Expect to pay any current liabilities within the same period. The current ratio gives a good idea of how able you are to meet current obligations with current assets.

The *quick assets ratio*, or the "acid test" as it is also called, is similar to the current ratio. It is also expressed as x to 1, and involves the same accounts as are used in computing the current ratio. A minimum of 1 to 1 is considered healthy for the quick assets ratio. Compute this ratio as follows.

$$\frac{\text{Current assets without inventories}}{\text{Current liabilities}} = \text{Quick assets ratio}$$

More conservative than the current ratio, the acid test measures the immediate ability to pay current debts. Since inventories are not necessarily available and may not be converted at all within one year, they are not considered as readily available.

The *capital to current liabilities ratio* is used to determine the owner's investment in a business versus the creditor's interests. Establish the relationship of debt to investment by comparing these two accounts.

$$\frac{\text{Capital}}{\text{Current liabilities}} = \text{Owner's investment versus creditors' interest}$$

The capital to current liabilities ratio is expressed as x to 1 or as a percentage. Review this ratio along with that of *capital to noncurrent assets,* or *funded debt* as below, to find the amount of fixed asset investments.

When the capital to current liabilities ratio is expressed as a percentage it indicates how much of the noncurrent assets have been supplied by investment and how much by outside funding. For example, a ratio of 100% means that the owner of the business has supplied all noncurrent assets. A ratio of more than 100% means that the owner is supplying a portion of current assets as well. A ratio of less than 100% means that the current assets and a portion of the long-term assets are supplied by creditors.

Working capital to funded debt indicates the ability of an operation to meet its obligations. It also indicates whether a business is heavily dependent on outside funding. It is expressed as:

$$\frac{\text{Working capital}}{\text{Funded debt}} = \text{\_\_\_\_}\%$$

Working capital is the net amount of current assets available; compute it as follows:

$$\text{Current assets } less \text{ current liabilities} = \text{Working capital}$$

The *specific current assets to total current assets ratio* shows you what percent of total assets the company's specific assets represent, such as inventory to total assets or cash to total assets.

$$\frac{\text{Specific current assets}}{\text{Total current assets}} = \text{Percentage}$$

Compare specific current assets to total current assets ratios over a period of time to come up with trends showing how your distribution of assets is changing. Figure 23-1 is an example of this ratio analysis for a five-year period. While accounts receivable have increased as a portion of total assets from 26.6% to 37.3% in five years, inventories have declined during the same period.

This type of analysis can be misleading, especially when it is reviewed by itself. Compare the figures with those of a similar analysis of sales in the sale period of time. While accounts receivable represent a larger share of total accounts classified as current assets, the actual amount in relation to total sales may have diminished. It is difficult to tell whether the trends are good or bad without a full knowledge of the rest of the operation's activities between the year this was drawn up and now.

The ratio of *net worth to assets* shows what share of assets is financed by owners and what share by creditors (represented by liabilities).

$$\frac{\text{Net worth}}{\text{Assets}} = \text{Owner financing versus debt financing}$$

A variation of this ratio is that of *net worth to current assets*. This shows what portion of working capital is financed by ownership, and what portion by credit. Use these ratios to analyze the relative investments in your assets and liabilities accounts.

COMBINED RATIOS

Ratios that compare accounts on the balance sheet to accounts on the income statement are called *combined ratios*. When you develop ratios for financial analysis, remember that the two accounts in the ratio should have a logical relationship to one another. For example, it

PETERSON CONSTRUCTION COMPANY
ANALYSIS OF CURRENT ASSETS

	Year -4	Year -3	Year -2	Year -1	Current Year
Cash	2.9%	2.6%	3.4%	1.9%	2.6%
Accounts Receivable	26.6	27.2	27.7	34.6	37.3
Bad Debts Reserve	(0.2)	(0.2)	(0.3)	(0.7)	(0.9)
Retainages	18.4	18.2	17.3	19.8	14.6
Inventory	52.3	52.2	52.0	44.4	46.4
Total	100.0%	100.0%	100.0%	100.0%	100.0%

Figure 23-1

makes no sense to compare Accounts Receivable to Payroll. Receivables and payrolls have nothing in common. At the same time, there are accounts which, although they are related, simply do not lend themselves to comparative analysis. For example, inventories are related to sales but should not be compared in a ratio because the two accounts are presented on a different accounting basis. Inventories are kept at cost while sales are designed for a margin of profit. When you work with combined ratios, make sure the balance sheet accounts and the profit and loss (income) accounts used are accounted for on a similar basis and bear a logical relationship to one another. Otherwise, the ratio you calculate will be misleading.

Use the *sales to accounts receivable ratio* to analyze a trend in your charge sales policy. This ratio is normally expressed as x to 1.

$$\frac{\text{Sales}}{\text{Accounts Receivable}} = \text{Trend in charge sales}$$

Don't conclude anything from this ratio unless the month-end activity for charge sales is consistent each month. Increased charges toward the end of a month will throw comparative ratios out of line if those month-end events belong properly to the following 30-day cycle. Examine this ratio over a year to determine a fair portion of charge sales to accounts receivable and whether more stringent controls over charge sales are needed. Compare average sales to average receivables for the period under study. This allows large seasonal distortions to be absorbed and may indicate a fairer trend, especially if major variations in sales and account receivable balances occur from month to month.

The *real turnover* or *physical turnover ratio* is one of the most helpful ratios a builder can calculate. It was mentioned previously that the inventory turnover ratio is not reliable as it compares net sales on a mark-up basis to inventory on a cost basis. The two are not comparable. But the real turnover ratio takes this into account:

$$\frac{\text{Cost of sales}}{\text{Inventory at cost}} = \text{Real turnover}$$

This is normally expressed as x number of times: "Real turnover of inventory occurred seven times last year." The value of this ratio is that it indicates the number of times the inventory is replaced (sold or repurchased). The preferable inventory number to use is the average inventory, but an ending inventory figure can give you a good indication of turnover.

A good application of turnover information is to divide the number of working days by the turnover to determine the days of inventory available to the builder:

$$\frac{\text{Total working days in one year}}{\text{Turnover}} = \frac{\text{Number of days of}}{\text{inventory available}}$$

Be aware, of course, that the number of days of inventory changes with the inventory level. Take this into account along with any seasonal changes in business volume. These ratios can help you control the amount of dead stock on hand and the problems of under or overstocking.

The *maintenance to fixed assets ratio* provides you with a guideline to control maintenance expenses in relation to your investment in fixed assets.

$$\frac{\text{Maintenance expense}}{\text{Fixed Assets}} = \frac{\text{Control of expense}}{\text{relative to investment}}$$

Remember that fixed assets are usually a builder's largest investment. As more equipment is acquired and existing equipment ages, a higher maintenance cost can be expected. Large increases in this account from one year to the next do not necessarily mean that the expense is uncontrolled. There should be a relationship between maintenance costs and:

- New equipment purchases
- Old equipment disposals (which should decrease year-to-year expenses)
- Old equipment aging

Each piece of equipment reaches a point in its useful life when maintenance cost exceeds the investment value for productive use of that equipment. If you develop good systems for analyzing profitability from equipment use you can recognize this point when it occurs and dispose of the equipment. One of the economic signs that this point has been reached is a narrowing margin between maintenance expenses and fixed assets as maintenance rises.

When the overall ratio begins changing, perform the test on each piece of equipment even though you may already know which equipment is causing problems. You can calculate the ratio by equipment piece if each piece has a complete cost file, as shown in Chapter 13.

The ratio of *equipment depreciation expense to equipment* is a good means of checking your depreciation policy.

$$\frac{\text{Depreciation expense}}{\text{Equipment}} = \frac{\text{Depreciation}}{\text{policy ratio}}$$

This ratio shows a rapidly decreasing percentage if you have few pieces of equipment and use accelerated depreciation methods. You should be able to tell from experience what the average useful life of equipment is and when to plan for replacements. Include in your replacement estimates a careful study of maintenance expenses by piece of equipment.

Use the *net income to net worth ratio* to check the overall efficiency of your operation.

$$\frac{\text{Net income}}{\text{Net worth}} = \text{True investment yield}$$

Unlike a comparison between income and sales, this ratio provides you with a *true* investment yield because it shows how the net assets of the operation have been applied to produce real profits.

INCOME ACCOUNT RATIOS

Income account ratios are used to analyze the relationships between gross profit and sales, expenses and sales, and net profit and sales.

The *gross profit to sales ratio* is expressed as a percentage. The ratio indicates the degree of control over direct costs.

$$\frac{\text{Gross profit}}{\text{Sales}} = \text{Percentage of gross profit}$$

Builders often find that this ratio begins to move in an unhealthy direction with significant increases in volume. This results from seasonal changes and from sudden market increases or decreases. Businessmen tend to relax their controls when money is not tight, and this ratio is a good control guide.

Changes in market types and in the cost to the builder of his materials and labor also affect the gross profit ratio. The ratio acts as a warning signal. If costs are increasing, adjust your pricing policies to reflect those increases. Otherwise your markup will not be the yield producer that it was before the past increase, when it was adequate to cover expenses and allow profits.

Calculate the *operating expenses to sales ratio* when sales volume changes significantly from month to month and year to year.

$$\frac{\text{Operating expenses}}{\text{Sales volume}} = \text{Expense control ratio}$$

Expense controls need adequate monitoring in an environment of change. Calculate the ratio regularly to spot expense changes, then identify the causes. Changes can indicate either an increase in management control efficiency, or a reaction to a change in volume. Study the rates for a period of time to establish a trend. Then make your expense expectation for a sales volume goal.

The *net income to sales ratio* shows the margin of profit. This is usually expressed as a percentage.

$$\frac{\text{Net income}}{\text{Sales}} = \text{Margin of profit}$$

The net income to sales ratio is best used when applied to financial statements broken down by line of business. Statements prepared this way minimize but do not eliminate the difficulties of assigning expenses to lines of business. Assigning expenses is always a estimate at best. If you are satisfied that your assignments are fair and reliable, analyze the different yields on the product and service mix. You may find, for instance, that twenty percent of total production time is spent on high-yield jobs and gives fifty percent of profits—and that the remaining eighty percent of production time is spent on lower-yield but necessary jobs.

The *usage test* compares depreciation to sales. This ratio shows how much depreciation your invested assets are undergoing for the amount of sales they produce.

$$\frac{\text{Depreciation}}{\text{Sales}} = \text{Usage test}$$

An investment in equipment and machinery should theoretically bring about sales. In reality, depreciation costs are "part of the package" when you buy equipment; these costs exist even if they generate no sales at all, unless a special depreciation rate such as "by use hours" is used.

This ratio seems illogical at first because there is no direct relationship between the two accounts. But the ratio can be very revealing. Depreciation is the spreading of equipment costs over the life of the asset. But during that life, the equipment will likely vary in its value. Some months, it will be used nearly 100% of the time generating income. At other times it may sit idle because little work is being performed. Every builder should be aware of the seasonal consequences of owning large equipment. Knowing how this rate relates to your own operation makes future asset purchasing decision better informed.

Increases in equipment use should increase this ratio, even though your investment in equipment is unchanged. Ratio increases should be accompanied by proper choices of depreciation methods. New equipment purchases affect the ratio, but they can be identified specifically as new depreciation is added.

PRESENTING RATIOS

Presenting ratios to others is as important as calculating the ratio itself. To have meaning, the ratio should be expressed so that its significance is best understood. There are four ways of presenting ratios:

1) As x to 1

2) As a percentage

3) As x per y (number of days, sales dollars, labor manhour, etc.)

4) As a fraction

Samples of three ratios are presented in Figure 23-2. The current ratio is expressed as 2 to 1, the best medium for exhibiting this particular relationship. The days of inventory available are expressed in the x to y format—the number of days, or 38 (x) days (y). The gross margin is expressed as a percentage. This commonly-used ratio is almost always presented in this format, demonstrating the percentage of margin before expenses.

Many books have been written on the subject

SAMPLES OF RATIOS

CURRENT ASSETS	266,450	
CURRENT LIABILITY	133,225	
Current Ratio		2 to 1
YEARLY WORK DAYS	240	
REAL TURNOVER	6.3	
Days of inventory available		38 days
GROSS PROFIT	254,814	
SALES	637,035	
Gross Margin		40%

Figure 23-2

of ratios and financial analysis. Generally, these are intended for use by bankers, loan officers, accountants and financial analysts. Remember that management conclusions drawn from detailed analysis are not often expressed in ratios. But ratios can be used to support a management conclusion.

COMPARATIVE RATIO ANALYSIS

No ratio can be calculated only once and be expected to remain valid for long. Businesses constantly change. The same ratio must be computed at regular intervals and the data compiled to show the trend of the activity reflected by the ratio. The greater the number of times the ratio is computed and compared with existing ratio calculations, the more valid the data becomes and the better able you are to draw conclusions based on the trend.

Analyze activities in your business using the most complete, comprehensive data available. Include all data that affects the activity in your study before you draw conclusions. For example, you can analyze sales trends in two main ways. You can trace sales totals over a period of time. But this way, you lack data that significantly changes those figures. Analyze sales along with net profits for a full year to discover the *quality* of your volume changes, not simply the dollar amount. This gives you the most profitable volume range for your business, since seasonal volume fluctuations would be included in the data and compared to seasonal net profits.

Here are some useful reports comparing related activities. Many activities also lend themselves to comparative ratio analysis.

- Average accounts receivable compared to average total sales

- Inventory and inventory turnover ratios compared year to year and season to season

- Progressive repair and maintenance expenses (identified by piece of equipment over its life) compared to total fixed assets

- Sales volume compared to net profits

Comparing your own ratios to those of another operation has some value. But remember that no single company is "average" in any industry-wide survey of ratio. Every business is unique, having its own set of underlying conditions affecting its ratios. Concentrate on your own operation. You know the conditions that make your ratios what they are. Establish acceptable ratio relationships for your own set of business circumstances. Set a personalized standard for the financial health of your company. Establish high and low ratio ranges, then check the effectiveness of your controls with monthly ratio analyses.

A LIST OF COMMON RATIOS

Here is a checklist of the ratios discussed in this chapter.

Balance Sheet Ratios:

Current assets / Current liabilities = Current ratio

Current assets without inventories / Current liabilities = Quick assets ratio (acid test)

Capital / Current liabilities = Owner's investment versus creditor's interests

Working capital / Funded debt = Ability to meet obligations

Current assets less current liabilities = working capital

Specific current assets / Total current assets = Percentage

Financial Ratios

Net worth / Assets = Owner financing versus debt financing

Combined Ratios:

Sales / Accounts receivable = Trend in charge sales

Cost of sales / Inventory at cost = Real turnover

Total working days per year / Turnover = Days of inventory available

Maintenance expense / Fixed assets = Control of expense relative to investment

Depreciation expense / Equipment = Depreciation policy ratio

Net income / Net worth = True investment yield

Income Account Ratios:

Gross profit / Sales = Percentage of gross profit

Operating expenses / Sales = Expense control ratio

Net income / Sales = Margin of profit

Depreciation / Sales = Usage test

Chapter 24
Putting Together A Statement

You should be able to delegate the task of preparing financial statements to an employee, a bookkeeper or an accountant. But in order to supervise the bookkeeper you need a working knowledge of statement procedures and should even be able to prepare the statements yourself. A good knowledge of your own statement procedures makes you more aware of the whole operation and better able to translate financial data into added profits by the time next year's statements come out. This chapter take you through the steps of preparing a statement, from account coding, posting accounts and test balancing, making the trial balance and closing adjustments, to drawing up the statement itself.

ACCOUNT CODING

The most common headache for bookkeepers and builders is coding accounts properly. The problem is one of consistency—a poorly defined and inconsistent chart of accounts is open to interpretation by each person who uses it.

For example, the purchase of business cards could be assigned to any one of the following accounts: office supplies, printing, sales promotion, or advertising. Each employee who handles coding may have his own idea about the proper assignment of this and the other accounts on your books. Within a year similar or identical expenses could end up in several different accounts unless one coding concept prevailed. Financial statements would thus contain inconsistent and misleading details because the general ledger had not been maintained properly.

You may have to go back through the accounts for several months and analyze the treatment of various expenses if you suspect that your coding has been inconsistent. Account coding mistakes are usually discovered when actual expenses are significantly higher than budgeted expenses and when accounts are obviously incorrect. One of the benefits of using a good expense budget is that whenever the actual results wander too far from the budgeted amounts you can perform an analysis to find out why. Once the inconsistency is discovered, adjust the account assignment accordingly and prevent future miscodings with a good chart of accounts. Keep handy a written definition of each account category and what types of items should be coded to it. You or your bookkeeper can then refer to the definitions whenever you code items. See Appendix A for a complete chart of accounts for the general ledger.

THE GENERAL LEDGER

The general ledger is nothing more than a summary of business. Each asset, liability, net worth, income, cost and expense account is listed here. Some types of accounts, such as expenses, can be combined into one large account. A sub-account can then be kept for each expense under the main heading as a ledger account control.

This saves space and posting time. The real detail is in the specialized journals—sales journal, check register, payroll, accounts receivable, payable records and the like.

Keep the general ledger as brief as possible, with as few divisions as necessary to prepare a good financial statement. The general ledger is not the place for massive control listings, and it should not be used to support in detail the figures on a financial statement. A brief general ledger is more easily controlled, and produces statements more easily, than a bulky ledger that is full of detail and hard to balance. Remember the following points when you make entries in the general ledger:

- Keep the general ledger in balance.

- Keep the general ledger in a summary format—don't let it become too detailed.

- Entrust the general ledger only to employees who are competent to work in it.

- Refer to the general ledger for spot-checking and quick reference.

- Develop a full understanding of the workings of the general ledger.

Leave a good *audit trail*. This means that whoever reviews your statement should be able to trace all your account entries back to the original document which supports the final entry itself. An audit trail might start with an entry on a financial statement referring to a specific account in the general ledger. The entry in the general ledger account should, in turn, refer back to an entry in the check register (for checks), the receipt journal (for income and cash receipts), the payroll journal (for payroll and payroll taxes), or the general journal (for all adjustments and special journals). The general ledger may refer to any of these entries by a code. For example, the check register may be called "CD" (for "cash disbursement"), and the check register page may be identified as "80-06" (year 1980, page 6), as explained later.

From any of these entries you should be able to trace back to a cancelled check, a duplicate bank deposit slip, or a full explanation of the entry (for journals). The check register, for instance, lists the check number and vendor. The paid bills file should contain in alphabetical order a copy of the check request, a voucher document, or a carbon copy of the check along with the paid bill itself. Income account entries are further supported in the accounts receivable system. Payroll entries are traced back to the payroll summary and time cards. Special journal entries can be traced back to well-documented worksheets.

The figures on the statement are thus verified all the way back to the original document. The audit trail is complete when you can do this for every account on the entire financial statement.

GENERAL LEDGER POSTING

The general ledger is broken down into the following main categories:

- Assets

- Liabilities

- Net Worth

- Income

- Direct Costs

- General Expenses

These main categories may be further broken down in sub-categories. For example, some general ledgers distinguish between selling expenses and direct overhead, or carry additional sections for Other Income and Other Expenses.

Pages in the general ledger can be marked with index tabs for more convenient access, as shown in Figure 24-1.

Figure 24-2 shows a typical general ledger account page. The standard procedure for posting the general ledger is to cross-reference each item with a check mark. Each ledger page

Figure 24-1

also includes a posting reference. As each entry is made from the source document to the general ledger (check register, income journal, etc.) a check mark (√) is placed in the source document. Thus you know that all entries from that document have been made to the ledger. At the same time the general ledger shows a reference back to the source document in a column labeled "posting reference." The following posting references are similar to those in common use.

- CR-7907 (Cash receipts journal, page seven of 1979)

- CD-8004 (Cash disbursement journal—or check register—page four of 1980)

- J-42 (General journal, page 42)

When you have double-checked the math in the general ledger, placing a check in the check mark (√) column verifies that, through a given point, the math accuracy has been proven and the general ledger is balanced.

In Figure 24-2, space has been provided on the page for the date, a description of the entry, and the activity (debit or credit) and balance of the account. Each page should be identified with the name, code, and page reference, which would include the page number within each account as well as the page number in the ledger. The date should include the year as shown below:

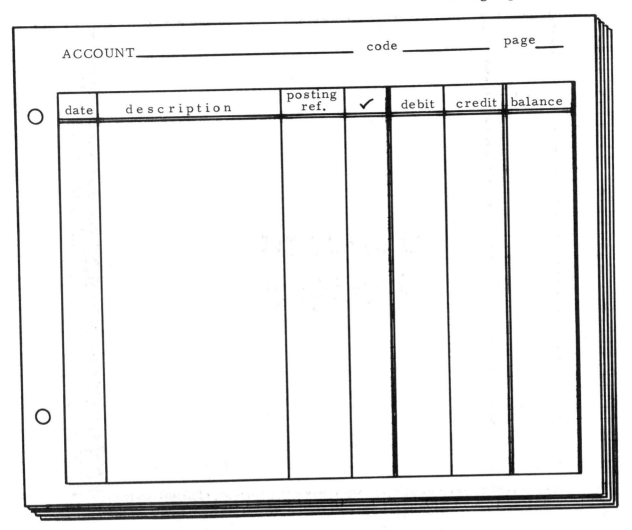

Ledger Page

Figure 24-2

The general ledger is a perpetual record. Pages are not replaced each year but are added instead, so identification by year is as important as by month.

The description column, often left blank, can be used for special notations and reminders pertinent to a specific account. It is a good idea in some accounts to identify the amounts in detail so that reversing or similar entries can be made in the correct account.

The balance of an account, plus the debits, less the credits, produces a new balance. In a credit-balance account, the negative balance forward will be decreased by (positive) debits and increased by (negative) credits. The double entry system was designed to assure that all entries are made correctly. If they are not, the general ledger will be out of balance. An exception to this, of course, is where an entry is correctly entered but in the wrong account. Only a careful procedure minimizes this problem, and you should be able to catch these mistakes while reviewing your trial balance.

Any time the general ledger is out of balance the error must be found. The first step is to check the source documents. These should be up to date and in balance. On the check register, the sum of all distributions (debits) is equal to the reduction of cash (a credit). On the receipts journal, the sum of all sales (credits) is equal to the accounts receivable total (a debit). All cash actually received (credits to accounts receivable) is equal to the increase (debit) in cash. By the same rule, journals must be equal in debits and credits.

Post the entries from the journals and source documents to the general ledger once you determine that they are all in balance. Many errors can occur in the posting procedure itself.

Here are some of the common posting errors:

- *Transposition* Two numericals within a number are reversed. For example, 2,146 is recorded as 2,416. In cases of transposition *the difference will always add up to nine.* In the example the difference this error creates is 270. Two plus seven is nine.

- *Decimal misplacement* A number is recorded with the decimal in the wrong place. For example, 2,146.00 is recorded as 21.46, which is difficult to find without examining every account.

- *Debit / credit error* A debit is entered in the credit column, or vice versa. The amount out of balance will be an *even* number, because this error doubles up the amount of the misplaced entry. To find such errors, divide the out-of-balance amount by two and search for an entry in the resulting amount.

- *Simple math error* An error is made 1) in adding the new balance in the general ledger, or 2) in adding one of the source documents. This can be discovered only through double-checking.

Keeping books in balance isn't always easy. Errors have a way of hiding so well that, occasionally, only a painful and time-consuming search will find them. But every accounting document can be balanced. Inexperienced bookkeepers often blame the system rather than their own lack of experience. If you find yourself with an employee who is having problems balancing the books, the best solution is patience and understanding. A motivated bookkeeper will learn quickly from his mistakes.

THE TRIAL BALANCE

The next step after posting the general ledger is to test the balance. Because all entries in the double-entry method consist of equal debits and credits, the *sum* of all accounts in the correctly-maintained general ledger will equal zero.

To test the balance, run a tape of the entire ledger. If the tape does not come out to zero, find and correct all the errors. A second tape should prove the balance by equalling zero. Then prepare a trial balance as explained in Chapter 21. The first two columns of Figure 24-3 are a trial balance from the general ledger, before adjustments. First, isolate the balance sheet accounts (assets, liabilities and net worth) on the top portion of the trial balance to get a year-to-date profit or loss total. All assets less all liabilities and net worth produce a difference. This difference should represent profit.

Secondly, isolate the income / cost / expense accounts in the lower portion of the trial balance. This gives the same difference as in the upper portion. Entries often consist of a debit to an asset account and a credit to an income account, or a debit to an expense account and a credit to an asset account. Thus, many entries are split with half going to the asset / liability / net worth side and half to the income side of the general ledger. The two sides will therefore not balance alone. But in a balanced ledger each side's difference will be the same.

ADJUSTING THE TRIAL BALANCE

Account balances must be adjusted when preparing financial statements. Record these adjustments only on a worksheet to produce statements during the year. But at the close of the year make the adjustment entries in the general ledger itself, as well. The second two columns in the worksheet in Figure 24-3 show the adjustments to the trial balance accounts in the first two columns. The following is an explanation of these adjustments.

- *Inventory adjustments* The year-end inventory, $1,600, is reversed out and replaced by the current inventory amount, $1,900. These adjustments are offset in the Materials Purchased cost account.

- *Depreciation* The Accumulated Depreciation fixed asset account is increased by a credit of $440.06. The offsetting debit is to the expense account, Depreciation.

Putting Together A Statement

MACDONALD CONSTRUCTION COMPANY
TRIAL BALANCE AND CLOSING ADJUSTMENTS
For the two month period ended February 28, 19___

	Trial Balance		Closing Adjustments		As Adjusted	
	Debit	Credit	Debit	Credit	Debit	Credit
Cash	1,460.18				1,460.18	
Accounts Receivable	12,780.00				12,780.00	
Reserve for bad debts		350.00				350.00
Inventory	1,600.00		1,900.00	1,600.00	1,900.00	
Fixed Assets	10,811.60				10,811.60	
Accumulated Depreciation		3,100.06		440.06		3,540.12
Accounts Payable		1,211.60	1,211.60	318.45		318.45
Payroll Taxes Payable		436.84				436.84
Current Note Payable		6,600.00				6,600.00
Long-term Note Payable		900.00				900.00
Net Worth		12,750.00				12,750.00
Total	26,651.78	25,348.50	3,111.60	2,358.51	26,951.78	24,895.41
Profit and Loss		1,303.28		753.09		2,056.37
Total		26,651.78		3,111.60		26,951.78
Income		10,418.40				10,418.40
Materials Purchased	5,216.00		318.45	1,211.60	4,022.85	
			1,600.00	1,900.00		
Direct Labor	1,400.00				1,400.00	
Office Salaries	800.00				800.00	
Payroll Taxes	204.00				204.00	
Depreciation	440.06		440.06		880.12	
Other Expenses	1,055.06				1,055.06	
Total	9,115.12	10,418.40	2,358.51	3,111.60	8,362.03	10,418.40
Profit and Loss	1,303.28		753.09		2,056.37	
Total	10,418.40		3,111.60		10,418.40	

Figure 24-3

- *Accounts Payable* The amount booked at year-end is reversed. The $1,211.60 is no longer truly payable, even though the amount is still listed in the general ledger. Actual booking of such entries is often made only a year, as was stated earlier. This reversal is replaced by the *current* Accounts Payable, $318.45. In this example, both past and current accounts payable are for materials, so the appropriate offsetting entries have been made to the Material Purchased account in the expense portion of the adjustments.

- *Other* No other adjustments are listed in this example. In a real situation, there may be numerous other adjusting entries.

Note that the debits and credits of the adjusting entries are equal, as they should be. Because the entries are split between the balance sheet accounts and the income accounts, the adjustment to increase income from adjustment entries is $753.09.

The final two columns represent the adjusted trial balance: in this example, the true current status of MacDonald Construction. Accurate financial statements can now be prepared from this adjusted trial balance.

Document all adjustment entries with notes to show why you have made the adjustments. For example, list the vendors and amounts purchased when making up the accounts payable adjustments. Specify what accounting method you used to arrive at the adjusted inventory total.

```
                MACDONALD CONSTRUCTION COMPANY
                         BALANCE SHEET
                       FEBRUARY 28, 19___

    Current Assets:
        Cash                                    1,460.18
        Accounts Receivable      12,780.00
        Less: Reserve for
            Bad Debts              (350.00)    12,430.00
        Inventory                                1,900.00
            Total Current Assets                            $15,790.18

    Fixed Assets:
        Equipment and Machinery              10,811.60
        Less: Accumulated Depreciation       (3,540.12)
            Net Fixed Assets                                  7,271.48

    Total Assets                                            $23,061.66


    Current Liabilities:
        Accounts Payable                        318.45
        Payroll Taxes Payable                   436.84
        Current Note Payable                  6,600.00
            Total Current Liabilities                       $ 7,355.29

    Long-Term Liability:
        Long-Term Note Payable                                  900.00

    Net Worth:
        Retained Earnings                    12,750.00
        Net Income year-to-date               2,056.37
            Total Net Worth                                  14,806.37

    Total Liabilities and Net Worth                         $23,061.66
```

Figure 24-4

BALANCE SHEET

Figure 24-4 shows the balance sheet for MacDonald Construction. The sheet has been prepared from the adjusted trial balance and shows balances as of February 28. All the pertinent information is available in a format designed for analysis. The current ratio is more than the 2 to 1 minimum standard, as can be seen at a glance from the Total Current Assets and the Total Current Liabilities columns. All other ratios prepared from balance sheet information can be calculated quickly and efficiently from the figures given.

The information presented in the balance sheet can be traced back through the trial balance to the unadjusted totals from the general ledger, to the original entry documents, and back still further to the duplicate deposits slips, cancelled checks, and paid invoices that prove the numbers.

Putting Together A Statement

```
              MACDONALD CONSTRUCTION COMPANY
                      INCOME STATEMENT
              For the two months ended February 28, 19___


    Income .........................................$10,418.40

    Cost of Goods Sold:
        Inventory  1-1 ................. 1,600.00
        Materials Purchased ............ 4,322.85
        Direct Labor ................... 1,400.00
            Total .................... 7,322.85
            Less: Inventory, 2-28 ...... 1,900.00
        Cost of Goods Sold......................   5,422.85

    Gross Profit.................................    4,995.55

    Operating Expenses:
        Office Salaries.................   800.00
        Payroll Taxes ..................   204.00
        Depreciation ...................   880.12
        Other Expenses .................. 1,055.06
            Total Expenses......................    2,939.18

    Net Income...................................  $ 2,056.37
```

Figure 24-5

INCOME STATEMENT

The income statement for MacDonald Construction is shown in Figure 24-5. Here again, all details are available for instant ratio analysis. Net profit is about 20% of gross sales, while gross profit is slightly less than 50% of gross sales. MacDonald should compare these figures to last month's and last year's to check the trend of his profits.

The change in inventories represents a $300.00 adjustment to the cost of goods sold. Both the beginning inventory (from year-end) and the ending inventory (from the current month) are shown in order to establish the method by which direct costs were computed. Showing both inventories is convenient because it allows MacDonald to see what his average inventory was during the period of the financial statement. Both beginning and ending balances are listed.

The flow of information from the original source documents through the financial statements is summarized in Figure 24-6.

CLOSING THE BOOKS

The books are *closed* at the end of each year. This means that the final adjustments are made to update all accounts and then recorded in the general ledger. Then all income, cost, and expense accounts are closed out. Do this by reversing the balance in each account to leave a zero balance. This must be at the close of each year to start the new year from scratch.

Since all entries must have offsetting entries, the net amount of profit or loss must be

Builder's Guide To Accounting

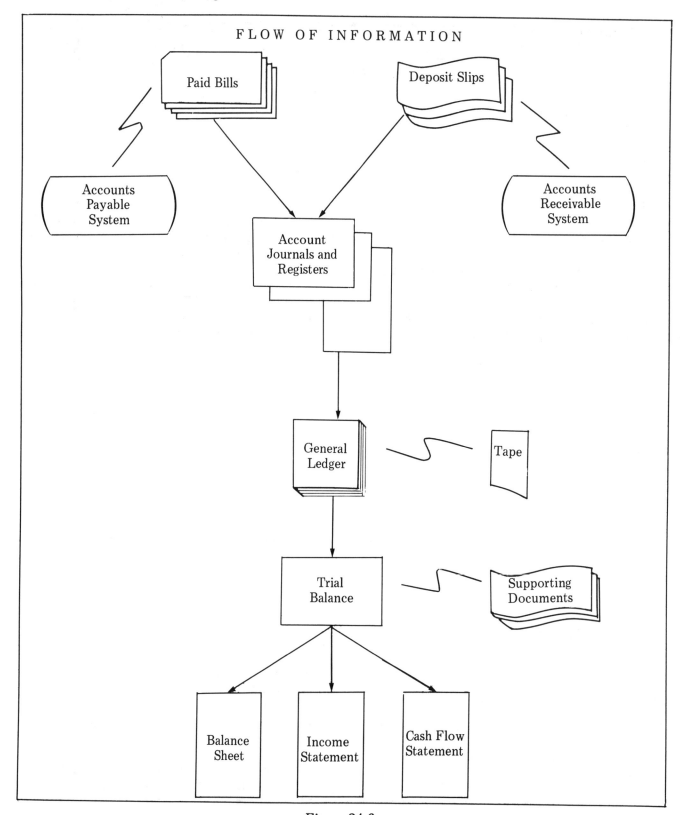

Figure 24-6

entered in the retained earnings account. Profit increases this account, and loss decreases it. On the balance sheet in Figure 24-4 notice that the Net Worth section consists of two parts: Retained Earnings and Net Income. The Net Income part represents a year-to-date total, while Retained Earnings is the *net sum* of all previous years' net profits and losses.

The following net worth accounts are commonly found in the general ledger itself.

- Capital stock accounts (corporations only)

- Partners' equity accounts (partnerships only)

- Partners' draw accounts (partnerships only)

- Retained earnings (corporations only)

- Profit and loss

Capital stock accounts change only when stockholders' investment balances change.

Partners' equity accounts, or owner's equity accounts for sole proprietorships, receive their share of all profits—otherwise called *retained earnings*—at the close of each year. They are also reduced by the amount of draw each year:

- Partners' Equity
  plus (Profit *less* Losses)
  less Draw
  equals:

- Adjusted Equity Accounts

A Retained earnings account, found exclusively on corporation books, is the sum of all prior years' profits and losses accumulated during the life of the corporation.

A Profit and Loss account is used to isolate the current year's results. This account is closed out at year-end into one of the other equity accounts described above.

Chapter 25

Comparative Period Statements

A ratio gains meaning when it is compared to others like it from past periods. Similarly, a financial statement becomes more valuable when you can compare it to those from past periods to spot overall trends. To do this, make up a *comparative period statement*, which lets you see progress in your financial condition from month to month or from year to year.

Comparative period statements are effective only when all statements are expressed on the same accounting basis. Coding procedures, account descriptions, statement formats, and valuation methods must be the same on each statement. If you have changed formats or accounting methods and want to include some statements in your comparison from both before and after the change, you must re-express on the current basis the statements prior to the change. Income statements, especially, must be figured the same way consistently from period to period. Hold to one accounting method: either percentage of completion or completed contract, or a consistent combination of the two. Otherwise, the income comparison will be meaningless.

COMPARATIVE BALANCE SHEETS

A comparative balance sheet lets you analyze on one sheet of paper the trend of your financial condition between the time one balance is taken and the time the next one is taken. Figure 25-1 shows a balance sheet comparing the two months following the close of one year and the two months following the close of the next. You might prepare this statement for a three-year period if significant growth is expected or has occurred recently.

MacDonald has drawn up his comparative balance sheet in February, giving him two month's operation after the first of the year to allow for full seasonal variations in his comparison. He has also made sure that the compared months are the same in each year; using different months invalidates the comparison.

The current ratio in the figure is better than 2 to 1 in both periods. MacDonald's financial relationship are basically well-controlled. Current debts are payable from current assets in both years. There has been a large increase in the amount of depreciation; this indicates that new equipment purchases are subject to larger-than-average depreciation in the early portion of the equipment's life. This affects income, of course, but *not* cash flow, because depreciation is not a cash exchange.

Comparative Period Statements

```
           MACDONALD CONSTRUCTION COMPANY
                    BALANCE SHEET
              FEBRUARY 28, 19__1 and 19__2
```

	2-28-__1	2-28-__2
Current Assets:		
Cash	$ 1,460.18	$ 2,260.00
Accounts Receivable	12,780.00	9,485.80
Reserve for Bad Debts	(350.00)	(125.00)
Inventory	1,900.00	1,200.00
Total Current Assets	$15,790.18	$12,820.80
Fixed Assets:		
Equipment and Machinery	$10,811.60	$ 7,811.60
Accumulated Depreciation	(3,540.12)	(223.65)
Net Fixed Assets	$ 7,271.48	$ 7,587.95
Total Assets	$23,061.66	$20,408.75
Current Liabilities:		
Accounts Payable	$ 318.45	$ 216.44
Payroll Taxes Payable	436.84	-0-
Current Note Payable	6,600.00	5,316.00
Total Current Liabilities	$ 7,355.29	$ 5,532.44
Long-Term Liability:		
Long-Term Note Payable	$ 900.00	$ 7,400.00
Net Worth:		
Retained Earnings	$12,750.00	$ 6,846.00
Net Income Year-to-date	2,056.37	630.31
Total Net Worth	$14,806.37	$ 7,476.31
Total Liabilities and Net Worth	$23,061.66	$20,408.75

Figure 25-1

A large long-term note, previously $7,400, is now only $900, raising the ratio of capital to liabilities and lowering the amount of outside funding. In the earlier period, assets were funded by nearly $13,000 in debts and only $7,000 in capital. In the later period, assets are funded only $8,000 by debts and $15,000 by capital. This shows that the owners have allowed their profits to remain in the business, or that additional capital has been fed into the company. Either way, the later statement appears to be much healthier.

MACDONALD CONSTRUCTION COMPANY
INCOME STATEMENT
For the two months ended February 28, 19___1 and 19___2

	2-28-_1	2-28-_2
Income	$10,418.40	$ 6,250.00
Cost of Goods Sold:		
Inventory 2-28	$ 1,600.00	$ 1,200.00
Materials Purchased	4,322.85	3,460.18
Direct Labor	1,400.00	600.00
Total	7,322.85	5,260.18
Less: Inventory 1-1	1,900.00	1,840.00
Cost of Goods Sold	$ 5,422.85	$ 3,420.18
Gross Profit	$ 4,995.55	$ 2,829.82
Operating Expenses:		
Office Salaries	$ 800.00	$ 300.00
Payroll Taxes	204.00	42.00
Depreciation	880.12	64.00
Other Expenses	1,055.06	1,793.51
Total Expenses	$ 2,939.18	$ 2,199.51
Net Income	$ 2,056.37	$ 630.31

Figure 25-2

Receivables make up the major increase in current assets. Inventories have not increased much in the second period and cash is lower. The owners should be able to establish that their receivables are collectible and that the reserve for bad debts is realistic. If the control policies in those areas are sound, the $3,000 increase in receivables does not seem unreasonable.

COMPARATIVE INCOME STATEMENTS

Macdonald Construction's comparative income statement in Figure 25-2 provides the same type and format of information as its annual income statement, but for two years. The information on the comparative statement was derived at the end of the same months for those years, as was done on the comparative balance statement.

Macdonald's volume of business increased tremendously during the year. Sales have increased by more than 65 percent. Yet gross profit has increased from 45 percent in the previous period to 48 percent in the current period. This indicates that controls have been enforced on direct costs, even with a large increase in volume. Net income has doubled from 10 percent to 20 percent.

General expenses have been summarized in Figure 25-2 in four categories—office salaries,

MACDONALD CONSTRUCTION COMPANY
COMPARATIVE PERCENTAGE INCOME STATEMENT
For the two months ended February 28, 19___1 and 19___2

	2-28-_1	2-28-_2
Income	100.0	100.0
Cost of Goods Sold:		
Inventory 1-1	15.4	19.2
Materials Purchased	41.5	55.4
Direct Labor	13.4	9.6
Total	70.3	84.2
Less: Inventory 2-28	18.2	29.5
Cost of Goods Sold	52.1	54.7
Gross Profit	47.9	45.3
Operating Expenses:		
Office Salaries	7.7	4.8
Payroll Taxes	2.0	0.7
Depreciation	8.4	1.0
Other Expenses	10.1	28.7
Total Expenses	28.2	35.2
Net Income	19.7	10.1

Figure 25-3

Payroll Taxes, Depreciation, and Other Expenses. This was done to keep comparative analysis to a minimum. You may wish to list every account in detail. But it is easy to get lost in a long column of numbers which lose much of their individual significance. By summarizing you can zero in on those accounts you want to keep a close eye on.

Builders often use an even more summarized format. The income statement is detailed using only the following categories: income, direct costs, gross profit, operating expenses, and net profit. These are the main sections of the income accounts in the general ledger. This format may be practical in some applications, but it does not allow statements to be used for detail analysis and control.

Because they are only calculated for two-month periods, comparative income statements are only indicators of future activity. Net profit to sales may change significantly beyond the statement period. But you can use other forecasting techniques and other ratios derived from the statement, such as the net to gross profit percentage, to help you estimate your income for the rest of the year. Compare last year's two-month net to gross profit figures with those from the annual statement at the end of

MACDONALD CONSTRUCTION COMPANY
MONTHLY INCOME STATEMENTS
For the year ended December 31, 19__1

	Jan	Feb	Mar	Apr	May	Jun	Jul	Aug	Sep	Oct	Nov	Dec	Total
Income	3318	2932	1766	3180	3366	3492	3440	6118	5766	4180	4764	5518	47840
Beginning Inventory	1760	1200	1375	1480	1492	1605	1866	1840	1756	1892	1910	1900	1900
Purchases	1682	1778	905	1060	1300	1210	1360	2418	2140	2004	1742	1810	19409
Direct Labor	300	300	300	300	300	450	450	800	800	800	700	700	6200
TOTAL	3742	3278	2580	2840	3092	3265	3676	5058	4696	4696	4352	4410	27509
Less: Ending Inventory	1840	1760	1200	1375	1480	1492	1605	1866	1840	1756	1892	1910	1840
COST OF SALES	1902	1518	1380	1465	1612	1773	2071	3192	2856	2940	2460	2500	25669
GROSS PROFIT	1416	1414	386	1715	1754	1719	1369	2926	2910	1240	2304	3018	22171
Salaries	150	150	150	150	150	150	150	400	400	400	400	400	3050
Payroll Taxes	21	21	21	21	21	21	21	83	83	83	32	32	460
Depreciation	32	32	32	32	32	440	440	440	440	440	440	440	3240
Other	963	831	342	425	480	510	543	622	601	549	562	515	6943
TOTAL EXPENSES	1166	1034	545	628	683	1121	1154	1545	1524	1472	1434	1387	13693
NET INCOME	250	380	(159)	1087	1071	598	215	1381	1386	(232)	870	1631	8478

Figure 25-4

last year. This shows how accurate the February income figures are at estimating for the rest of the year. You can then assume that this year's February figures have the same accuracy, give or take factors you expect to experience in the coming year which may affect the accuracy of the current February prediction.

Another comparative income analysis format is shown in Figure 25-3. Here, the dollar figures from the previous comparative income statement have been converted to percentages of gross income or sales. The most striking comparison in this statement is the change in net income from 10 percent to nearly 20 percent. But this is only a two-month statement. As mentioned earlier, do not place too much significance on these figures, except as they indicate a trend, until all outside factors have been considered.

Macdonald's comparative percentage income statement in Figure 25-3 shows clearly that Other Expenses should be broken out still further. This category comprises such a large portion of the total that the accounts making it up should be shown in detail.

Total Expenses decreased from 35.2 percent of sales to 28.2 percent. MacDonald seems to be controlling his sales expenses and overhead with increased volume. But without better detail of expenses he cannot know how well he is actually doing. He may be able to make further expense cuts, but only if he can locate the expense areas that need policing.

MONTHLY INCOME STATEMENTS

A complete twelve-month income history, as in Figure 25-4, helps plan future cash flow and expense budgets. This one-page summary can be prepared easily from monthly general ledger totals. Or summarize the data from monthly financial statements. Add each month's results to the worksheet as you gather them at the end of the month throughout the year, checking the results for that month with your expectations and past results. Trends in expenses can be seen at a glance, and the percentages of gross profit can be plotted by month through the year.

Comparative Period Statements

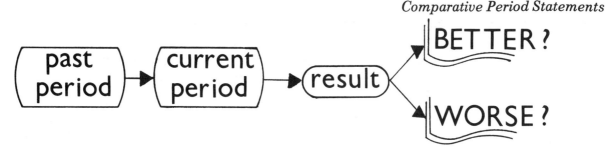

Figure 25-5

Note the seasonal changes in volume in MacDonald's monthly income statement. Fifty-five percent of the year's sales were accounted for in only the last five months. This same five-month period accounted for 59 percent of the year's profits.

Also note that the inventory figure is relatively stable throughout the year. This indicates that MacDonald maintains a stable level of materials on hand despite the rise in income.

USING COMPARATIVE DATA

Follow up on your comparative analyses by taking a course of action. First, compare actual results to the budgets established before the current period began. This lets you see how your original plan has turned out, as in Figure 25-5.

You should have established a high expectation and a low expectation. Do the results fall between these? Or are they below or above that range? Some type of control is needed if the plan was prepared realistically and the results still came in below budget. The comparative statement should show which current conditions are on a downward trend from last period. Then look at the statement along with the budget. This should show you where the actual results fell short of budgeted expectations. Decide what controls are needed to raise the down trend conditions.

If the results were better than the high expectation, you probably set your sights too low. Or, something unusual occurred in the current period that should not be counted on to repeat itself. Study the comparative statement carefully and define exactly what went right. How can this trend be repeated? Choose a future course of action that encourages the profitable trend.

OTHER COMPARATIVE APPLICATIONS

Comparative analysis has applications all through the builder's books and records. Comparative cash flow statements can be prepared to discover increases or decreases in cash efficiency. Or you can compare expense budgets from one year to the next to judge your estimating ability. And a comparison of contract estimates to total actual contract costs helps in bidding and controlling current and future jobs.

You are constantly making comparative studies without realizing it. For example, you compare similar contracts for expense, inventories, manpower and the like. You also compare your operation to other construction businesses. It may not be possible to compare your financial statements with another builder's. But you may regularly check the similarities in two operations to get a competitive edge. For instance, you may have more modern equipment, offices, and retail yard than another builder you know, and you are thus able to afford a higher volume of business due to greater capital potentials.

An important type of comparison within your own operation involves labor and payroll. You can quickly prepare a payroll cost accounting and by-the-job summary when each payroll is made up. Compare current payroll to (1) previous payrolls on the same project, (2) payrolls on other projects, and (3) progress billings and work in progress.

You also cannot help but compare the men making up your labor force to one another. Some men turn out to be more valuable, since they are more efficient, more dependable, and better able to produce a better product faster.

The comparative analyses you perform from the books and records are only a small part of the total comparative function involved in running a construction business. But these comparisons are the only firm basis you have on which to make your business judgments. Knowing how to compare factors—and knowing which of the many factors in your operation to choose from—is a daily practical skill you need to direct your operation.

COMPARISONS WITH BUDGETS

You may want to compare actual results to budget expectations. This type of comparative statement has three columns, each for a different set of figures:

1) Actual results

2) Budgeted amounts

3) Variance from budgeted amounts, over or under.

This kind of statement is often misused, since its purpose is often overlooked. The statement allows the builder to see in *summary* format how well he has budgeted and what kinds of controls are needed to bring accounts back into line. But too often, the statement is drawn up complete with pages of narrative explanation to document every variance from the budget. This is a time-consuming but thorough job requiring much research for each answer. Often the research is wasted. The detailed answers are satisfactory, yet no action is taken to control conditions or reverse trends. It's almost as though the variance is acceptable as long as everyone knows the reason for it. Use this type of comparative statement most effectively to stress *overall* changes from budgeted amounts. You only gain with this format if you look for the variances and bring them back into line, rather than take comfort in the good documentation.

Budgets are made to set a standard. Therefore, they should be used as tools of enforcement. If the standard is realistic, the budget is controllable—but only if general areas of variance are controlled and brought into line.

Some builders pay close attention to unfavorable variances (budget overruns) but ignore favorable variances, assuming that they are money ahead by overbudgeting. But be cautious of favorable variances. 1) If you are overbudgeting, your cash flow projections are too conservative and you won't be able to capitalize on available cash since there is no full plan for cash utilization. 2) Favorable variances can result merely from delayed payments. Make certain that you understand why the favorable variance occurs. If the favorable trend were to reverse in the future, would funds be available to pay the bills? Budget analysis includes being able to interpret data and make decisions with an awareness of the practical consequences.

Keep in mind that all comparative data is much more valuable than data on single-period statements. While two statements can be reviewed separately and some information gained from each of them, your most significant conclusions by far come from comparing trends and conditions.

A word of caution when preparing comparative statements: First make certain that your information is truly comparative. This means ensuring that you are looking at periods of the same length — 12-month years as compared to short years, for example. It may also mean making some adjustments if you've put some entries in different classifications than you did the previous year. Finally, take out any extraordinary items, such as write-offs of inventory, exceptionally high bad debts, or a heavy casualty loss.

Chapter 26

Restatements By Accounting Methods

You maintain the books using one of several accounting methods. Each method has its own advantages and disadvantages and presents data in a different way. Income, costs and expenses are treated similarly each method. Percentage of completion accounting requires that all booked items reflect the degree of job completion. This means that items may be either accrued or deferred to bring the books into line with job status. Completed contract accounting requires that all income, costs and expenses be kept out of the income statement until a job is finished. This method of accounting assumes that calculating any percentage of completion is at best an estimate. Because the two methods are so different from each other, their resulting income statements will be different. Percentage of completion accounting gives the builder a better idea of his true income status, since it provides current information. But keeping records for completed contract accounting is easier in many ways.

From time to time, the most practical way to keep the books is to use a combination of the two methods. Maintain percentage of completion records for long-term contracts and completed contract records for short-term contracts of less than thirty days.

From time to time, analyze your financial statements by examining results using both methods. This requires that you prepare a restated income statement. First, produce an income statement entirely by one of the two methods. This sets up one of the bases for reporting. Then make the adjustments to arrive at results by the second method.

Figure 26-1 is a restated income statement for Beacon Construction Company. Beacon uses the percentage of completon method and has restated to the completed contract method. The following steps must be taken when you restate in this way. Beginning with the percentage of completion column:

1) Remove uncompleted contracts and all related costs and expenses from the percentage of completion figure.

2) Add contracts completed in previous periods and previously reported as percentage of completion contracts. You are adding the

BEACON CONSTRUCTION COMPANY
RESTATED INCOME STATEMENT
For the four month period ended April 30, 19___

	Percentage of Completion	Less: Uncompleted	Plus: Contracts Completed	Completed Contract
Income	$ 286,400	$ 210,680	$ 184,010	$ 259,730
Cost of Goods Sold:				
Inventory 4-30	10,416	-0-	-0-	10,416
Purchases	132,680	124,441	109,616	117,855
Subcontractors	13,000	13,000	9,000	9,000
Direct Labor	44,500	37,650	27,600	34,450
Total	200,596	175,091	146,216	171,721
Less: Inventory 1-1	18,600	-0-	-0-	18,600
Total Costs	181,996	175,091	146,216	153,121
Gross Profit	104,404	35,589	37,794	106,609
Operating Expenses	77,380	12,402	15,980	80,958
Net Profit	27,024	23,187	21,814	25,651

Figure 26-1

portions of jobs that have been included in prior accounting periods.

The second adjustment above is difficult to arrive at in some circumstances, depending on how accurately you have kept your books. Determine which contracts are completed in the current period. Then find out how much income, costs, and expenses were booked in previous periods for those contracts.

Note that the following factors will distort the restatement somewhat: any interim adjustment to the treatment of accounts, major changes in receivable balances, or large changes in inventory. Point out these changes in the restatement presentation, or it will appear that one or the other method is not represented.

The restated income statement can not be used as a direct comparison document. Total costs and operating expenses are out of proportion to relative total income. But it is not known from the statement how many months or years are involved in recapturing old completed contract basis items. The company's controls may have improved significantly in the past year, or prices may have changed dramatically. But the statement contains no job schedule or any way of keying income to the calendar to allow for comparative analysis.

Restated income statements are most valuable as support documents when trying to obtain credit or demonstrate solvency to lenders. Being able to produce a statement that compares results by different accounting methods may show the lender your willingness to cooperate. Additionally, you can tell from a restated income statement how your tax results would differ under another accounting method.

CASH ACCOUNTING

Both accounting methods discussed above are accrual methods. This means that some items (such as accounts receivable, accounts payable, and accrued taxes) are included in balance sheets to show profit and loss even though cash has not exchanged hands.

The cash accounting method is also acceptable for income tax purposes. In this method, no funds are assigned to accounts and reported as

BEACON CONSTRUCTION COMPANY
COMPARATIVE ACCOUNTING BALANCE SHEET
April 30, 19____

	Accrual Basis	Cash Basis
Current Assets:		
Cash in Bank	1,201	1,201
Accounts Receivable	16,204	-0-
Retainages	1,600	-0-
Inventory	10,416	10,416
Total Current Assets	29,421	11,617
Fixed Assets:		
Furniture and Equipment	57,000	57,000
Less: Accumulated Depreciation	(2,945)	(2,945)
Total Fixed Assets	54,055	54,055
Other Assets:		
Prepaid Insurance	1,900	-0-
Accrued Contract Income	19,460	-0-
Total Other Assets	21,360	-0-
Total Assets	104,836	65,672
Current Liabilities:		
Accounts Payable	10,380	-0-
Taxes Payable	451	-0-
Current Notes Payable	3,600	3,600
Total Current Liabilities	14,431	3,600
Long-Term Liabilities:		
Long-Term Note Payable	7,200	7,200
Other Liabilities:		
Deferred Contract Income	21,630	-0-
Total Liabilities	43,261	10,800
Net Worth:		
Retained Earnings	34,551	34,551
Net Income, Current Year	27,024	20,321
Total Net Worth	61,575	54,872
Total Liabilities and Net Worth	104,836	65,672

Figure 26-2

BEACON CONSTRUCTION COMPANY
COMPARATIVE ACCOUNTING INCOME STATEMENT
For the four months ended April 30, 19___

	Accrual Basis	Cash Basis
Income	$ 286,400	$ 270,766
Cost of Goods Sold:		
Inventory 1-1	10,416	10,416
Purchases	132,680	122,300
Subcontractors	13,000	13,000
Direct Labor	44,500	44,500
Total	200,596	190,216
Less: Inventory 4-30	18,600	18,600
Total Costs	181,996	171,616
Gross Profit	104,404	99,150
Operating Expenses	77,380	78,829
Net Profit	27,024	20,321

Figure 26-3

profit or loss until cash changes hands. This is an easy way to keep books, because only cash exchanges are reported. In other words, all bank deposits are equal to income and all checks are equal to expenses, costs, and other payments (such as equipment purchases and note payments). Any funds kept out of the checking account would have to be included as income or expense as well.

The big disadvantage of the cash method is that it is far from accurate reporting. Accounts receivable are truly income accounts, and accounts payable are truly costs and expenses. Reporting on a cash basis in a charge society is not realistic. The builder interested in producing accurate reports for lenders and tax records, and having accurate information on hand for his own cost accounting and analysis, must use the accrual method.

We can compare cash basis accounting and accrual basis accounting on a balance sheet, as in Figure 26-2, to show the inadequacy of the cash system. The comparison is for a single month. The cash basis statement does not reflect receivables, so accounts receivable and retainage are removed from the cash side of the comparison. Income is reduced by the same amount.

Prepaid insurance is removed from the accrual statement as well, and the insurance expense is increased by that amount. In the cash accounting method, expenses are booked when the check is written, regardless of the period of coverage.

Accrued contract income, representing the portion of percentage of completion income not yet billed, cannot be reflected on a cash basis statement, so the $19,460 asset is removed. Income is reduced accordingly.

Accounts payable is removed totally, and material purchases are decreased by the same amount. Taxes payable are also moved out, and general expenses are decreased correspondingly.

Deferred contract income (representing either overbilled percentage of completion income or the total of uncompleted contracts for the completed contract income) must also be removed. This increases income by the same amount.

The cash-basis results shows that Beacon has no reliable information upon which to draw conclusions. None of the standard ratios produce true relationships between accounts. The cash basis statement is nearly useless to the builder who wants sound financial information to make informed business decisions. The resulting effect on Beacon's income by using cash basis accounting is shown in Figure 26-3. Income decreases nearly $16,000 dollars, while costs decrease only $10,000. Expenses increase more than a thousand dollars and show a much lower net profit.

Chapter 27

Statements By Job

Financial statements by job provide you with a logical basis for analyzing and controlling your profits and losses. Look at profits by job, not just in total. Otherwise, low-yield job types remain hidden and effective control over individual jobs is impossible.

A well-maintained, detailed record system is the foundation for accurate by-job statements. Begin your by-job breakdown with a detailed income statement, and remember the following two rules for breaking down income and expense data by job. See Figure 27-1.

1) All parts of the income statement should be broken down by job or category.

2) The sum of all jobs or categories must equal the total income statement.

All income-producing areas of the operation shown on a detailed income statement must be broken down into job categories. Below are typical categories into which each job you do can be accounted for:

1) Percentage of completion contracts

2) Contracts under the completed contract method finished in the current period

3) Completed short-term contracts (other than those in 2 above)

4) Short-term contracts in progress (other than those in 2 above)

5) Other lines of business, one category for each (retail sales, wholesale material sales, etc.)

Breaking out the income from different types of contracts lets you look at income and expenses more carefully. A statement including both percentage of completion and completed contract items could be misleading unless the distinction between them is pointed out.

Short-term contracts can sometimes differ from contracts subject to completed contract accounting. It may be that you wish to define different types of work in different ways. For example, you could consider all short-term contracts as a special category or as exclusively completed contracts. Some builders don't exclude short-term contracts from the general ledger until the end of the year because too much recordkeeping is involved. A special category for short-term contracts lets you divide up your contracts and cut down on bookkeeping.

Income statements broken down by project

Statements By Job

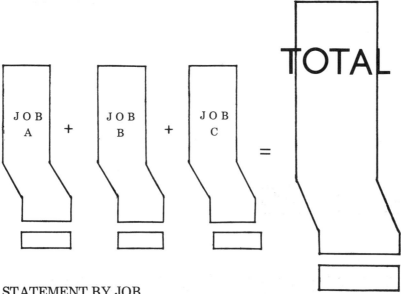

STATEMENT BY JOB

All parts of the income statement are broken down by job.

The sum of the jobs must equal the statement.

Figure 27-1

require complete by-job records. The job cost card discussed is an important feature of good accounting controls. Keep one for each major contract and one for each broad category of work as well. This way, all income, costs, and expenses are assigned to specific cost centers. The control document for the entire by-job system would then be the income statement itself. All cost items are posted to a job cost cards as follows:

- Income can be posted from the sales register.

- Direct costs can be posted from the check register.

- Expenses are posted from a special analysis done once each month.

- Payroll can be posted from the analysis done once each payroll period.

At the end of the month, each month's current job card postings should equal the current month's profit or loss (total income less total costs and expense).

PLOTTING JOB PROGRESS

Visual presentations of data can be more revealing than lists of numbers, as in Figure 27-2. Portraying upcoming jobs on a graph lets you plot your profits and compare them month by month with actual results from your monthly job summary. Job 426 is expected to last 17 months. Within that time, it is expected to yield about $27,800 in net profits. This profit is plotted to proceed along the curve in the figure, based on estimates and previous experience with similar work.

As large variances begin to occur between plotted and actual figures, the trend can be identified and controlled. No job will follow this curve exactly. But remember that most profits are produced in the middle part of the job. If the curve of actual profits starts upward *too early* in the job schedule, it indicates that you will:

- Finish early

- Make a higher overall job profit

- Need to reschedule labor and material for earlier dates

243

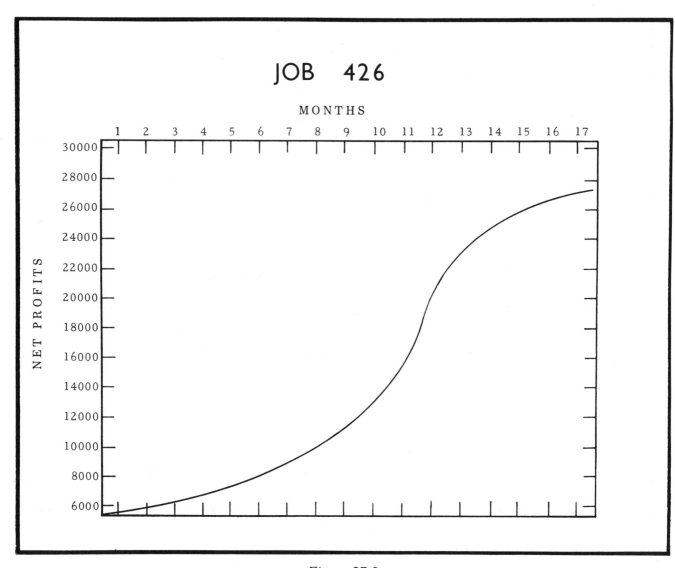

Figure 27-2

- Need to examine the estimate to determine why you were off.

But if the profit curve does *not* swing up *soon enough* or *fast enough*, it can mean that you will:

- Finish late

- Make a lower overall job profit

- Need to delay scheduled labor and material until later in the job

- Need to examine the estimate and the actual supervision of the job to determine why the job is off schedule.

A graph like the one in Figure 27-2 can be prepared for each major contract. Profits determined by cost center and then added together into one figure can be entered on the graph as the job progresses.

Figure 27-3 shows another type of statement by job expressed as a graph. This give you simultaneous profits data for any job you currently work on. From it, future cash flow can be estimated, because the rise and fall of profits on a job can be estimated compared to profits on other jobs.

But while it provides a visual summary of the year's profits, this graph is not necessarily the best way of presenting results outside the office to lenders, investors, and others. Develop your own graphs for in-house use, or show office employees the proper way to summarize profits

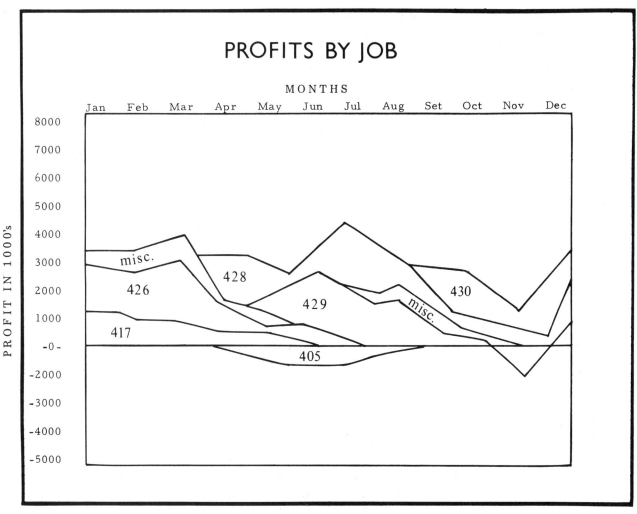

Figure 27-3

by job. You might want to see the volume of income broken down by job. Or you might wish to compare the volume to profits by job. The latter would provide the most meaningful visual data—a comparison of volume by degrees of success. After all, the percentage of yield on gross volume determines whether the estimate is a good one, and whether the expected return on investment will be realized. Most important, comparison between volume and profits indicates what lines of business or types of projects are profitable and which are not. You obviously want to avoid costly, time-consuming work that does not contribute to the overall profit standard. Without a basis for comparison, you can't tell how much or how little your jobs yield for the amount of work and expense you put in.

LONG RANGE JOB PLANNING

Part of a thorough by-job analysis of your operation's profitability includes deciding on which jobs to take and which to pass up. You think about this every time you bid on a job. But here is a list of factors that affect your ability to do various kinds of work. Refer to it along with your by-job profit analyses and your own best judgment to plan up-coming work.

- *Personality* This is the most important factor in any builder's success. How you approach problems and handle people contributes heavily to the kinds of work you can do successfully.

- *Capital* The degree of investment capacity usually dictates the limit of market involvement. Some kinds of work require heavy equipment investments, bonding qualification, and cash that many builders don't have.

- *Management ability* A well-organized and experienced builder knows his management limits and can work within them. Entering into new ventures depends on having a knack for knowing how much time, resources, manpower, and data you need and can handle well. Your ability to visualize all the organizational problems in taking on a new job can determine whether or not you bring that job in successfully, on time, and within budget.

- *Work force* The mix of specialists to general laborers, the types of specialists you have now and would have to hire for a new job, the size of the labor force, and the duration of average employment all affect the jobs you can do.

- *Competition* How do companies bidding on the same job compare to your own operation in capital, experience, and past success.

- *Experience with similar jobs* Use job records of past jobs and your own knowledge of your trade to assess your capability to do the new job. The less experience you have at the new task or job the longer the job schedule should be, within acceptable limits. But a longer than average schedule means higher than average costs and expenses. Know that you can finish the job or task using your usual economical methods of time, manpower and expense, or build some allowances for schedule and budget overruns into your plans for unfamiliar work.

Use statements by job throughout the control process, as illustrated by Figure 27-4. Plan your new projects using records from previous jobs. Use job status reports to help control current job progress, and compare reports from various jobs to control your operation's overall status. Use final profit and expense results to plan future work and improve your efficiency.

LEVELS OF CONTROL AND JOB PLANNING

Think of the control process as having three stages. The first stage is a foundation of solid, well-designed and efficient books and records. All the information you need is prepared and available in the shortest possible time. The second stage is made up of reports and worksheets related to the broad business goal—cost accounting reports, graphs, statements, and analyses. The third stage comprises the builder's broad business goals, which are

STATEMENTS AND THE CONTROL CYCLE

Figure 27-4

realized through the first and second level controls.

Statements by job are useful immediately in short-range planning. The builder uses them to control business on this level from day to day. Money must be available for weekly payrolls, rent payments, and material purchases. Equipment must be kept running. Controls have to be enforced: idle time must be kept to a minimum by careful daily scheduling, receivables must be kept collectible through vigorous collection procedures, the levels of inventory must be kept reasonable. Statements by job help indicate immediate needs in each area of the operation and help cut bad trends before they become drastic losses. The long-range goals of a building operation are made up of short-term realities. Each job the builder begins and ends contributes to or detracts from the overall success of the operation. Statements by job are the best indicators of how well you are controlling short-term realities. They help you discover what areas of business are not producing profits within your range of acceptability.

Faced with unacceptably low profits on some of your regular jobs, you can do one of the following:

- Institute new controls to increase productivity on those jobs

- Get out of those markets

- Accept low profits as inevitable in some

Statements By Job

THE LEVELS OF CONTROL

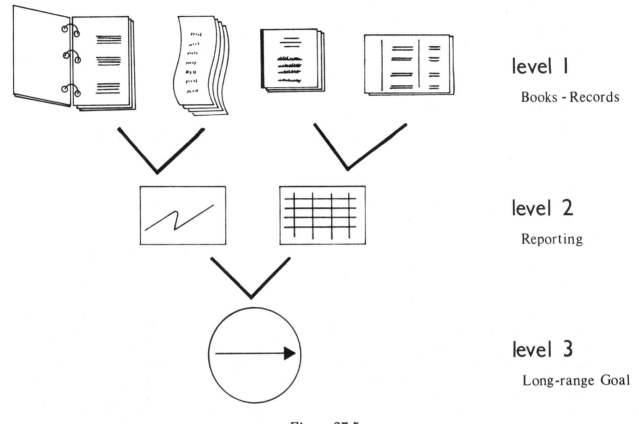

Figure 27-5

seasons and with some jobs and try to offset their effect by increasing productivity in other markets.

Monitor controls continuously that are designed to bring about increases in productivity. This requires timely, accurate statements by job. Take positive action to tighten up costs and expenses or find some way to cut the cost of business. You may have to increase the markup to offset higher costs for a particular type of project, thus possibly diminishing your market.

Eliminate a specific kind of job or task only if you understand the consequences of the action. Some portion of the overhead assigned to that task or job will exist whether you are in that market or not. Make sure you can actually increase yield by dropping a type of job. Otherwise, you may discover that overall profits fall because they have absorbed a portion of the fixed overhead previously included in high-volume, low-yield work.

The only way to retain a low-yield job type and still bring about overall profit goals is to increase productivity in other areas. Make this decision in view of all the consequences. A move to improve yields requires tight control of costs and expenses or you must increase income by markup. Total cash flow suffers when volume is reduced, so make sure that you will not be handicapped by a lower cash flow.

Short-range results add up to long-term results. Analyze short-range by-job data carefully to decide on events for next year and beyond. What you see this week in one of your statements by job may not show up except as relatively small and unnoticeable trend in your yearly and quarterly statements. But correcting the error in upcoming jobs and in other current work may significantly affect how soon you realize your overall goal.

Chapter 28
Statements For Loan Applications

Many builders don't understand loan procedures and how to produce statements for lenders. So financing is often impossible to obtain. This chapter shows how to prepare for obtaining credit, what documents you need when you go to a banker for a loan, and crucial lending procedures every builder should know. In addition, this chapter introduces the Small Business Administration and its services for builders. Loans are obtainable if your approach is correct and if you can document your need for the loan and your ability to repay it.

WHAT IS A LOAN?

You are buying money when you take out a loan. Your payments are broken down into two parts: 1) principal, or the money borrowed, and 2) interest, the cost of borrowing the money. Because different banks have different lending policies, the cost of money varies from bank to bank. This variation is not necessarily wide, but it can make a difference of hundreds or thousands of dollars. For example, the difference of one-half of one percent on a thirty-six month loan for three thousand dollars is about eighty dollars. For that reason it pays to shop around for a loan.

The availability of money affects the rate of interest. Money is more available at some times than at others. This availability is affected by the economy, inflation, unemployment, government programs, and the bank's investment markets. If the bank can invest its money in such a way that it will make 12 percent interest, it won't be as willing to loan the builder that same money for nine percent. But if the going rate for consumer loans is higher than any other readily available investment, more money will be available.

LOANS AS INVESTMENTS

A bank views loans as investments. It allows consumers (among them builders) to use its money for a certain length of time. The builder pays interest on that loan, producing a yield to the bank. Of course, some of the bank's loans are not collected and become losses. The amount of interest charged on loans takes these losses into account.

Builders, especially modest-size operations, have trouble obtaining loans because contractors fail in business at a higher rate than many other businesses. These failures are usually a combination of poor management and misuse of capital. For the builder to obtain a loan, he must do more than show the lender he is the best craftsman in town.

He must convince the bank that he is able to control costs and expenses, produce a profit, pay back the loan with minimum strain, and maintain a healthy business status with the proceeds from the loan. The loan will not be granted if the bank concludes that it won't help the builder. After all, the bank must have the highest confidence that its customers can keep up their monthly loan payments. Here are some of the personal and business characteristics you must be able to demonstrate to the bank if you want to be considered a good risk:

- Your business is healthy.

- There is no management problem, and funds are controlled constantly.

- The loan is needed.

- The loan payments will not cause undue financial strain.

- The loan itself will generate higher cash flow and increase profits.

All of the above make a good impression on banks and help show that you should get the loan you want. Specifically identify the purpose of the loan. "Working Capital" is often given as the purpose. But this doesn't really spell out what use the proceeds of the loan will be put to. More accurate descriptions might include "To purchase heavy equipment" or "To prepay materials for upcoming contracts." Make your explanation complete. Demonstrate how the loan will help you generate income. Below are two typical kinds of explanations:

- *For heavy equipment purchase:* "The company currently charges $17.50 per hour in estimates with old equipment. The current vehicle produces about 25 usable hours of work per week, as it is often out of commission for repairs. With the new equipment work can proceed at about twice the present speed, and the equipment will be usable for over 30 hours per week. This should enable us to reduce our hourly charge and compete for contracts at greater savings."

- *For prepaying contract materials:* "We will lose on an upcoming contract about eight percent of our material costs due to limited cash flow from lost discounts and quantity price breaks not currently affordable by the operation. A well-timed series of materials purchases would produce materials to be used over the next four months of construction. The interest on such a loan would equal about one-half of our savings from discounts and price reductions for quantity purchases. The loan can be repaid within ninety days from the installment payments on the same contracts."

Explanations should be in English. Use common-sense reasoning. Vague explanations do not describe the exact purpose of the loan or the way in which money is to be saved. Such applications are likely to be rejected.

Some loans are not needed, and the bank does the builder a favor by turning him down. The builder who would have trouble making loan payments doesn't need a loan. First he should tighten up controls over cash flow and improve his management techniques.

LOAN APPLICATIONS

Most banks have their own forms for loan applications. The two types of forms are *personal* and *business*. The builder's business statement must be completed, although the bank may additionally require that a personal statement be included.

Most bankers require all or most of the following information in the application package:

- The loan application form

- An income statement for one full year

- A current balance sheet

- A projection for cash flow for one year

- A federal income tax return for the previous year.

Explain as much of the financial statements as necessary to provide complete information. A typical explanation would include the following:

- Amount of current accounts receivable (an aging list would be helpful)

- The last physical count of inventory and the method used to value it

- Due dates on retainage

- The accounting method used for sales and costs/expenses, the anticipated profits on contracts in progress and the duration of those contracts, and the expected new business during the coming year

- Details on all accounts such as "Prepaid Assets," "Refundable Deposits," and "Organizational Costs"

- Details and aging of accounts and taxes payable

- Monthly payments and pay-off dates of all existing loans

- Form of organization (sole ownership, partnership, or corporation) and identity of owners or principal stockholders

- Growth in sales over the past three years, as well as gross and net profits in the same periods

- Depreciation methods used and the remaining useful life of equipment

- Explanations of any unusually large cost or expense items

- Comparative cash flow and net profits, with and without expected loan proceeds

- Equipment use and estimated hourly cost

- Discount and price break policies of major material suppliers

- Labor force breakdown—hourly rates by job classifications, numbers of foremen, journeymen, apprentices, and so on.

Any additional financial information is helpful. If the builder has had an audited statement in the past year, you may wish to include the accountant's opinion and financial statements even though they may not be the latest ones available.

There is no such thing as giving too much information when you apply for a loan. The banker who reviews your application can not possibly learn about a business in his short analysis, so give him as much valid information as is available.

PRESENTING DATA TO LENDERS

Imagine being a loan officer at the local bank. A builder approaches you with a loan application. The financial statements have been scribbled on a yellow pad, no explanations are attached, and the application is incomplete. Your first reaction, based entirely on the format of the presentation, is to reject his request for a loan. Obviously, little care was taken in preparing for the application. How can you judge this person as a good credit risk?

But assume that the application and the statements are complete, neatly typed, and supplemented with several pages of concise, readable notes. These notes are cross-referenced to the statements themselves. A cash flow projection and a short summary of sales and profits for the last three years are included.

You can now tell a lot about this man's operation. More important, he has taken some time, made your job easier, and has ended up with a concise summary of facts. You know from experience that he is likely to be as well-organized in all his affairs. You might conclude that he must have a well-organized shop and management to produce such well-defined material.

You must deal with individuals in banks, and final decisions are made not only on the material presented, but in the manner of the approach—and even on personal appearance. One banker may be totally objective and look at the documented facts only. Another banker may turn down a loan application because the builder didn't shave that morning. Banks insist that all loans are given a fair review. But they also admit that their loan officers use a good amount of judgment and weigh all factors in considering loans and loan applicants.

Most banks will gladly assist the builder in completing the application correctly. Each bank asks its own questions, and these are not always easily understood. Most applications require a financial statement to be completed on the form itself. The best approach to this is to fill in the form completely and attach a typed financial statement with notes. Very often, the format of a bank's loan application is not as informative as the builder's own statement, and there may not be space enough to include full notes. There is no room for a comprehensive cash flow statement on an application, or for any extra worksheets or historical data in the detail necessary.

Prepare a good list of credit references, complete with addresses and phone numbers, and a short description of your credit history with each person or firm.

CO-SIGNERS AND GUARANTORS

Banks are sometimes willing to grant loans on the condition that the applicant get a co-signer or a guarantor.

A *co-signer* is a person whose credit is well established and who can vouch for the builder's ability and reputation. The bank usually requires co-signers for young adults or minors and for people without a long business record or their own credit references. A *guarantor* is similar to a co-signer, except that a guarantor actually promises to pay back the loan if the builder does not. This person will be liable for the balance of the loan if the builder defaults, and the bank would expect payment from the guarantor in such cases.

The largest guarantor of loans in the United States is the Small Business Administration. The SBA was formed to encourage free enterprise, and it acts to help small businesses obtain financing.

SMALL BUSINESS ADMINISTRATION

The SBA itself cannot lend money until the builder has established that he is unable to get a loan from his local bank. In cities with populations of 200,000 or more the builder must be turned down by two banks before he can apply to the SBA.

The SBA considers a small business to be one that is independently owned and operated. Annual sales for retail and service organizations must be under $2 million to $8 million, depending on the industry. Other standards are sometimes used to define a small business; consult the SBA in your area. The credit requirements of the SBA include that an applicant:

1) Be of good character

2) Show that he can operate successfully in his business

3) Be well capitalized so that operations will be sound with the SBA loan

4) Prove the value of the loan requested

5) Prove his ability to repay the loan

6) Be able to put up enough of his own funds to capitalize the venture (if a new business).

The SBA does not lend money directly to businessmen, but guarantees loans from private banks. Where this is not feasible, the SBA itself advances funds as a last resort to those they consider qualified borrowers. Guarantees can be as high as 90% or $350,000, whichever is less, and in some cases the guarantee can be as high as half a million dollars. These loans may be for terms as long as ten years, or 20 years for new construction loans. Working capital loans are usually limited to seven years.

Interest on SBA loans ranges between six and eight percent. Collateral must consist of one or more of the following.

- Mortgage on land or equipment

- Assignment of receipts

- Guarantees or assignment of current receivables

The SBA does not grant loans if funds can be obtained elsewhere, so you must have been turned down by banks before applying to the SBA. SBA loans are not approved to pay off other loans, to pay business owners, or to pay the business back for funds used to pay up previous loans. Loans are not granted for speculative investments or to finance real property held for sale or investment (such as speculative building projects).

The Small Business Administration requires that loan applicants follow its procedure exactly:

1) Prepare a current balance sheet.

2) Prepare a one-year income statement and an income statement for the current year to date.

3) Prepare personal balance sheets for each

business owner or principal stockholder (those owning 20% or more of stock).

4) List collateral to be offered as security.

5) State the amount and exact purpose of the loan.

6) Take this material to the bank. Ask for a direct loan, and if refused, ask whether the bank will participate under the SBA's Loan Guaranty Plan. If the bank is interested, ask that they contact the SBA. Banks usually deal directly with the SBA in such cases.

7) If the bank does not wish to participate in the SBA's Loan Guaranty Plan, contact the nearest SBA office yourself for an appointment.

SBA field offices are located throughout the United States. These offices are listed below:

Small Business Administration Field Offices

Alabama
    Birmingham

Alaska
    Anchorage
    Fairbanks

Arizona
    Phoenix

Arkansas
    Little Rock

California
    Fresno
    Los Angeles
    Sacramento
    San Diego
    San Francisco

Colorado
    Denver

Connecticut
    Hartford

Delaware
    Wilmington

District of Columbia
    Washington

Florida
    Coral Gables
    Jacksonville
    Tampa
    West Palm Beach

Georgia
    Atlanta

Guam
    Agana

Hawaii
    Honolulu

Idaho
    Boise

Illinois
    Chicago
    Springfield

Indiana
    Indianapolis

Iowa
    Des Moines

Kansas
    Wichita

Kentucky
    Louisville

Louisiana
    New Orleans

Maine
    Augusta

Maryland
    Baltimore

Massachusetts
    Boston
    Holyoke

Michigan
    Detroit
    Marquette

Minnesota
    Minneapolis

Mississippi
    Biloxi
    Jackson

Missouri
    Kansas City
    St. Louis

Montana
    Helena

Nebraska
    Omaha

Nevada
    Las Vegas

New Hampshire
    Concord

New Jersey
    Newark

New Mexico
    Albuquerque

New York
    Albany
    Buffalo
    Elmira
    New York
    Rochester
    Syracuse

North Carolina
    Charlotte
    Greenville

North Dakota
    Fargo

Ohio
    Cincinnati
    Cleveland
    Columbus

Oklahoma
    Oklahoma City

Oregon
    Portland

Pennsylvania
    Harrisburg
    Philadelphia
    Pittsburgh
    Wilkes-Barre

Puerto Rico
    Hato Rey

Rhode Island
    Providence

Statements For Loan Applications

Small Business Administration Field Offices

South Carolina
   Columbia
	Texas
	   Corpus Christi
	   Dallas
	   El Paso
		Vermont
		   Montpelier
			West Virginia
			   Charleston
			   Clarksburg

South Dakota
   Rapid City
   Sioux Falls
	Houston
	   Lower Rio Grande Valley
	   Lubbock
	   Marshall
		Virginia
		   Richmond
			Wisconsin
			   Eau Claire
			   Madison
			   Milwaukee

Tennessee
   Knoxville
   Memphis
   Nashville
	San Antonio
	
	Utah
	   Salt Lake City
		Washington
		   Seattle
		   Spokane
			Wyoming
			   Casper

LOANS FOR NEW BUSINESSES

New businesses always have difficulty obtaining loans. They have no history to prove a track record of success in business. A large number of businesses fail within the first year, and banks are often not willing to take a chance. This doesn't mean it is impossible or not advised for new businesses to get loans. It can be a wise decision, but is harder to accomplish.

The SBA advice to new businessmen seeking loans is sound whether or not an SBA loan is requested. If you are a new business, the SBA require that you do the following before filling out a loan application.

1) Describe in detail the kind of business you operate.

2) Profile your business experience and capability.

3) Estimate the amount of your own investment and the amount of the loan to be requested.

4) Prepare a current personal financial statement.

5) Estimate the first year's earnings.

6) List any collateral and its approximate market value.

7) Take the material to the bank and request a loan.

The procedure from this point on for obtaining an SBA loan is the same as described previously. Application processing for SBA loans takes between three and eight weeks.

OTHER SBA SERVICES

The Small Business Association encourages free enterprise through loans and loan support, but the SBA also offers additional valuable benefits to the builder. Special programs have been developed for veterans, the handicapped, and the general businessman. Free counseling is provided by specialists. Builders can receive free advice on management problems, budgeting, business planning, and all other business matters from another builder working voluntarily with the SBA.

These services are provided through two SBA-associated groups, SCORE (Service Corps of Retired Executives) and ACE (Active Corps of Executives). These groups recognize that good management skills are the single most important factor in the successful operation of business. SCORE/ACE organizes talks with individual builders and various special management workshops of interest to builders.

The SBA publishes a 19-page booklet, "Business Plan for Small Construction Firms." This booklet summarizes many of the areas covered in detail by this book, and it is available free from any SBA field office; it is publication number 221. Publication 221 can serve as a guideline for general matters for the builder, and it may act as a basis for more specific SCOPE/ACE questions.

Publications 115A and 115B, also available from the SBA, list additional references available to the businessman. Builders who wish to study a specific area should obtain these, as they point toward the specific management problem you may have in mind.

You can obtain many services from the Small Business Administration in addition to loan counseling. The SBA is available to all businessmen to help with every conceivable management problem. Apply for literature or visit the nearest SBA office for more information.

THE PROSPECTUS FORMAT

Narrative explanations of your operation's history, markets, recent profits, and cash flow can be highly effective presentations to lenders. This type of factual, descriptive presentation is called a *prospectus format*. The prospectus should be brief and it should be written honestly and without overstating the facts. Don't ignore the bad to point out the good in your business. Bankers appreciate a totally honest presentation when they review a loan application. The sales promotional, public relations approach to presenting financial information is not appropriate for loan applications, as bankers are not sold by glossy, attractive, and slickly written reports. The prospectus is less formal than the columns of figures and pure accounting data that goes in the financial statements. Its purpose is to describe the operation and its functions through brief educational sketches and, in some cases, examples. This format is often used by large corporations when they present their annual statement to stockholders. These elaborate and often expensive projects include a letter from the board of directors, financial statements and notes, and a large amount of text and photographs about the products and the year's successes.

But the "annual statement" type of presentation is not appropriate for loan applications for small business. These are sales presentations that are carefully prepared to highlight good news and downplay bad news. Although issued by corporations, they are normally prepared by public relations firms who write the letter from the chairman of the board, all the product explanations, and the picture captions.

The modest prospectus format, on the other hand, is a simple three- or four-page summary of the company's markets and history. It is informative and meaningful to a banker. It quickly familiarizes him with the operation in a way that affects his lending decision. An expensive, hard-sell presentation may make him suspicious: why is the builder trying so hard? Forget about elaborate, impressive presentations for loan applications. The most effective presentation combines the best qualities of the accounting and the prospectus methods:

1) Complete set of financial statements—balance sheet, income statement, and cash flow statement.

2) Detailed notes explaining any special circumstances, accounting methods, inventory valuations, accounts receivable aging, sales and profits for specified periods, and the like. Make sure your notes are fully cross-referenced to the financial statements.

3) Narrative description of the company and its markets. Be brief, but include solid, straightforward information. Make the description honest and well-written. Mention both good and bad aspects of your operation's condition.

It is important to mention both the good and the bad. The accounting approach and the public relations approach to the same narrative description or note may be very different. The accounting approach tends to weigh negative conditions just as heavily as positive financial conditions, while the public relations approach is very often more optimistic. Find a reasonable explanation somewhere in between which reflects a businesslike approach to solving problems and changing negative trends.

For example, assume that sales for the past year fell far below previous years and below current expectations. The three approaches that could be used are given below.

- *The accounting approach:* Sales were 25 percent lower than the previous year. The entire drop in volume was in the bid contract market for new homes.

- *The public relations approach:* Sales were lower than in previous periods; however,

the estimates for next year are promising, as the company is in a better competitive position than ever before.
- *The honest compromise:* Sales were 25 percent lower than the previous year, due to less activity in the new homes market. This was caused by faulty estimating procedures. A more organized method is now being used.

Notice that in the honest compromise good and bad news are placed together to offset one another. The lender can thus make his judgments for himself with all the information he needs at hand. He isn't forced to hunt elsewhere for negative conditions or positive trends that offset the note under examination.

WHERE TO WRITE

Contact the Small Business Administration by telephone at 800-368-5855 (or 202-653-7561 in Washington D.C.). You may also want to call a local Small Business Development Center (SBDC) for counseling.

You can write to the Small Business Administration as well:

National Headquarters
1441 L Street, NW, Room 317
Washington, DC 20416   Tel: 202-653-6365

Region I
60 Batterymarch Street, 10th Floor
Boston, MA 02110   Tel: 617-223-3204

Region II
26 Federal Plaza, Room 29-118
New York, NY 10278   Tel: 212-264-7772

Region III
One Bala Cynwyd Plaza, West Lobby
Bala Cynwyd, PA 19004   Tel: 215-596-5901

Region IV
1375 Peachtree Street, NE
Atlanta, GA 30367   Tel: 404-347-4999

Region V
230 South Dearborn Street
Chicago, IL 60604   Tel: 312-353-0359

Region VI
8625 King George Drive, Building C
Dallas, TX 75235   Tel: 214-767-7643

Region VII
911 Walnut Street, 13th Floor
Kansas City, MO 64106   Tel: 816-374-3163

Region VIII
1405 Curtis Street, 22nd Floor
Denver, CO 80202   Tel: 303-844-5441

Region IX
450 Golden Gate Avenue
San Francisco, CA 94102   Tel: 415-556-7487

Region X
2615 Fourth Avenue, Room 440
Seattle, WA 98121   Tel: 206-442-5676

You may also request booklets, self-study courses, and pamphlets from the SBA:

Booklets
15 Handbook of Small Business Finances
Stock No. 045-000-00208-0 $4.50

25 Guidelines for Profit Planning
Stock No. 045-000-00137-7 $4.50

44 Financial Management: How to Make a Go of Your Business
Stock No. 045-000-00233-1 $2.50

103 Small Business Incubator Handbook: A Guide for Start-Up and Management
Stock No. 045-000-00237-3 $8.50

Self-study series
1001 The Profit Plan
Stock No. 045-000-00192-0 $4.50

1002 Capital Planning
Stock No. 045-000-00193-8 $4.50

1003 Understanding Money Sources
Stock No. 045-000-00194-6 $4.75

1004 Evaluating Money Sources
Stock No. 045-000-00174-1 $5.00

Pamphlets (free)
MA 1.001 The ABC's of Borrowing
MA 1.004 Basic Budgets for Profit Planning
MA 1.016 Sound Cash Management and Borrowing
MA 2.008 Business Plans for Small Construction Firms

Appendix

The Chart Of Accounts
Johnson Construction Company–Complete Financial Statements
Expanding The Accounting System
Automating Your Accounting
Income Tax Planning
Blank Forms

Appendix A
The Chart Of Accounts

A good chart of accounts is easy to work with, forces a logically ordered general ledger, and allows enough room for flexibility. Almost every business needs to add additional accounts from time to time. The well-designed chart of accounts makes it easy to add additional accounts when they are needed.

The descriptions of many accounts are long and space-consuming so it is convenient to assign a numerical code to each account. This can be a three-digit code or higher. Each number in the code should have some meaning, so that the logic of the chart is consistent throughout the general ledger.

Your operation may put its books on a small computer system some time in the future. Therefore, set up your chart of accounts so that an existing manual system can be fed into the computer's memory with as little conversion or programming time as possible. Yet the chart and its numbering system should still be convenient enough for manual referencing. This appendix can be used as a guide to creating your own chart of accounts that is adequate for both manual and computer coding.

The first step toward designing a chart of accounts is to define the broad classifications of the general ledger.

1. Assets
2. Liabilities
3. Net Worth
4. Income
5. Direct Costs
6. Selling Expenses
7. Fixed Overhead
8. Other Income and Expense
9. Tax Provisions

Other divisions are possible. The expense categories could be lumped together, or other income and other expenses could be separated. But assume that the designated classifications are as listed. They are numbered from one to nine. These numbers are the first digits of the accounts in each category. This level of identity, the *classification* of accounts into categories, is called *level one*. All other levels, up to level five, add numbers to this single digit and further differentiate accounts.

Level two is called the *sub-classification* of accounts. There are two digits in the sub-classification. The additional digit greatly increases the possible number of accounts that can be classified in each first level category. The second level accounts listed below have been spaced within the range of each first level account to allow for more second level accounts to be added later. For example, in category 1, Assets, four accounts in level two have been listed, but an additional five level two accounts can be coded between those numbers: 11, 13, 15, 17, and 19. Each new level expands the number of possible accounts that can be coded to the next lowest level.

- 12 Current Assets
- 14 Long-term Assets
- 16 Other Assets
- 18 Intangible Assets
- 22 Current Liabilities
- 24 Long-term Liabilities
- 32 Net Worth Accounts (for corporations)
- 34 Net Worth Accounts (for partnerships)
- 36 Net Worth Accounts (for sole ownerships)
- 42 Income Accounts
- 48 Returns and Allowances
- 52 Direct Cost Accounts
- 62 Selling Expenses
- 72 Fixed Overhead Expenses
- 82 Other Income Accounts
- 86 Other Expense Accounts
- 92 Income Tax Provision Accounts

All asset accounts begin with the asset classification number 1. All liabilities begin with 2, and so forth. Notice that the sub-classifications are also interrelated. For example, current assets are sub-classified as "12" and current liabilities are sub-classified as "22."

Level three of the account code consists of three digits, the two above plus one more. This is further extension of the classification, and allows for many more accounts to be coded to each original single-digit category and double-digit sub-classification. Below is a complete level three chart of accounts. Note that there is enough room for additional accounts throughout.

- 121 Petty Cash Funds
- 122 Cash in Banks
- 123 Cash in Escrow Accounts
- 124 Notes Receivable
- 125 Accounts Receivable
- 126 Reserve for Bad Debts
- 127 Retainage
- 128 Inventories

- 141 Land
- 142 Building
- 143 Furniture and Fixtures
- 144 Autos and Trucks
- 145 Equipment
- 146 Machinery
- 147 Small Tools
- 148 Improvements

- 152 Accumulated Depreciation - Building
- 153 Accumulated Depreciation - Furniture and Fixtures
- 154 Accumulated Depreciation - Autos and Trucks
- 155 Accumulated Depreciation - Equipment
- 156 Accumulated Depreciation - Machinery

The Chart Of Accounts

157 Accumulated Depreciation - Small Tools

158 Accumulated Amortization - Improvements

161 Prepaid Assets

162 Deferred Assets

163 Deposits

164 Organizational Cost (net of amortization)

165 Suspense Account

166 Accrued Construction-in-Progress Debits (income)

167 Deferred Construction-in-Progress Debits (expenses)

181 Covenants not to Compete

182 Goodwill

221 Accounts Payable

222 Payroll Taxes Payable

223 Sales Taxes Payable

224 Income Taxes Payable

225 Other Taxes Payable

226 Accrued Expenses Payable

227 Notes Payable - Current Portion

241 Notes Payable - Long-term Portion

261 Deferred Construction-in-progress credits

265 Suspense Account

321 Capital Stock

322 Retained Earnings

328 Profit and Loss - Corporations

342 Owners' Equity Accounts

343 Owner's Drawing Accounts

348 Profit and Loss - Partnerships

362 Owner's Equity Account

363 Owner's Drawing Account

368 Profit and Loss - Sole Ownerships

421 Sales - Percentage-of-Completion Contracts

423 Sales - Completed Contracts

425 Short-term Contracts

428 Adjustments to Gross Income for Sales Tax Included (clearing account)

481 Sales Returned - Percentage-of-Completion

483 Sales Returned - Completed Contracts

485 Sales Returned - Short-term Contracts

487 Discounts Allowed on Sales

521 Materials Purchased

523 Direct Labor

525 Subcontractors

527 Freight and Delivery

621 Payroll Taxes on Direct Labor

622 Travel

623 Entertainment

624 Auto and Truck - Gas & Oil

625 Auto and Truck - Repairs and Maintenance

626 Depreciation - Autos and Trucks

627 Depreciation - Equipment

628 Depreciation - Machinery

629 Depreciation - Small Tools

630 Other Repairs and Maintenance

631 Advertising

699 Selling Expenses - Cost Accounting Control

721 Salaries and Wages

722 Payroll Taxes

723 Rent

724 Utilities

725 Telephone

726 Insurance

727 Office Supplies

728 Bonds, Licenses and Fees

731 Property Taxes

732 Other Taxes

735 Printing

736 Postage and Delivery

737 Legal and Accounting

738 Outside Services

741 Dues and Subscriptions

742 Donations and Contributions

745 Union Welfare

751 Depreciation - Building

752 Depreciation - Furniture and Fixtures

756 Amortization - Improvements

757 Amortization - Organizational Costs

761 Building Maintenance

766 Interst Expense

768 Bad Debts

771 Miscellaneous

799 Administrative Expenses - Cost Accounting Control

821 Interest Income

822 Capital Gains / Losses

823 Miscellaneous Other Income

861 Miscellaneous Other Expense

899 Other Income / Other Expense - Cost Accounting Control

921 Federal Income Tax Provision

925 State Income Tax Provision

928 Local Income Tax Provision

999 Tax Provision - Cost Accounting Control

Level three accounts are those most often used for posting the general ledger. But occasionally, additional account information is required in your books and analyses. This is best done by a further breakdown of the accounts on the chart into *level four* sub-accounts. Some level four accounts are given below.

122.01 Cash in Bank - General Account

122.02 Cash in Bank - Payroll Account

122.03 Cash in Bank - Tax Account

128.01 Inventory - Raw Materials

128.02 Inventory - Finished Goods

222.01 Payroll Taxes: State Disability accrued

222.02 Payroll Taxes: F.I.C.A. accrued

222.03 Payroll Taxes: Federal withholding

The Chart Of Accounts

222.04 Payroll Taxes: State withholding

222.05 Payroll Taxes: F.U.T.A. accrued

Level five is the cost accounting control level. Here, the specific cost centers (projects, markets, etc.) are coded and assigned exact costs and expenses in addition to income. This level five code may correspond to a job number or general category, as below:

100 through 599	Project numbers
600	Retail sales
700	Home repairs and maintenance
800	Various commercial projects
900	All other income / cost centers

You may choose not to use level five accounts. The complete account numbering method is illustrated in Figure A-1.

Classification	X	level one
Sub-Classification	X X	level two
Account	X X X	level three
Sub-Account	X X X . X X	level four
Cost Account	X X X . X X . X X X	level five

The Complete Account Numbering Method
Figure A-1

Some controls can be built into the chart of accounts so that both manual handling and any eventual automation are equally efficient. Level five should only be used on income, cost and expense accounts. One rule might be, "Accounts beginning with the digits '4' and '5' must include a cost account. No cost account is to be used for any other accounts." Another rule helps you assign expenses, other items, and tax provisions for final cost accounting break-out: "Accounts with '99' as the second and third digits are clearing accounts, to be used only for assigning costs." Thus, the accounts 699, 799, 899, and 999 would comprise the total control for your cost estimate accounting system. Income and direct costs do not need this feature, of course, as they can be coded to specific jobs throughout the posting and recording process.

Computerized accounting systems can utilize coding rules for programming logic. These rules give good coding control. Some of the typical rules for computerized systems are listed below:

1) No account is accepted unless it is already on the computer's file. This prevents many coding errors.

2) All accounts must be the same numeric length within each level. This helps make the posting process uniform. Accounts which have no sub-accounts should use "00." An exception to this rule is in (3) below.

3) Income and direct cost accounts (as identified by the first digit, or level one) must always be included with a cost accounting code (level 5).

4) In computerized posting, all checks in a numerical sequence must be posted on the same date. The batch total is thus accurate for each day. Most computerized systems edit each day's input and allow you to check the work before it is fully integrated with the stored general ledger data.

5) The data provided to the computer must always be in balance.

6) All data needed for each program (such as payroll, check-writing, or recording income) must be completed on a uniform input form. This assures that the data is entered in a dependable format which can catch errors and avoid inputting delays.

7) Before integrating a daily batch of data, either the computer must be instructed that the information is correct or corrections must be entered and verified.

8) Users of the computer and the general ledger program must gain access to the computer through a password or a numeric series. This assures that only those employees trusted with the password can work on the general ledger—minimizing the chances for embezzlement or fraud. This security feature can be built in to most business computer systems, and is a necessary part of a good computerized general ledger system.

An accounting document can be reviewed and errors in coding spotted more readily with a good chart of accounts. Say you see a common expense coded to a 400-series account. You can quickly ask yourself why that expense has been coded to income (in the 400 series) rather than to its proper expense account.

You can't use any other builder's chart of accounts without modification, and even the one included here may require some work to adapt it to your needs. But the chart of accounts format here outlined can systematically accomodate the large number of new accounts you need for any future expansion, and the numeric system can be easily programmed for computerized applications.

Appendix B
Johnson Construction Company – Complete Financial Statements

Remember the following points in presenting financial statements to lenders and others.

- The format should be consistent. It is very difficult to compare statements from one period to another when the presentation and order of data are not the same in both.

- The format should be the one most informative to you. Be able to zero in on important information yet perform detail analysis.

- No unnecessary data should be included. You only burden the preparer with more material than he can handle, and you delay the time it takes to produce the finished statement package.

- The statement must be produced efficiently. Otherwise, it will not be ready for review when the builder most needs it—at the earliest possible moment after each month's closing.

- Distinguish between a closing statement, where notes and full explanations are included, and an interim statement, which summarizes monthly data.

Most builders can benefit from a comparative statement, especially if it compares current information with that from the previous year. Of course, the books and records must be maintained on the same basis in both periods for any comparative statement to be meaningful.

Because this is often a problem, many builders have to wait for a full year before comparative statements are available to them. This unfortunate delay is caused by inconsistency in bookkeeping methods. Very likely, the process of "reconstructing" a previous year's books onto the current year's format would be more work than the builder could justify; therefore, he must wait until the following year. Reconstructing a previous year's books would not only involve a careful study of the methods for recording sales and accounts payable, but a detailed audit of all coding to determine whether expenses are being coded in the same way as the previous year.

Figures B-1, B-2, B-3, and B-4 are the

JOHNSON CONSTRUCTION COMPANY
BALANCE SHEET
DECEMBER 31, 19_2 AND 19_1

ASSETS

	12/31/-2	12/31/-1
Current Assets:		
Petty Cash Funds	$ 75.00	$ 75.00
Cash in Banks	5,216.41	2,960.18
Cash in Escrow Accounts	18,000.00	-0-
Notes Receivable	-0-	4,000.00
Accounts Receivable (Note 3)	24,415.60	19,942.08
Reserve for Bad Debts	(885.00)	(840.00)
Retainages	10,008.50	4,586.00
Inventories (Note 8)	16,400.00	9,460.50
Total Current Assets	$ 73,230.51	$ 40,183.76
Long-Term Assets (Note 2a,2b)	$113,199.00	$108,990.00
Other Assets:		
Prepaid Assets	$ 1,400.00	$ 4,218.66
Deferred Assets	800.00	-0-
Deposits	2,000.50	2,000.50
Organizational Costs	1,213.25	4,150.00
Suspense Accounts	300.00	662.18
Accrued Construction	1,607.35	4,000.00
Deferred Construction	300.60	2,806.62
Total Other Assets	$ 7,621.70	$ 17,837.96
Intangible Assets:		
Covenants Not to Compete	$ 1,600.00	$ 3,200.00
Goodwill	5,000.00	5,000.00
Total Intangible Assets	$ 6,600.00	$ 8,200.00
Total Assets	$200,651.21	$175,211.72

Figure B-1

JOHNSON CONSTRUCTION COMPANY
BALANCE SHEET
Page Two

LIABILITIES AND NET WORTH

	12/31/-2	12/31/-1
Current Liabilities:		
Accounts Payable	$ 12,366.16	$ 9,400.07
Payroll Taxes Payable	1,606.00	909.60
Sales Taxes Payable	1,016.33	1,104.00
Notes Payable, Current	18,000.00	14,500.00
Total Current Liabilities	$ 32,988.49	$ 25,913.67
Long-Term Liability:		
Notes Payable, Long-Term	$ 46,606.00	$ 61,400.03
Deferred Credits:		
Deferred Construction	$ 2,604.00	$ 5,900.50
Suspense Accounts	66.50	207.00
Total Deferred Credits	$ 2,670.50	$ 6,107.50
Total Liabilities	$ 82,264.99	$ 93,421.20
Net Worth:		
Capital Stock		
authorized 100,000 shares		
issued 4,600 shares	$ 46,000.00	$ 46,000.00
Retained Earnings	35,790.52	2,576.46
Net Income, current year	36,595.70	33,214.06
Total Net Worth	$118,386.22	$ 81,790.52
Total Liabilities and Net Worth	$200,651.21	$175,211.72

Figure B-2

JOHNSON CONSTRUCTION COMPANY
INCOME STATEMENT
For the years —2 and —1

	12/31/-2	12/31/-1
Earned Income (Note 1)	$524,680.41	$537,980.65
Returns and Allowances	1,400.00	322.60
Net Sales	523,280.41	537,658.05
Cost of Goods Sold (Note 5)	277,585.00	294,927.66
Gross Profit	245,695.41	242,730.39
Selling Expenses (Note 6)	82,796.68	76,441.60
Profit before Fixed Overhead	162,898.73	166,288.79
Fixed Overhead (Note 7)	126,303.03	133,074.73
Net Income	$ 36,595.70	$ 33,214.06

Figure B-3

complete financial statements for Johnson Construction Company. The format calls for consistently comparing current year-end results with the same information in the previous year. For easier analysis, maintain the same format throughout the presentation, as was done here, in all detailed financial statements.

It is apparent that Johnson Construction has been able to maintain nearly all of its profits in the business. Fixed assets increased with the net purchase during the year, and current assets have increased more than $30,000.00.

The current ratio of 7.3 to 3.3 is better than two to one, although last year's current ratio was below that standard at 4.0 to 2.6. Interestingly, the volume of sales dropped more than $13,000.00, but gross and net profits increased. This indicates that the company is better equipped to control costs as well as seek out better quality volume.

JOHNSON CONSTRUCTION COMPANY
STATEMENT OF CASH FLOWS
COMPARATIVE TWO YEARS

	12/31/-2	12/31/-1
Funds were provided from:		
Net Profits	$ 36,595.70	$ 33,214.06
Plus: Non-cash items:		
Depreciation	10,031.00	7,717.00
Amortization	4,736.75	4,004.16
Decrease in Prepaid Assets	2,818.66	1,412.60
Funds provided by operations	54,182.11	46,347.82
Sales of Fixed Assets:		
Autos and Trucks	6,500.00	-0-
Decrease in Suspense Accounts	221.68	18.12
Decrease in other assets	4,898.67	7,206.41
Total Provisions	$ 65,802.46	$ 53,572.35
Funds were used for:		
Purchase of Fixed Assets:		
Autos and Trucks	$ 14,940.00	$ 21,400.00
Machinery	6,000.00	3,000.00
Increase in Deferred Assets	800.00	-0-
Decrease in Long-Term Notes Payable	14,794.03	12,600.00
Decrease in other liabilities	3,296.50	1,806.10
Total Uses	$ 39,830.53	$ 38,806.10
Increase in Funds (Note 9)	$ 25,971.93	$ 14,766.25

Figure B-4

> NOTE 1 – Accounting Basis for Contracts
>
> Long-term contracts are accounted for and reported by means of the percentage-of-completion method. The income, costs and expenses as reported represent estimates of the degree of completion for each contract and have been developed based on past experience. Revisions to the degree of recognized income (and applicable costs and expenses) are made when facts requiring revisions become known.
>
> Short-term contracts are accounted for and reported by means of the completed contract method. Income, costs and expenses for each contract are recognized upon completion of the total project.
>
> Contracts that have durations of less than one month are reported as follows: all income is recognized at the time of final billing whether or not the contract has been completed; costs and expenses are recognized at the time of payment or accrual.
>
> Losses on contracts are accrued at the time that such losses become known.

Figure B-5

The statement of cash flows shows a substantial increase in the availability of funds, proving that the company has managed its cash well. Control of receivables and bad debts, inventories, outside financing, and costs and expenses has paid off this year.

Figure B-5 explains the accounting basis used to record income and recognize costs, expenses and profits. Include an explanation of your accounting basis with all detailed financial statements.

Figure B-6 and Figure B-7 detail the current and past year balances of long-term assets. The increase of $15,000.00 in the gross assets is moderate, considering the size of the business and its existing level of fixed asset investment.

Figure B-8 (Note 3) provides an aging of accounts receivable and an explanation of the method used to calculate bad debts. Many published financial statements exclude this information. Yet it is a good way to prove that receivables are controlled and that the bad debts reserve is large enough to cover any uncollectible accounts.

NOTE 2a – Long-Term Assets, Current Year

	Gross	Depreciation - Amortization	Net
Land	$45,000.00	$ -0-	$45,000.00
Building	62,000.00	13,527.00	48,473.00
Furniture and Fixtures	2,400.00	960.00	1,440.00
Autos and Trucks	46,000.00	37,650.00	8,350.00
Equipment	4,400.00	2,904.00	1,496.00
Machinery	9,000.00	1,260.00	7,740.00
Small Tools	800.00	700.00	100.00
Improvements	1,600.00	1,000.00	600.00
Total	171,200.00	58,001.00	113,199.00

Figure B-6

NOTE 2b – Long-Term Assets, Past Year

	Gross	Depreciation - Amortization	Net
Land	$45,000.00	-0-	$45,000.00
Building	62,000.00	11,460.00	50,540.00
Furniture and Fixture	2,400.00	720.00	1,680.00
Autos and Trucks	37,560.00	31,440.00	6,120.00
Equipment	4,400.00	2,460.00	1,940.00
Machinery	3,000.00	290.00	2,710.00
Small Tools	800.00	600.00	200.00
Improvements	1,600.00	800.00	800.00
Total	156,760.00	47,770.00	108,990.00

Figure B-7

NOTE 3 – Accounts Receivable

An aging of Accounts Receivable follows:

	12/31/-2	12/31/-1
0-30 Days	$12,699.79	$ 7,745.82
31-60 Days	9,486.63	6,100.03
61-90 Days	1,684.18	4,260.50
Over 90 Days	545.00	1,835.73
Total	$24,415.60	$19,942.08

Bad Debts are estimated each month, based on a long-term analysis of actual bad debts, on varying levels of charge sales over a twelve-month period. Actual bad debts are written off against the established Bad Debts Reserve.

Figure B-8

NOTE 4 – Contingent Liabilities

No substantial claims against the company's assets are known. Two suits were filed in the past five years, neither resulting in a judgement against the company. Contingent losses on those claims were under $50,000.00, and both cases involved claims related to contract performance.

Current claims involve three cases of minor complaint. The total contingency amounts to under $4,000.00, and it appears that the claims can be satisfied under the terms of existing contracts without the need for further litigation.

The company has never experienced losses for non-performance of contracts nor for non-payment of legitimate debts. No such claims are anticipated in the future.

Contingent Liability - $4,000.00

Figure B-9

NOTE 5 – Cost of goods sold

	12/31/-2	12/31/-1
Inventory January 1	$ 9,460.50	$ 8,215.06
Materials Purchased	162,418.00	164,200.99
Direct Labor	88,300.00	91,412.06
Sub-Contractors	32,000.00	37,460.00
Freight and Delivery	1,806.50	3,100.05
Total	293,985.00	304,388.16
Less: Inventory Dec. 31	16,400.00	9,460.50
Cost of Goods Sold	$277,585.00	$294,927.66

Figure B-10

NOTE 6 – Selling expenses

	12/31/-2	12/31/-1
Payroll Taxes on Direct Labor	$ 9,106.16	$ 9,418.60
Travel	1,604.00	806.16
Entertainment	405.90	216.50
Gas and Oil	20,416.82	11,994.66
Repairs and Maintenance	26,419.80	35,607.62
Depreciation - Autos, Trucks	6,210.00	4,650.00
Depreciation - Equipment	444.00	444.00
Depreciation - Machinery	970.00	216.00
Depreciation - Small Tools	100.00	100.00
Other Repairs and Maintenance	15,400.00	11,388.06
Advertising	1,720.00	1,600.00
Total Selling Expenses	$82,796.68	$76,441.60

Figure B-11

Figure B-9 explains the company's experience with contingent liabilities. Much detail is provided to establish that the company has not experienced losses from contract performance complaints. This supports the contention that current contingent liabilities are unlikely to be realized. $4,000.00 is listed as a conservative contingent amount.

Figures B-10, B-11, and B-12 are included to remove as much detail as possible from the main income statement. The summarized statement is available as well as the details of direct costs, selling expenses, and fixed overhead.

Figure B-13 (Note 8) explains the valuation of inventories and the policy of Johnson Construction Company regarding physical counts. This is an important note to include with any financial statement, as it lends credibility to the claimed levels of inventory. A reader who is unfamiliar with your operation may assume that inventories are estimated in many cases; but builders who can state honestly that they have counted their stock have stronger financial statements.

Figure B-14 summarizes the components of the increase in cash flow. Specific provisions and uses of funds are detailed on the cash flow statement itself, but this note helps to point out where the increases came from through the year's change in current assets and liabilities.

NOTE 7-Fixed overhead

	12/31/_2	12/31/_1
Salaries and Wages	$ 34,625.00	$ 31,008.60
Payroll Taxes	3,866.10	3,304.00
Rent (Warehouses)	5,400.00	4,800.00
Utilities	2,206.16	2,988.42
Telephone	2,160.00	2,719.04
Insurance	15,900.00	23,204.66
Office Supplies	702.14	796.80
Bonds, Licenses and Fees	12,408.50	10,900.66
Property Taxes	4,812.66	4,590.14
Other Taxes	1,440.22	3,606.12
Printing	1,220.63	1,109.80
Postage and Delivery	1,760.16	790.04
Legal and Accounting	3,650.00	4,725.00
Outside Services	1,200.00	755.50
Dues and Subscriptions	280.00	375.00
Donations and Contributions	1,300.00	250.00
Union Welfare	12,600.06	14,090.50
Depreciation - Building	2,067.00	2,067.00
Depreciation - Furniture	240.00	240.00
Amortization - Improvements	200.00	200.00
Amortization - Organizational	2,936.75	2,204.16
Amortization - Covenants	1,600.00	1,600.00
Building Maintenance	1,366.05	1;712.60
Interest Expense	6,218.60	7,400.98
Bad Debts	1,495.00	1,712.50
Miscellaneous	4,647.20	5,923.21
Total Fixed Overhead	$126,303.03	$133,074.73

Figure B-12

NOTE 8 - Valuation of Inventories

Inventories are valued at cost. All stock is specifically identified at the time of purchase.

The inventory is kept in three sections-wood products, metals and all other. Each of these sections is counted physically on a rotating schedule every three months. A complete physical count is taken every December 31.

Figure B-13

NOTE 9 - Components of Increase in Cash Flow

	12/31/-2	12/31/-1
Cash in Banks	$ 2,256.23	$ (118.36)
Cash in Escrow Accounts	18,000.00	-0-
Notes Receivable	(4,000.00)	4,000.00
Accounts Receivable	4,428.52	12,180.82
Retainages	5,422.50	2,650.00
Inventories	6,939.50	4,245.44
Accounts Payable	(2,966.09)	(4,866.23)
Payroll Taxes Payable	(696.40)	(216.45)
Sales Taxes Payable	87.67	41.53
Notes Payable, Current	(3,500.00)	(3,150.50)
Increase in Funds	$25,971.93	$14,766.25

Figure B-14

Appendix C

Expanding The Accounting System

An internal accounting system is easier to understand when books and records are thought of as belonging to several distinct categories:

- *Books and records or original entry* These are supported by a document from outside the company, such as a check, a bill, or a purchase order received.

- *Ledgers* These include the general ledger and the records for any subsidiary systems, such as accounts receivable records.

- *Support systems* These include the records that are maintained to help in documenting information, such as purchase orders, sales invoices, bank account forms, and petty cash slips.

- *Cost and control reports* These include the reports and worksheets needed to determine the trends and conditions of business, the cost of equipment, maintenance, bad debts or delinquent accounts, and the ways in which profits can be maintained or improved.

Train yourself to know at a glance where all parts of the system fit within these four categories. Your accounting work takes less time and effort when you can see each report and ledger in relation to the overall system.

Develop this overall understanding by tracing a specific number up through the stages of records or a final number backwards to source documents. Here are a few of these steps:

1. Sales:
   Sales invoice
   Accounts receivable control
   Check and bank deposit
   Sales journal
   General ledger entries:
      Increases to cash, accounts receivable, sales

2. Material purchases:
   Purchase order
   Invoice
   Cancelled check
   Check register
   Bank statement
   General ledger entries:
      Increases to materials and decreases to cash

3. Payroll:

Assignment sheet
Time card
Payroll register
Individual payroll earnings card
General ledger entries:
   Increases to Direct Labor or Salaries accounts, payroll tax accruals and expense, decrease to Payroll Account (cash)

4. Petty cash expense:
   Petty cash voucher slips
   Reimbursement check
   Imprest fund summary
   General ledger entries:
      Increase to expense accounts, decrease to cash by way of check register

5. Selling expense or fixed overhead:
   Invoice or schedule of recurring payments
   Check register
   Bank statement
   General ledger entries:
      Increases to specific accounts for expenses, decrease to cash

6. Note payments:
   Original note agreement
   Check register
   General ledger entries:
      Increase to interest expense, decrease to note balance and cash

7. Depreciation:
   Equipment and depreciation record
   General journal
   General ledger entries:
      Increases to depreciation expense and Accumulated Depreciation

8. Prepaid insurance:
   Original worksheet to set up the asset
   General journal
   General ledger entries:
      Increase to insurance expense, decrease to prepaid asset.

Accounting information must flow so that you can track all entries from original documents to summary totals. The bookkeeping system must be able to prove all entries made to it anywhere. Yet it must be an efficient, smoothly running procedure that can be serviced with a minimum investment of time. Finding this middle ground depends on the builder's ability to pick the best method for his particular operation. No one system will work for everyone.

COMPUTERIZED SYSTEMS

Manual systems are workable in most cases for new businesses. This is because the volume has not yet built up enough to indicate how much efficiency will be needed. It is often a good idea not to establish long-range systems until you know your future volume goals. An expensive computer installation, programming, and training period could become obsolete before it is fully utilized if your volume levels off rather than follows your expected trend.

Most modest-size operations can benefit from one of the many pegboard methods when volume can justify the changeover. A pegboard system cuts the number of steps from three or four down to one, in many cases, saving time and effort and increasing your direct control over the books.

But any system has a limit to its capacity. Do not install a system that is larger than your forseeable needs. But you can justify computerized accounting systems when the volume of recordkeeping becomes greater than your pegboard system can handle.

You may start out by contracting with a data service bureau to handle a specific phase of the operation, such as accounts receivable. Charges and payments on account are coded on a weekly or daily basis, creating daily or weekly batches of information. These batches are delivered to a central processing location and entered into a memory unit. Batch delivery can be by hand, or batches can be entered directly into the computer from a remote terminal. Remote terminals use telephone lines and can be located hundreds of miles from a central computer. At the end of each month, the computer automatically issues statements and bills each customer.

This system has several drawbacks. Often a large amount of information is flowing into the computer from several users at one time, forcing you to "wait in line" for access to the computer. This means that entries must be made at the computer's convenience, not yours. Another disadvantage is the lack of direct control. Data service bureaus are often slow in returning or processing information, and daily or weekly analyses can be delayed unless you can work out a schedule with the bureau. The main advantage of these arrangements is the savings in time versus the processing cost. Computerized processing through a data bureau is a cheaper way to do business when a large volume of processing is involved.

The data service bureau is intended for businesses that are too large for manual processing and too small for their own in-house computer. But recognize that in a growing, healthy operation, the outside service arrangement is primarily a transitional service. Both the continual drop in costs for small computers and the expansion of your building volume can justify installing your own computer at some point.

Many computer systems allow you to do data processing without extensive knowledge of computer programming and operation and without on-the-spot trained computer professionals. The market for computers is changing so rapidly that the specifications, sizes, capacities, and prices of small business computers are likely to be obsolete within a few months. But mini-computers or micro-computers are currently well adapted for use by small to mid-size builders, and are expected to be more than adequate for some time to come. These models are small, usually small enough to take up less than the space of a small office desk. The cost of the computer itself, a display terminal, and a printer lowers each year as new technology makes better equipment available at lower manufacturer cost. This equipment is called *hardware*. The programming that enables a machine to function is called *software*.

The cost of software seems to increase each year as more highly specialized programs are developed. But many computer companies sell and install the hardware, provide complete instructions to the user, and sell package deals. These deals involve ready-to-operate programs designed to do a specific chore such as accounts receivable record-keeping and monthly billing. The program is contained on a small disk or cassette, and is usually sold at a fixed price. The program can be modified to fit the builder's exact needs. For example, a builder would want his name and address to appear on a blank form statement, and he may want special features such as an aging list or a separate system to control and bill retainage. Such features can be added to ready-made programs at small cost.

Before you install any system—manual, pegboard, or small computer—research the market thoroughly and make a decision based on the experience of other businessmen and builders. Small computers can be valuable if the investment in such equipment is justified by the amount of time and effort it saves and by the degree of control and efficiency it provides.

Here are some questions you should ask yourself before you turn to computerization.

- What data will be fed into the system to achieve what result?

- What will be expected from the system? Can the proposed computer or service bureau deliver this result?

- Since this will probably be a long-term commitment, how adaptable is the proposed system? Can it handle more than you need now?

- Who will program the computer for you? How expensive will that be?

- Does the contract come with a maintenance agreement? How much will it cost?

It pays to shop around for computers. Compurization is an expensive move, so get exactly what you want at a price you can afford. Make sure the system is flexible enough to allow you to tailor it to your individual needs, and find out if it can be expanded to grow with you. Make sure that sales people fully explain their systems to you. While there are many good inexpensive systems available, there are also many that would unnecessarily complicate your operation.

To evaluate a system's value, compare the cost, efficiency, and practicality of each alternative. Figure C-1 shows one comparative method for judging systems. Prepare evaluation sheets for each system under consideration. Assign or deduct points from each system, depending on its positive aspects and drawbacks.

Or list the pros and cons of each system, assigning a point value to each feature, and find out which system has the highest total point value. The following should be considered as good or bad points:

- The total cost

- The need for special training

SYSTEM EVALUATION

Description of system _____
_____

COST
Total cost of the system

$ _____

LIFE
Estimated useful life   Cost per year

_____ years $ _____

SAVINGS
Savings, in hours, per month

hours _____

SUMMARY
Add the following:

    One point for each year of useful life   _____

    One point for each hour saved each month   _____

    Less: one point for each $100 of cost   _____

    Total points   _____

NOTES
Other points to consider:

    What special training is required with this system?

    What tax benefits will result from depreciation and investment credits?

    Will more employees be needed to operate?

    How much processing time will there be to obtain information from this system?

    How readily will this system be converted into even better systems later on?

    What hidden costs will come with this system—maintenance, special supplies, etc.?

Figure C-1

Expanding The Accounting System

- Savings in overhead

- Special maintenance expenses

- Access of materials

- Limitations caused by possible breakdowns

- Actual time savings

- Tax benefits of purchase

- Feasibility of leasing versus buying

- Degree of increased efficiency expected

- Expandability of the system

- Life value of the system

- Maintenance agreements that come with with system, if any

- Availability of replacement parts or needed supplies (thus, time required for maintenance and repairs)

- Cost of specialized supplies that must be used within the system (special paper, forms, and so forth)

- Multi-uses possible with the system once it is installed

- The degree of work, time, and expense required to convert the current system

- The effect of lapses in efficiency because of conversion time

- Financing arrangements and the effect on cash flow

- The effect on overhead expenses—decreases in clerical time due to higher efficiency or increases due to greater office specialization

Changing the system totally is always a major business decision. Review current needs in view of the present system and draw up a proposal for an improvement. The cost of a new system must be justified by the savings created. You will probably not realize a savings advantage immediately. But compare the life of a new system with the expected cost recovery period.

To calculate this, amortize your expected cost over the useful life of the system, much as you take depreciation over the useful life of a new piece of equipment. But the difference here is that you do this to judge the feasibility of the system itself and to get an idea of its annual cost.

How does this cost compare with the estimated annual savings in clerical time, the value of statements produced more quickly, the quality of records, and so forth? This may be difficult to determine, as you must assign a dollar value to an intangible factor such as the value of reports that will be available sooner by computer.

In some cases this will not be possible, and known costs must be compared to the proposed costs of the new system. For example, the following comparisons can be made:

- Current versus expected bookkeeper's time

- Monthly supply costs of current and proposed systems

- Monthly or annual maintenance costs of current and proposed systems

- Tax benefits of current system depreciation, if any, versus investment credit and depreciation of new system

- Reduced accountant billings from higher efficiency under a new system, compared with current expense

Comparing actual or known expenses rather than intangible benefits lets you obtain comparison of current and proposed systems based on dollar value. The intangible benefits will simply add to the value of the conversion.

Some examples of accounting improvements without total conversion would be:

- Expansion of a pegboard system

- Modification to partial automation

- Changing from a batch-input computer service to the purchase of a mini-computer

- Upgrading an existing mini-computer by adding to its memory capacity, its processing speed, or adding to its functions.

You need enough knowledge about the proposed system to judge its strong and weak points before going ahead with conversion or improvement. You must first be close enough to the books and records to see how the work could be done better, faster, and cheaper without losing existing efficiency. Then, if you decide to convert or upgrade, become totally familiar with the new procedures before trying to assess how well the new system performs. The builder who is in touch with his business systems will usually have a higher degree of success than one who isolates himself from his own data.

Appendix D
Automating Your Accounting

"My bookkeeping system is a mess. I don't understand accounting. It's too complex. I guess I need a computer."

That statement is made far too often, and it's based on a false assumption. Yes, accounting is complex. And yes, computers do process information more efficiently than humans can. But a computer is only a powerful adding machine. If you buy a computer to set up and organize your bookkeeping system, you're going to be disappointed. If you already have a good bookkeeping system, a computer may save some time and eliminate some clerical errors. But it won't organize what is already a mess. More likely it will just compound the mess.

Computers process information and follow simple commands. They have no knowledge or skill.

When should you switch to a computerized accounting system? Under these conditions:

1) When the volume of transactions is so heavy that you're wasting too much valuable time just entering numbers — especially numbers treated the same way over and over again (like accounts receivable for many customers).

2) When your business and the number of transactions is growing rapidly.

3) When one or more routines, such as payroll or job costing, takes way too much processing time.

4) When you know exactly how to get information from your records to your books, but it's taking too much time and there are too many mistakes.

When should you *not* automate?

1) When you don't understand how your system runs today.

2) When the number of transactions is low.

3) When you haven't fully explored other ways of doing things more efficiently. For example, don't try to do payroll on your own computer if your bank will handle the chore for $20 or $30 per pay period.

4) When there's no time available to learn the new system and not enough money on hand to buy the hardware, software and peripherals that will be needed. Moving your bookkeeping routines to a computer will take more time, not

less, for the first few months, and will add a lot to expenses without showing any additional income.

RESEARCH

If you're determined to use a computer to keep your books, do some investigating and thinking before buying.

Start with a candid evaluation of yourself and your staff. Do you have people on the payroll who can learn to run the computer and keep it running? Or will you end up doing it yourself? Don't assume that you can rely on outside experts to set up the system and keep it running. You have to accept responsibility for learning to use the computer and computer programs and for keeping the whole thing running even when nothing goes right. Calling in experts to solve every little problem will be far too expensive.

The cost of the computer and the computer programs won't be the most expensive part of converting to automated bookkeeping. The most expensive part will be the value of your time and the time of your staff in learning to use the computer and the computer programs. Figure that learning time will be more than half the costs on nearly every installation. If you don't have the patience to puzzle over manuals until late at night, maybe a computer isn't right for you.

Second, take a look at your transaction load. If you run a clean, neat, efficient construction company (no more than 2 or 3 jobs at a time, nearly all work subcontracted, only 3 or 4 employees, write no more than 100 checks a month), you may not need a computer. One of the most common errors builders make is to believe that an automated general ledger is somehow "better" than one kept manually. It isn't true.

Third, review what's on the market. Software (computer programs) come in all sizes. You'll find some fairly comprehensive packages for less than $50. One that comes to mind is DacEasy Accounting by Dac Software, 4801 Spring Valley Road, Suite 110B, Dallas, Texas 75244 (214-458-0038). The largest builders can spend $50,000 or more for a comprehensive package. Most likely, you'll spend between $300 and $800 for a professional accounting system. Look for these features:

1) Ease of use. Don't buy any system based on a quick demonstration or promises made by a salesperson. Many computer salespeople have one goal — to get you out the door with a computer in your arms. Make sure you have the chance to sit at a terminal, open the user's manual, and experiment with the program yourself before buying. A reputable computer store will not only allow you to do this, they'll encourage it. Check it out thoroughly before you buy.

One of the most important features you can find in an accounting package is flexibility. A flexible program can be changed to fit your needs. Don't get stuck with a program that makes you change your whole bookkeeping system. Most small, inexpensive programs tend to be fairly rigid. Look for a program that gives you freedom to set report formats your own way.

2) Check the manual. Make sure it's easy to understand. It should take you through each procedure in the right order, explaining each step with examples and alternatives. Even the best program will be worthless if the manual is poorly written, incomplete, overly technical, and vague on details. Generally, shoddy programs come with shoddy operating manuals. Judge the quality of the programs by the quality of the manual.

3) Find out about support before buying. You'll have questions. You'll get stuck on some step and won't be able to continue without help. What then? Who can you call on? The selling price of some software includes free phone support for several months. Other vendors require that you pay for support on a monthly basis or by the minute of consultation. More expensive packages include a specified number of free hours of consultation over the phone. Try calling the support number before you buy. Test how responsive and knowledgeable your vendor is going to be.

4) Once you find software you like, find the computer you need to run it. Don't make the mistake of buying a computer first and then discovering that the software you want won't work with that computer. Most microcomputer programs written for business use will run on IBM computers and "compatibles." But there are degrees of compatibility. Some IBM clones are more compatible than others. It's always best to try before you buy.

5) Check computer and printer prices. Be sure you're comparing apples to apples. Each computer you're comparing should have the same number and capacity of floppy and hard disks, the same video boards, similar terminals, and the same memory capacity. The computer you buy should have at least 512K units of Random Access Memory (RAM). Accounting records tend to take up a lot of space on your disk, so you'll probably want a hard disk. Twenty megabytes is probably enough capacity for most builders.

6) Maintenance will always be a problem. Eventually everything mechanical breaks. Hard disks can be problem prone. You need either a spare computer always on hand and available when needed, or someone you can call on to make repairs and supply replacement parts. Don't buy any hardware without a clear understanding of where repair service will be available.

On-site maintenance is most convenient. But the cost of even a simple repair on site will usually be at least several hundred dollars. Drive-in service will usually cost less. But how long will the computer sit on a bench at the repair shop before it's ready for return? You may not be able to get by without a computer for a week or two while your technician works off his backlog.

An annual maintenance contract will usually cost about 25% of the original cost of the equipment maintained. If you can afford it and can't be without your computer for very long, buy a maintenance contract. If you insist on going first class, get on-site service. If you have to hand-carry or ship the computer back to the store, will you get one-day service unless parts aren't immediately available? Most repair shops and some manufacturers will make a commitment to repair broken hardware promptly.

THE CONVERSION

Even if you make all the right choices and end up with software and hardware that will do the job, plan on a gradual conversion. You won't simply go back to the office, plug it in, and start doing work. Plan on spending a week or two learning to use the computer and programs. You'll want to experiment quite a bit before actual processing begins.

Continue keeping your books on the old system for a month or two until all the kinks are worked out. Find the errors and shortcomings in the new system before they do any permanent damage to your company. Plan to run two sets of books for at least one full month.

If it's practical and your timing is right, try to install your new system around October or November (assuming your tax year ends in December). That way, you can use the last two months of the year as a test period. If your fiscal year does happen to end in December, chances are it's a slow period for you — perfect for a conversion.

Try to start using your new system at the beginning of the year. That will usually reduce the number of records that have to be entered twice.

If you plan to have employees involved in the conversion, follow these guidelines:

1) Get them involved in decision-making early. Set aside time for training. But don't set deadlines. That makes mistakes, overtime and half-measures more likely.

2) Expect some resistance to change. Deal with it directly but with understanding. Some people simply can't work with computers. Personnel changes may be necessary. But don't force the issue.

3) Be sure every step of the process is fully understood.

The cost of staff turnover (or the need to train someone to run the system when you're too busy) can be one of the highest hidden costs of automating.

In most construction companies, the owner will have to learn to use the computer first. No one else has the incentive or interest in making a successful conversion. Then the owner will train others to do the work that needs to be done. The first person to learn the programs will always take the longest. Everyone else gets a shortcut; they can ask questions and get a demonstration. They don't have to dig it all out of the manual.

You'll probably want to write up your own summary of how the program is to be used and set up a processing procedure and set of monthly deadlines.

ALTERNATIVES TO A COMPUTER

You can use a computer without buying one, of course. Here are three alternatives to consider:

1) Batch processing services (or service bureau). These are companies that specialize in doing some form of automated reporting. For example, you could hire a batch processing service to send out your bills each month. These services offer little or no flexibility in statement format or billing dates. But the advantage is a fairly low cost service if you handle a lot of transactions each month.

Many companies offer to handle bookkeeping services on computer. The Yellow Pages in your phone book will list several in your community. The cost will probably be modest — usually less than $50 per month, plus a set charge per record processed. The disadvantage is that the processing service will probably take a week or more to process your records and then return them to you. That's not too bad, unless there's a mistake in the figures you sent to the service and the mistake has to be corrected. Then you'll have to wait another week and pay the minimum processing charge again. If you did the work in your office with your own computer, correcting a mistake might take only a few minutes and cost next to nothing.

2) A payroll service. Many specialized companies and even banks offer payroll processing services. They prepare payroll checks, federal and state deposits, quarterly and annual reports, and information forms you're required to file. In addition, they'll supply you with a summary, breaking down payroll by department or job. Considering the paperwork involved in payroll and the frequency with which rates change, using an outside service is a good choice for most builders. Your bank may even waive the monthly fees if you maintain some specified balance in your checking account.

3) Time sharing. You can buy time on a large mainframe computer through one of the large national firms specializing in this service. These include ADP, Tymshare, and Control Data. You invest in little or no hardware beyond renting or buying a terminal. In exchange, you're given free instruction and training. In some cases, specialized modifications can be made to some programs or routines at no additional cost. Also, you have access to a large library of existing programs.

The cost of time sharing is based on the amount of storage, processing, and sign-on time you actually use. Larger computers running more complex programs will cost more. Time sharing probably isn't a good choice for bookkeeping unless you're handling a heavy volume of checks every day.

I can't end this short section without emphasizing that there is no easy solution, no one choice that fits every construction company. The best alternative for you depends on your budget, your preferences, you willingness to deal with computers and service bureaus, and the time you can devote to making the system work. My advice is to get the recommendation of your accountant and other contractors. Find someone who is using a computer and likes it. Nearly everyone who has a computer is anxious to talk about it.

The next time you get together with other contractors, get them talking about how they do their bookkeeping. You're sure to learn something. In fact, you may discover that everyone else is having more trouble with accounting than you are!

Appendix E

Income Tax Planning

Every business needs a bookkeeping system to prove the income and deductions you claim on your income tax returns. The tax law requires no specific method for keeping records, only that your records are adequate and complete. This requires that:

1) Business records must be kept separate from personal accounts.

2) All vouchers and receipts should be clearly marked and cross-referenced.

3) Income should be traced from bank deposits to entries in your books. It's important to deposit all income in your business account, and to explain fully any other money you deposit (such as proceeds from loans, capital you contribute, or deposits held for customers).

4) All business expenses should be paid with a business check. If you have cash expenses, keep the receipts and write a reimbursement check from your business account.

5) Don't use the business account for payment of personal bills.

6) Include with your tax return extra documentation or letters explaining highly unusual deductions.

REDUCING YOUR TAXES

It's perfectly legal to pay the minimum tax required by law. You're allowed to arrange your financial affairs to reduce your tax burden. Techniques used to minimize taxes include shifting, conversion, deferral, and sheltering.

SHIFTING

If you expect a greater tax burden this year than next, you can shift income to the following year and increase expenses this year. This technique is only effective if you report on the cash basis (since accrued income and expenses must be reported in the period incurred).

An example: You're on the cash basis, and need to reduce this year's tax liability. You can tell your customers to delay making payments due in December. Have them pay you in January instead. As long as the customer agrees and doesn't issue a check until January, that income is shifted.

If a check is available to you in December and you simply defer depositing it until

January, you can't defer reporting it. The money must not be available to you. For example, if you don't pick up your mail from a post office box until January 2, but a large payment on account had been placed in the box on December 31, it's reportable as income during December.

On the other hand, a check issued in December but not *received* until January is deferred.

CONVERSION

Up until 1986, it was possible to convert fully taxable gains to long-term capital gains by keeping them longer than six months. Under rules in effect at that time, only 40% of your gain was taxable. For 1987 and thereafter, it's no longer possible to convert ordinary gains to long-term capital gains.

It's still possible to convert income. You can, for example, pay one of your children for labor he or she performs in your yard, shop, or on a crew. Within limits, the income earned by a child will be taxed at that child's lower rate.

DEFERRAL

You can defer taxes in several ways. One is to use the completed contract accounting method to the extent allowed under the law. Another is to delay, until the following year, the sale of equipment on which you expect to profit.

If you have discretion in the timing of completion on part of a job, delaying work until the following year will effectively defer the recognition of gain, even on the accrual basis. Assume, for example, that you report on the percentage-of-completion basis. You have promised to complete a major section of a job no later than January 15. Crews are available to work between December 20 and January 10. While you could complete the work before year-end, that will mean reporting the gain during this tax year. If you delay completion until January 10, you'll be able to defer that income.

You can also defer tax on income by placing up to $2,000 per year in an Individual Retirement Account (IRA). While you may not qualify for a deduction for contributions made, the earnings on an IRA are not taxed until withdrawn.

SHELTERING

You can shelter portions of your income, even after the 1986 tax reform act. If you're self-employed, for example, you can set up a Keogh plan. You can place 25% of your net earnings in the account and deduct those contributions from your adjusted gross income. This 25% limit refers to income *after* computing your Keogh contribution, so it's actually only 20%. For example, if you earn $100,000 net income, your Keogh contribution could be $20,000, which is 20% of the actual net or 25% of the $80,000 net remaining *after* the Keogh is subtracted.

Like an IRA, earnings in a Keogh account are deferred for tax purposes until money is withdrawn. So a Keogh account provides you with two tax advantages. First, you shelter the money you put away. Second, earnings are not taxed until you begin withdrawals. You can withdraw money at any time after you reach age 59½. If you take the money out before that date, it's taxed immediately, and you're charged an early withdrawal penalty (as a form of additional tax).

OTHER TAX PLANNING

As the owner and manager of your own business, you can and should plan for taxes all year, not just at year end. Once the end of the year has passed, it's too late to reduce taxes.

Here are some other ways to reduce your tax burden:

Timing asset purchases to maximize the depreciation and expensing provisions under the law— For example: You have already purchased equipment this year and plan to write off the maximum $10,000 through the expensing allowance. You plan to purchase another piece of equipment, but could wait until next January. That would allow you an additional expensing allowance next year.

Timing asset sales to delay recognition of a gain— For example: You want to sell a heavy truck, on which you'll report a profit. If you do so this year, it will increase your profit beyond the maximum tax bracket you currently expect. By waiting two months, you can delay reporting that income until next year, when you expect net profits to be lower.

Making contributions— For example: You have about $2,500 in useless, obsolete inventory. You have been offered $250 by another contractor. That's probably the most you could get. However, by contributing the inventory to a charity, you'll be allowed to write off the *market value* as a contribution. If your tax bracket is 15%, this will represent a savings of at least $375. Making the contribution will be more profitable than selling the obsolete stock. In addition, you can time your contribution to gain maximum advantage, based on your expected profits this year and next.

Periodic reviews— Meet with your accountant or tax advisor well before year end. Review income to date and estimate your tax liability for the year. Devise ways to shift, convert, defer or shelter income to reduce the liability.

Create an accounting system that supports your tax awareness. You don't keep financial records just to comply with the law. Use your accounting records to reduce the tax burden within the law. Make your accounting system a source for profits.

Your books aren't working for you unless they provide the information you need when you need it. The more knowledge you gain from an efficient and informative accounting system, the more valuable that system is. Your accounting system should be a source for knowledge, and not just as a necessary part of operating a business. Here are some specific ideas:

1) Keep your record keeping system as simple as possible. The easier to record information, the easier it is to compile and report.

2) Train employees to give you summarized information. The sooner you get what you need, the more able you will be to take action quickly.

3) Design every report, record and file to facilitate your tax planning and cost control procedures. Devise methods for compiling information, either manually or through an automated system, so that you can review your bottom line whenever you want.

4) Check profit and loss throughout the month as well as monthly and quarterly. Estimate net profit and tax liability frequently. And stay in touch with the levels of income and debt. When you see the numbers wandering away from your original estimates, consult with your accountant and take necessary actions. This may include tax reduction planning or revision of estimated payments to avoid underpayment penalties.

5) Keep yourself aware of changing rules, either by subscribing to a tax update newsletter, or by speaking regularly with your accountant or tax advisor. Remember that the complex system of rules under which you operate may change through one or more methods (new tax legislation, technical corrections, or tax court interpretations of the tax rules).

Appendix F

Blank Forms

The following blank accounting forms can be copied for use in your own operation. They are general forms of the types used in most businesses. Try to develop a set of forms that works best for you. No system or series of preprinted forms can ever take the place of a specially designed procedure that gives you exactly what you want. Blank forms are also available in stationery stores, and these are just as adaptable to a variety of uses as the ones included here. Specially designed forms can also be custom designed at relatively low cost. A graphic artist can do this for you, and nicely finished proofs can then be printed at a corner printshop for much less than columnar blank paper costs.

Remember that blank forms must fit into a more or less uniform size of binders and files. So forms should be the same size whenever possible. Call the printshop to find out what paper sizes are the cheapest and most practical to print.

Figure F-1 is a petty cash voucher. Each cash item can be listed on this form. The voucher then acts as a summary support document for the check issued to pay back the imprest fund. Since payments bring the account back up to its original balance, the beginning and ending totals should be identical. Copy and use this form as is, or use it as a guide in developing your own.

Forms for balancing a bank account are shown in Figures F-2, F-3, F-4 and F-5. Document the bank account balancing fully, not only for your own information, but to speed balancing in the following month. Documentation assures that all previous adjustments are cleared. Follow an established balancing procedure. Outstanding check lists can be long, depending on the timing of check writing, the number of checks, and the average length of time they are outstanding. Several pages are needed for listing checks on large-volume accounts. When checks clear in following periods, place a check mark (✓) next to each line. This helps pinpoint remaining outstanding amounts. In most cases, few if any deposits in transit need to be listed. Figure F-4 may be used to list outstanding deposits for several months in a row.

List adjustments and explain them fully each month on a form like Figure F-5. Because of the nature of most adjustments, you often need space for a written explanation as provided in the figure.

Figure F-6 is a form for keeping track of a bank account's balance. This form is especially useful in pegboard or other manual systems. You can use the form in the figure, but a bank balance form of one type or another should be kept. You should be able to verify the accuracy of the math by applying the following formula each month before balancing the bank account: Balance forward *plus* deposits and *less* checks *equals* ending balance.

A customer ledger as in Figure F-7 helps maintain accurate records of customer accounts receivable. Pre-printed forms can also be purchased in stationery stores for pegboard systems. The form in the figure is useful only with manually-kept procedures for Accounts Receivable.

Keep an equipment record as in Figure F-8 for each piece of equipment. Fill it out when you purchase it and keep it on file after you sell it. This gives you a complete record of your fixed asset investments.

A record of depreciation form goes hand in hand with an equipment record. See Figure F-9. The builder needs a written source to justify monthly (or annual) depreciation totals, and a quick way to compute the gain or loss of an asset when it is sold. The equipment and depreciation records should agree with the balances in the general ledger for those accounts.

A check request form makes issuing checks a uniform procedure. See Figure F-10. You make checks out for invoices and statements you receive from those you owe. But you must also be able to document check approvals, delivery instructions, and account coding. Not all statements allow room for these items, so you need a single form of your own. The best are three-part forms on carbonless paper. The first copy, the original should be filed on top of the check copy and invoice/statement in an alphabetical paid bills file. Maintain the second copy in numerical order by check for easy reference. The third copy can be a multi-purpose form.

Every builder needs a general journal. See Figure F-11. Items recorded on general journals include all recurring entries (depreciation, amortization), adjustments, and year-end accruals. Have a standardized form for this and all other permanent accounting documents.

Builder's Guide To Accounting

PETTY CASH VOUCHER

BEGINNING BALANCE

LESS-Payments [vouchers attached]

Description/Code	Amount

TOTAL PAYMENTS

Cash over or short

TOTAL NEEDED TO BALANCE FUND

ENDING BALANCE [same as beginning balance]

prepared by _____ date paid _____

approved by _____ check number _____

Figure F-1

Blank Forms

BANK RECONCILIATION
month of _____

Balance per Bank	Ledger Balance
_____	_____
_____	- - - -
_____	- - - -

BALANCE

ADD: deposits in transit

LESS: outstanding checks

Other Adjustments

ADJUSTED BALANCE

Notes _____
_____
_____
_____
_____

Prepared by: _____ date _____

Approved by: _____ date _____

Figure F-2

OUTSTANDING CHECKS
month of _____

page ____

Number	Amount

Number	Amount

Number	Amount

Total, this page ☐
Grand Total ☐

Figure F-3

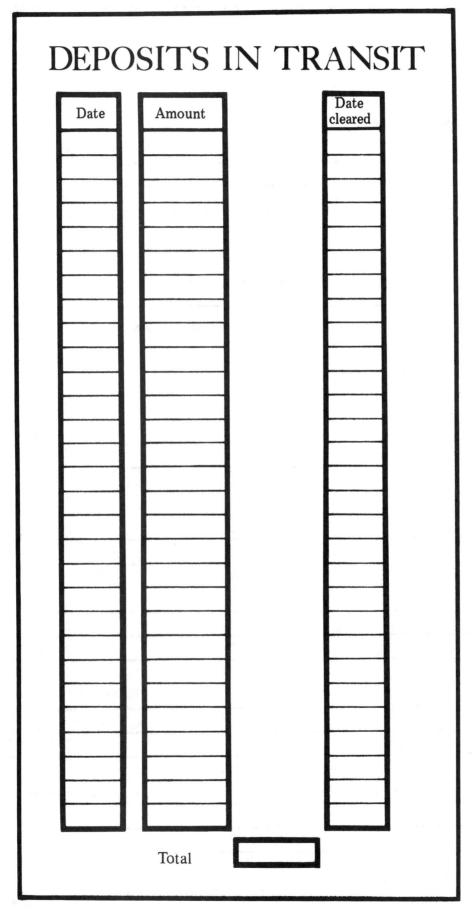

Figure F-4

Builder's Guide To Accounting

ADJUSTMENTS

BANK ERRORS | amount

1) _____
   _____
   _____

2) _____
   _____
   _____

Total bank errors

CHECK BOOK ERRORS

1) _____
   _____
   _____

2) _____
   _____
   _____

3) _____
   _____
   _____

4) _____
   _____
   _____

5) _____
   _____
   _____

Total check book errors

Figure F-5

RECORD OF BANK BALANCE

BANK _____

ACCOUNT _____   MONTH _____

Date	check numbers	−CHECKS	+DEPOSITS	BALANCE
1				
2				
3				
4				
5				
6				
7				
8				
9				
10				
11				
12				
13				
14				
15				
16				
17				
18				
19				
20				
21				
22				
23				
24				
25				
26				
27				
28				
29				
30				
31				
Totals				

Figure F-6

Builder's Guide To Accounting

CUSTOMER LEDGER

Name _____ Phone _____
Address _____ Account # _____
City _____ Discount terms _____
                                        Credit limit _____

Date	Description	✓	Charges	Payments	Balance

Figure F-7

RECORD OF EQUIPMENT

Description of Asset _____

Company Number _____ Classification _____

Size _____ Model _____ Style _____

New ☐   Used ☐   Serial # _____

Purchased from _____ Terms _____

Location of Property _____

Cost information
- Price
- Tax
- Delivery charges
- Installation costs
- Other _____
- Total

Date sold _____ Sold to: _____

Sales price:
- Gross
- Tax
- Total

Figure F-8

RECORD OF DEPRECIATION

Description of Asset _____ Company Number _____
Classification of Asset _____ Estimated useful life _____ years
Method of Depreciation _____ Salvage Value _____

| Year/Month | Month | Amount of Depreciation ||| Date depreciation booked | Date asset sold |
		Year-to-date	History to date	Total to date		

Figure F-9

CHECK REQUEST

Date _____

Pay to: _____
_____
_____

For invoice(s) _____
_____

Delivery instructions
    mail ☐   give to _____
    hold for _____
    other _____

Coding

code	amount		code	amount

Total

Requested by _____ Approved by _____

Date paid _____ Check # _____
Processed by _____

Figure F-10

page____	GENERAL JOURNAL			
Date	Explanation	✓	Debit	Credit

Figure F-11

Index

A
Account coding, 220
Accounting methods, 18-27
Accounts payable, 87-91
    double register method, 88
Accounts receivable, 33-39
    average length study, 43
Accounts management, 42-48
Accruals, 193-198
    controls, 198
    receivables, 194
Accrued income, 19
Amortization, 144, 197
Application for credit, 45
Asset, definition, 14
Asset ratios, 214, 218
Automated accounting, 10-11, 279-282
Average length study, 43

B
Bad debts, 49-55
    bad debt trend, 52
    estimating, 50-55
    journal entries, 50
    ratio summary, 54
    recognizing, 49
    writing off, 49
Bad debt study, 50
Balance sheet, 199, 202-205, 226
Balance sheet ratios, 213-214
Bank account, 178-185
    balancing entries, 181-182
    cancelled checks, 178
    checklist, 182-184
    constructing a balance, 184
    outstanding checks, 184-185
    reconciliation forms, 179, 289-293
Bank service charges, 80
Breakeven point, 152-153

Budgeting expenses, 120-122
Building account, 131

C
Capital stock, 205
Cash accounting, 27, 198, 238
Cash budgeting, 147-157
    breakeven point, 152-153
    cash movement, 149-152
    cash plan, 148
    checklist, 147
    current ratio, 148-149
    debt to capitalization ratio, 149
    expense ratio, 149
    gross profit, 149
    income to net worth ratio, 149
    preparation, 149-152
    source and application of funds, 149-152
Cash clearing, 90
Cash disbursements journal, 13
Cash flow statement, 15, 199, 201
    controls, 210-211
Cash movement budgeting, 149-152
Cash on hand. See Petty Cash.
Cash Sales, 28-29
Charge sales, 29-33
Chart of accounts, 11, 220, 257-262
Check request, 297
Checks
    accounts payable, 87-91
    computerized, 92
    double register, 88
    general account, 89
    maintenance, 77-93
    multiple businesses, 91
    payroll account, 89-90
    payroll accounting, 102-117
    records, 6-7
    register, 82-83

returned, 80
tax account, 90
voided, 85
voucher checks, 93
written to "cash," 83
Closing the books, 227-229
Collection letters, 46-48
Combined ratios, 214-216
Comparative statements, 230-236
Completed contract accounting, 19
Computer accounting, 10-11, 279-282
Contingent liabilities, 204-205
Control account, 34
Co-signers, 251
Cost accounting, 158-173
    deferrals, 160
    direct labor, 167-169
    job cost ledger cards, 170-171
    labor performance, 162-163
    materials, 169
    overhead, 169-170
    subcontracts, 169
Credit, definition, 12
Credit application, 45
Current assets, 202-204
Current liabilities, 204
Customer ledger, 294

D
Debit, definition, 12
Debit/equity ratios, 214-218
Declining balance depreciation, 141-142
Deferred expenses, 127, 197
Deferred income, 18
Depreciation, 137-144
    Accelerated Cost Recovery System, 137-139
    class lives, 143
    declining balance, 141-142
    form, 140, 296
    obsolescense, 138
    straight line, 139-141
    sum of the years' digits, 141-142
Depreciation ratio, 217, 219
Direct costs
    labor, 167-169
    materials, 94-101, 169
    subcontracts, 169
Disbursements journal, 13
Double-entry system, 12-16

E
Earned income, 18
Earnings record, 109
Equipment, 130-146
    amortization, 144
    capitalized, 130
    classifying, 131
    controls, 134
    cost assignment, 136-137
    cost summary, 136
    depreciation, 131-132, 137-145
    form, 133, 295
    leasing, 144-145
    sales, 145
    time schedule, 135
    trade-ins, 145
    unit cost, 134
Equity. See Net Worth.
Estimating, 186-191
    equipment costs, 189
    job completion schedules, 186-187
Expense control ratio, 217-219

F
Financial statements, 199-206
    balance sheet, 199, 202-205
    capital stock, 205
    contingent liabilities, 204-205
    current assets, 202-204
    current liabilities, 204
    direct costs, 206
    equity account, 205
    fixed assets, 204
    gross sales, 205-206
    income statement, 199, 205-206
    long-term liabilities, 204
    net profit, 205
    operating expenses, 206
    other assets and liabilities, 204
    retained earnings, 205
    returns and allowances, 206
    supplementary schedules, 202
    tax provision, 206
    trial balance, 199, 202
Finished goods inventory, 99-100
Fixed assets, 204
Fixed expenses, 118-120
Forms, 12

G
General account, 89
General journal, 298
General ledger, 13, 221-224
Gross profit ratio, 216-217, 219
Gross sales, 205-206
Guarantors, 251

I
Imprest system, 174-175
Improvements, 131

Index

Income
    accrued, 19
    completed contract, 20
    deferred, 18
    earned, 18
    percentage of completion, 18-19
Income account ratios, 216-217
Inventory statement, 199, 205-206, 227
Income tax planning, 283-285
Inventory, 98-101
Inventory ratios, 215-216, 219
Investment yield, 217, 219

L
Labor, 167-169
Land, 131
Leasing equipment, 144-145
Ledger sheet, 13
Liabilities, 14
Loan applications, 248-255
Long-term assets. See Equipment.
Long-term liabilities, 204

M
Machinery and equipment, 131
Maintenance cost ratio, 217, 219
Margin of profit, 74, 217, 219
Materials, 94-101
    delivery terms, 95
    inventory, 98-101
    purchase journal, 97
    purchase order system, 95
    volume buying, 95

N
Net profit, 205
Net worth, 14, 205

O
Office furniture and equipment, 131
Organizational expenses, 127
Other assets and liabilities, 204
Overdraft charges, 80
Overhead expenses, 118-129
    budgeting, 120
    deferred, 127
    fixed, 118-120
    general ledger accounts, 122-124
    organizational, 127
    prepaid, 127
    recording, 126-127
    selling profit, 122
    variable, 118-119
Overhead ratios, 217, 219
Owner's equity, 205

P
Payroll account, 89-90, 104
Payroll accounting, 102-117
    checks, 106-108
    earnings record, 108-109
    forms, 115-117
    payroll taxes, 104-105, 110-115
    register, 109-110
    time card, 105-106
Pegboard systems, 37, 85-87
Percentage of completion accounting, 18-19
Perpetual inventory, 100
Petty cash, 174-177
    imprest system, 174-175
    voucher, 175-176, 288
Prepaid expenses, 127, 194-197
Profit planning, 68-75
    management guidelines, 74-75
    margin of profit, 74
    successful volume, 71
Provision accounts, 197-198, 206
Purchase journal, 97
Purchase order system, 95

Q
Quick assets ratio, 214, 218

R
Ratios, 212-219
    asset, 214, 218
    balance sheet, 213-214
    combined, 214-216
    comparative analysis, 218
    current, 212-213, 218
    debt/equity, 214, 218
    depreciation policy, 217, 219
    expense control, 217, 219
    gross profit, 216-217, 219
    income account, 216-217
    inventory, 215-216, 219
    investment yield, 217, 219
    maintenance cost, 217, 219
    margin of profit, 217, 219
    presenting, 217-218
    quick assets, 214, 218
    real turnover, 215-216, 219
    sales trend, 216, 219
    true investment yield, 217, 219
    turnover, 215-216, 219
    usage test, 217, 219
    working capital, 214, 218
Raw materials inventory, 99
Real turnover ratio, 215-216, 219
Receivables. See Accounts Receivable.
Reconciliation forms, 179, 289-293

Restatements, 237-241
Retained earnings, 205
Returned checks, 80
Returns and allowances, 206

S
Salaries. See Payroll accounting.
Sales accounting, 18-27
Sales budgeting, 56-62
    cash projection, 60
    cash receipts summary, 59
    profits and sales summary, 61
Sales, equipment, 145
Sales planning, 63-67
    checklist, 63
    historical income statement, 65-66
    market assumptions, 66-67
    yield/risk standards, 64
Sales records, 5-6
Sales trends and ratios, 42, 216, 219
Selling profit, 122
Small Business Administration, 251-254
Small tools, 131
Source and application of funds
    budgeting, 149-152
Standards, 208-210
Statements by job, 242-247
    control cycle, 246
    levels of control, 246-247
    long-range planning, 245-246
    plotting jobs, 243-245
Statements for loan applications, 248-255
    co-signers, 251
    guarantors, 251
    loans, 248
    presenting data, 250
    prospectus format, 254
    Small Business Administration, 251-254
Stop payments, 80
Straight line depreciation, 138-141
Subcontracts, 169
Sum of the years' digits depreciation, 141-142
Supplementary schedules, 202
Supplies inventory, 99

T
"T" account, 13
Tax account, 90
Tax planning, 283-285
Tax provision, 206
Time card, 105-106
Trade-ins, equipment, 145
Trends, 42
Trial balance, 199, 202, 224-225
Trucks and autos, 131
True investment yield, 217, 219
Turnover ratios, 215-216, 219

U
Uncollectible accounts. See Bad Debts.
Unearned income, 18
Usage test, 217, 219

V
Variable expenses, 118-119
Voiding checks, 80, 85
Voucher checks, 93

W
Wages. See Payroll accounting.
Work in process inventory, 99
Working capital ratio, 214, 218

Other Practical References

Contractor's Survival Manual

How to survive hard times and succeed during the up cycles. Shows what to do when the bills can't be paid, finding money and buying time, transferring debt, and all the alternatives to bankruptcy. Explains how to build profits, avoid problems in zoning and permits, taxes, time-keeping, and payroll. Unconventional advice on how to invest in inflation, get high appraisals, trade and postpone income, and stay hip-deep in profitable work. 160 pages, 8½ x 11, $16.75

Builder's Office Manual Revised

Explains how to create routine ways of doing all the things that must be done in every construction office — in the minimum time, at the lowest cost, and with the least supervision possible: organizing the office space, establishing effective procedures and forms, setting priorities and goals, finding and keeping an effective staff, getting the most from your record-keeping system (whether manual or computerized), and more. Loaded with practical tips, charts, and sample forms for your use. 192 pages, 8½ x 11, $15.50

Blueprint Reading for the Building Trades

How to read and understand construction documents, blueprints, and schedules. Includes layouts of structural, mechanical, HVAC and electrical drawings. Shows how to interpret sectional views, follow diagrams and schematics, and covers common problems with construction specifications. 192 pages, 5½ x 8½, $11.25

Cost Records for Construction Estimating

How to organize and use cost information from jobs just completed to make more accurate estimates in the future. Explains how to keep the records you need to track costs for sitework, footings, foundations, framing, interior finish, siding and trim, masonry, and subcontract expense. Provides sample forms. 208 pages, 8½ x 11, $15.75

Building Cost Manual

Square foot costs for residential, commercial, industrial, and farm buildings. Quickly work up a reliable budget estimate based on actual materials and design features, area, shape, wall height, number of floors, and support requirements. Includes all the important variables that can make any building unique from a cost standpoint. 240 pages, 8½ x 11, $16.50. Revised annually

National Construction Estimator

Current building costs for residential, commercial, and industrial construction. Estimated prices for every common building material. Manhours, recommended crew, and labor cost for installation. Includes *Estimate Writer*, an electronic version of the book on computer disk, with stand-alone estimating program — free on 5¼" high density (1.2Mb) disk. The National Construction Estimator and *Estimate Writer* on 1.2Mb disk cost $26.50. (Add $10 if you want *Estimate Writer* on 5¼" double density 360K disks or 3½" 720K disks.) 592 pages, 8½ x 11, $26.50. Revised annually

How to Succeed With Your Own Const. Business

Everything you need to start your own construction business: setting up the paperwork, finding the work, advertising, using contracts, dealing with lenders, estimating, scheduling, finding and keeping good employees, keeping the books, and coping with success. If you're considering starting your own construction business, all the knowledge, tips, and blank forms you need are here. 336 pages, 8½ x 11, $19.50

Roof Framing

Shows how to frame any type of roof in common use today, even if you've never framed a roof before. Includes using a pocket calculator to figure any common, hip, valley, or jack rafter length in seconds. Over 400 illustrations cover every measurement and every cut on each type of roof: gable, hip, Dutch, Tudor, gambrel, shed, gazebo, and more. 480 pages, 5½ x 8½, $22.00

National Repair & Remodeling Estimator

The complete pricing guide for dwelling reconstruction costs. Reliable, specific data you can apply on every repair and remodeling job. Up-to-date material costs and labor figures based on thousands of jobs across the country. Provides recommended crew sizes; average production rates; exact material, equipment, and labor costs; a total unit cost and a total price including overhead and profit. Separate listings for high- and low-volume builders, so prices shown are accurate for any size business. Estimating tips specific to repair and remodeling work to make your bids complete, realistic, and profitable. *Repair & Remodeling Estimate Writer* FREE on a 5¼" high-density (1.2 Mb) disk when you buy the book. (Add $10 for *Repair & Remodeling Estimate Writer* on extra 5¼" double density 360K disks or 3½" 720K disks.) 320 pages, 11 x 8½, $29.50. Revised annually

Builder's Comprehensive Dictionary

Never let a construction term stump you again. Here you'll find almost 10,000 construction term definitions, over 1,000 detailed illustrations of tools, techniques, and systems, and a separate section of common legal, real estate, and management terms. 532 pages, 8½ x 11, $24.95

Handbook of Construction Contracting

Volume 1: Everything you need to know to start and run your construction business; the pros and cons of each type of contracting, the records you'll need to keep, and how to read and understand house plans and specs so you find any problems before the actual work begins. All aspects of construction are covered in detail, including all-weather wood foundations, practical math for the job site, and elementary surveying. 416 pages, 8½ x 11, $24.75

Volume 2: Everything you need to know to keep your construction business profitable; different methods of estimating, keeping and controlling costs, estimating excavation, concrete, masonry, rough carpentry, roof covering, insulation, doors and windows, exterior finishes, specialty finishes, scheduling work flow, managing workers, advertising and sales, spec building and land development, and selecting the best legal structure for your business. 320 pages, 8½ x 11, $24.75

Wood-Frame House Construction

Step-by-step construction details, from the layout of the outer walls, excavation and formwork, to finish carpentry and painting, with clear illustrations and explanations. Everything you need to know about framing, roofing, siding, insulation and vapor barrier, interior finishing, floor coverings, and stairs — complete step-by-step "how to" information on building a frame house. 240 pages, 8½ x 11, $14.25. Revised edition

Carpentry Estimating

Simple, clear instructions on how to take off quantities and figure costs for all rough and finish carpentry. Shows how to convert piece prices to MBF prices or linear foot prices, use the extensive manhour tables included to quickly estimate labor costs, and how much overhead and profit to add. All carpentry is covered; floor joists, exterior and interior walls and finishes, ceiling joists and rafters, stairs, trim, windows, doors, and much more. Includes sample forms, checklists, and the author's factor worksheets. 320 pages, 8½ x 11, $25.50

Rough Carpentry

All rough carpentry is covered in detail: sills, girders, columns, joists, sheathing, ceiling, roof and wall framing, roof trusses, dormers, bay windows, furring and grounds, stairs, and insulation. Explains practical code-approved methods for saving lumber and time. Chapters on columns, headers, rafters, joists, and girders show how to use simple engineering principles to select the right lumber dimension for any species and grade. **288 pages, 8½ x 11, $17.00**

Illustrated Guide to the National Electrical Code

This fully-illustrated guide offers a quick and easy visual reference for installing electrical systems. Whether you're installing a new system or repairing an old one, you'll appreciate the simple explanations written by a code expert, and the detailed, intricately-drawn and labeled diagrams. A real time-saver when it comes to deciphering the current NEC. **256 pages, 8½ x 11, $24.00**

Estimating Home Building Costs

Estimate every phase of residential construction from site costs to the profit margin you include in your bid. Shows how to keep track of manhours and make accurate labor cost estimates for footings, foundations, framing and sheathing finishes, electrical, plumbing, and more. Provides and explains sample cost estimate worksheets with complete instructions for each job phase. **320 pages, 5½ x 8½, $17.00**

Electrical Construction Estimator

This year's prices for installation of all common electrical work: conduit, wire, boxes, fixtures, switches, outlets, loadcenters, panelboards, raceway, duct, signal systems, and more. Provides material costs, manhours per unit, and total installed cost. Explains what you should know to estimate each part of an electrical system. *Electrical Estimate Writer is included FREE with the book on a 5¼" high-density (1.2 Mb) disk.* (Add $10 for extra 5¼" double-density 360K disks or 3½" 720K disks.) **416 pages, 8½ x 11, $28.50. Revised annually**

Bookkeeping for Builders

Shows simple, practical instructions for setting up and keeping accurate records — with a minimum of effort and frustration. Explains the essentials of a record-keeping system: the payment, income, and general journals, and records for fixed assets, accounts receivable, payables and purchases, petty cash, and job costs. Shows how to keep I.R.S. records and accurate, organized business records for your own use. **208 pages, 8½ x 11, $19.75**

Contractor's Year-Round Tax Guide

How to set up and run your construction business to minimize taxes using corporate tax strategy to your advantage, why you should consider incorporating, and what to look for in your contracts with others, including sample contracts. Covers tax shelters for builders, write-offs and investments that'll reduce your tax liability, what the I.R.S. allows and what often questions. Explains how to keep records and protect your company from tax traps. **192 pages, 8½ x 11, $16.50**

Contractor's Guide to the Building Code Revised

This completely revised edition explains in plain English exactly what the Uniform Building Code requires. Based on the 1991 code, the most recent, it covers many changes made since then. Also covers the Uniform Mechanical Code and the Uniform Plumbing Code. Shows how to design and construct residential and light commercial buildings that pass inspection the first time. Suggests how to work with an inspector to minimize construction costs, what common building shortcuts are likely to be cited, and where exceptions are granted. **544 pages, 5½ x 8½, $28.00**

Contractor's Growth and Profit Guide

Step-by-step instructions for planning growth and prosperity in a construction contracting or subcontracting company. Explains how to prepare a business plan: select reasonable goals, draft a market expansion plan, make income forecasts and expense budgets, and project cash flow. You'll learn everything that most lenders and investors require, as well as the best way to organize your business. **336 pages, 5½ x 8½, $19.00**

---

10 Day Money Back GUARANTEE

- ☐ 11.25 Blueprint Reading for the Building Trades
- ☐ 19.75 Bookkeeping for Builders
- ☐ 24.95 Builder's Comprehensive Dictionary
- ☐ 15.50 Builder's Office Manual Revised
- ☐ 16.50 Building Cost Manual
- ☐ 25.50 Carpentry Estimating
- ☐ 19.00 Contractor's Growth and Profit Guide
- ☐ 28.00 Contractor's Guide to the Building Code Revised
- ☐ 16.75 Contractor's Survival Manual
- ☐ 16.50 Contractor's Year-Round Tax Guide
- ☐ 15.75 Cost Records for Construction Estimating
- ☐ 28.50 Electrical Construction Estimator with
  free Electrical Estimate Writer
  on 5¼" (1.2Mb) disk. *Add $10 for extra*
  ☐ 5¼" (360K) or ☐ 3½" (720K) disks.
- ☐ 17.00 Estimating Home Building Costs
- ☐ 24.75 Handbook of Construction Contracting Vol. 1
- ☐ 24.75 Handbook of Construction Contracting Vol. 2
- ☐ 19.50 How to Succeed With Your Own Const. Business
- ☐ 24.00 Illustrated Guide to the National Electrical Code
- ☐ 26.50 National Construction Estimator with
  free Estimate Writer on 5¼" (1.2Mb) disk.
  Add $10 for extra Estimate Writer on either
  ☐ 5¼" (360K) or ☐ 3½" (720K).
- ☐ 29.50 National Repair & Remodeling Estimator with
  free Repair & Remodel Estimate Writer
  on 5¼ (1.2Mb) disk. *Add $10 for extra*
  ☐ 5¼" (360K) or ☐ 3½" (720K).
- ☐ 22.00 Roof Framing
- ☐ 17.00 Rough Carpentry
- ☐ 14.25 Wood-Frame House Construction
- ☐ 22.50 Builder's Guide to Accounting Revised

Craftsman Book Company

6058 Corte del Cedro, P. O. Box 6500
Carlsbad, CA 92018

In a hurry?

We accept phone orders
charged to your MasterCard,
Visa or American Express

Call 1-800-829-8123
FAX (619) 438-0398

Include a check with your order and we pay shipping

Name (Please print) _____

Company _____

Address _____

City/State/Zip _____

Send check or money order

Total enclosed _____ (In California add 7.25% tax)

If you prefer, use your ☐ Visa, ☐ MasterCard or ☐ American Express

Card number _____

Expiration date _____ Initial _____

Call 1-800-829-8123 for a **FREE** Full Color Catalog with over 100 Titles

Craftsman Book Company

6058 Corte del Cedro
Carlsbad, CA 92018

☎

In a hurry?
We accept phone orders charged to
your MasterCard, Visa or American Express

Call (619) 438-7828

Name (Please print) _____
Company _____
Address _____
City/State/Zip _____
Total Enclosed _____ (In CA add 7.25% tax)
Use your ☐ Visa ☐ MasterCard ☐ Amer.Exp.
Card # _____
Exp. date _____ Initials _____

10 Day Money Back GUARANTEE

- ☐ 11.25 Blueprint Reading for Blding Trades
- ☐ 19.75 Bookkeeping for Builders
- ☐ 24.95 Builder's Comprehensive Dictionary
- ☐ 15.50 Builder's Office Manual Revised
- ☐ 16.50 Building Cost Manual
- ☐ 25.50 Carpentry Estimating
- ☐ 19.00 Contractor's Growth & Profit Guide
- ☐ 28.00 Cont Guide to Blding Code Rev
- ☐ 16.50 Contractor's Year-Round Tax Guide
- ☐ 15.75 Cost Records for Const Estimating
- ☐ 28.50 Electrical Const. Estimator with
  free *Electrical Estimate Writer* on
  5¼" (1.2Mb) disk. *Add $10 for extra*
  ☐ 5¼" (360K) or ☐ 3½" (720K) disks.
- ☐ 17.00 Estimating Home Building Costs
- ☐ 24.75 Handbook of Const Cont Vol. 1
- ☐ 24.75 Handbook of Const Cont Vol. 2
- ☐ 19.50 How to Succeed w/ Own Const Bus.
- ☐ 24.00 Illust. Guide to Nat Electrical Code
- ☐ 26.50 National Construction Estimator with
  free *Estimate Writer* on 5¼" (1.2Mb) disk.
  Add $10 for extra *Estimate Writer* on either
  ☐ 5¼" (360K) or ☐ 3½" (720K) disks.
- ☐ 29.50 Nat. Repair & Remodeling Est. with
  free Repair & Remodeling Estimate Writer
  on 5¼" (1.2Mb) disk. *Add $10 for extra*
  ☐ 5¼" (360K) or ☐ 3½" (720K) disks.
- ☐ 22.00 Roof Framing
- ☐ 17.00 Rough Carpentry
- ☐ 14.25 Wood-Frame House Construction
- ☐ 22.50 Builder's Guide to Accounting
- ☐ Free Full Color Catalog

Craftsman Book Company

6058 Corte del Cedro
Carlsbad, CA 92018

☎

In a hurry?
We accept phone orders charged to
your MasterCard, Visa or American Express

Call (619) 438-7828

Name (Please print) _____
Company _____
Address _____
City/State/Zip _____
Total Enclosed _____ (In CA add 7.25% tax)
Use your ☐ Visa ☐ MasterCard ☐ Amer.Exp.
Card # _____
Exp. date _____ Initials _____

10 Day Money Back GUARANTEE

- ☐ 11.25 Blueprint Reading for Blding Trades
- ☐ 19.75 Bookkeeping for Builders
- ☐ 24.95 Builder's Comprehensive Dictionary
- ☐ 15.50 Builder's Office Manual Revised
- ☐ 16.50 Building Cost Manual
- ☐ 25.50 Carpentry Estimating
- ☐ 19.00 Contractor's Growth & Profit Guide
- ☐ 28.00 Cont Guide to Blding Code Rev
- ☐ 16.50 Contractor's Year-Round Tax Guide
- ☐ 15.75 Cost Records for Const Estimating
- ☐ 28.50 Electrical Const. Estimator with
  free *Electrical Estimate Writer* on
  5¼" (1.2Mb) disk. *Add $10 for extra*
  ☐ 5¼" (360K) or ☐ 3½" (720K) disks.
- ☐ 17.00 Estimating Home Building Costs
- ☐ 24.75 Handbook of Const Cont Vol. 1
- ☐ 24.75 Handbook of Const Cont Vol. 2
- ☐ 19.50 How to Succeed w/ Own Const Bus.
- ☐ 24.00 Illust. Guide to Nat Electrical Code
- ☐ 26.50 National Construction Estimator with
  free *Estimate Writer* on 5¼" (1.2Mb) disk.
  Add $10 for extra *Estimate Writer* on either
  ☐ 5¼" (360K) or ☐ 3½" (720K) disks.
- ☐ 29.50 Nat. Repair & Remodeling Est. with
  free Repair & Remodeling Estimate Writer
  on 5¼" (1.2Mb) disk. *Add $10 for extra*
  ☐ 5¼" (360K) or ☐ 3½" (720K) disks.
- ☐ 22.00 Roof Framing
- ☐ 17.00 Rough Carpentry
- ☐ 14.25 Wood-Frame House Construction
- ☐ 22.50 Builder's Guide to Accounting
- ☐ Free Full Color Catalog

Mail This Card Today For A FREE Full Color Catalog

Craftsman Book Company / 6058 Corte del Cedro / P.O. Box 6500 / Carlsbad, CA 92018

Over 100 books, videos, and audios at your fingertips with information that can save you time and money. Here you'll find information on carpentry, contracting, estimating, remodeling, electrical work, painting, and plumbing.

All items come with an unconditional 10-day money-back guarantee. If they don't save you money, mail them back for a full refund.

Name _____
Company _____
Address _____
City/State/Zip _____